New Horizons in Evolution

New Horizons in Evolution

EDITED BY

SOLOMON P. WASSER
Haifa University, Israel

MILANA FRENKEL-MORGENSTERN
Bar-Ilan University, Israel

ELSEVIER

ACADEMIC PRESS
An imprint of Elsevier

Academic Press is an imprint of Elsevier
125 London Wall, London EC2Y 5AS, United Kingdom
525 B Street, Suite 1650, San Diego, CA 92101, United States
50 Hampshire Street, 5th Floor, Cambridge, MA 02139, United States
The Boulevard, Langford Lane, Kidlington, Oxford OX5 1GB, United Kingdom

British Library Cataloguing-in-Publication Data
A catalogue record for this book is available from the British Library

Library of Congress Cataloging-in-Publication Data
A catalog record for this book is available from the Library of Congress

ISBN: 978-0-323-90752-1

Publisher: Charlotte Cockle
Acquisitions Editor: Anna Valutkevich
Editorial Project Manager: Lena Sparks
Production Project Manager: Selvaraj Raviraj
Cover Designer: Miles Hitchen

Typeset by MPS Limited, Chennai, India

Contents

List of contributors

Walter F. Bodmer
Cancer and Immunogenetics Laboratory, Weatherall Institute of Molecular Medicine, John Radcliffe Hospital, University of Oxford, Oxford, United Kingdom

Hynek Burda
Department of Game Management and Wildlife Biology, Faculty of Forestry and Wood Sciences, Czech University of Life Sciences, Praha, Czech Republic

Gon Carmi
Cancer Genomics and BioComputing of Complex Diseases Lab, The Azrieli Faculty of Medicine, Bar-Ilan University, Safed, Israel

Amy K. Conley
Department of Biology, Washington University, St. Louis, MO, United States; New York Natural Heritage Program, College of Environmental Science and Forestry, State University of New York, Albany, NY, United States

Daniel J.M. Crouch
JDRF/Wellcome Diabetes and Inflammation Laboratory, Wellcome Centre for Human Genetics, University of Oxford, Old Road Campus, Oxford, United Kingdom

Milana Frenkel-Morgenstern
Cancer Genomics and BioComputing of Complex Diseases Lab, The Azrieli Faculty of Medicine, Bar-Ilan University, Safed, Israel

Alessandro Gorohovski
Cancer Genomics and BioComputing of Complex Diseases Lab, The Azrieli Faculty of Medicine, Bar-Ilan University, Safed, Israel

Mikhail I. Katsnelson
Institute for Molecules and Materials, Radboud University, Nijmegen, The Netherlands

Eugene V. Koonin
National Center for Biotechnology Information, National Library of Medicine, National Institutes of Health, Bethesda, MD, United States

Etty Kruzel-Davila
Department of Nephrology, Rappaport Faculty of Medicine and Research Institute, Technion-Israel Institute of Technology, Rambam Health Care Campus, Haifa, Israel

Xionglun Liu
Southern Regional Collaborative Innovation Center for Grain and Oil Crops in China, Hunan Agricultural University, Changsha, P.R. China

Zhiyong Liu
State Key Laboratory of Plant Cell and Chromosome Engineering, Institute of Genetics and Developmental Biology, The Innovative Academy of Seed Design, Chinese Academy of Sciences, Beijing, P.R. China

Jennifer L. Neuwald
Department of Biology, Washington University, St. Louis, MO, United States; Department of Biology and Graduate Degree Program in Ecology, Colorado State University, Fort Collins, CO, United States

Eviatar Nevo
Institute of Evolution, University of Haifa, Haifa, Israel

Junhua Peng
Center of Crop Germplasm Enhancement and Utilization, Huazhi Bio-Tech Co., Ltd., Changsha, P.R. China

James A. Shapiro
Department of Biochemistry and Molecular Biology, University of Chicago, Gordon Center for Integrative Science, Chicago, IL, United States

Karl Skorecki
Department of Nephrology, Rappaport Faculty of Medicine and Research Institute, Technion-Israel Institute of Technology, Rambam Health Care Campus, Haifa, Israel; Azrieli Faculty of Medicine, Bar-Ilan University, Safed, Israel

Dongfa Sun
College of Plant Science and Technology, Huazhong Agricultural University, Wuhan, P.R. China

Moshe Szyf
Department of Pharmacology and Therapeutics, Faculty of Medicine, McGill University, Montreal, QC, Canada

Alan R. Templeton
Department of Biology, Washington University, St. Louis, MO, United States

Solomon P. Wasser
Institute of Evolution, Department of Evolutionary and Environmental Biology, University of Haifa, Haifa, Israel

Yuri I. Wolf
National Center for Biotechnology Information, National Library of Medicine, National Institutes of Health, Bethesda, MD, United States

Jun Yan
Key Laboratory of Coarse Cereal Processing, Ministry of Agriculture Rural Affairs, School of Pharmacy and Bioengineering, Chengdu University, Chengdu, P.R. China

Foreword

The book you are holding in your hands contains the proceedings of an international conference on "New Horizons in Evolution" which was dedicated to the 90th birthday of Prof. Eviatar Nevo, a prominent scientist in the field of evolutionary biology.

On February 2, 2019, Prof. Nevo (Eibi to his many friends), celebrated his 90th birthday. We began our celebration of Prof. Nevo's career in May 2019, by organizing the "New Horizons in Evolution" international conference. This 2-day event (May 27–28, 2019) was dedicated to new horizons in evolutionary theory and processes. This volume recounts the proceedings of the conference and presentations delivered by leading scientists in genomics (J. Shapiro, United States), epigenetics (M. Szyf, Canada), evolutionary theory (E. Koonin, United States), cancer evolution and evolutionary medicine (W. Bodmer, United Kingdom and K. Skorecki, Israel), evolutionary biology networking (A. Templeton, United States), the evolution of sensing underground (H. Burda, Czech Republic/Germany), evolutionary agricultural domestication of wheat (J. Peng, China), the evolution of protein domains (M. Frenkel-Morgenstern, Israel), and sympatric speciation based on the evolution canyon model (E. Nevo, Israel). These figures include naturalists, experimentalists, and theoreticians, all contributing new insight to evolutionary theories of speciation, adaptation, and regulation, together with reports on their application in medicine and agriculture.

Evolution, the constant agent of change in nature, is one of the deepest of human ideas. It is found at the core of physical, biological, social, and humanistic sciences, linking energy, matter, life, and consciousness. It is also a dynamically changing theory. The dramatic genomic revolution of recent decades has revealed new informative horizons of unparalleled combinatorial diversity and exposed the innovation of nature as one moves from the simple to the complex, all through insight into the dynamic and universal language of life that is DNA. Much of our understanding of genome complex adaptive architecture, symbiosis, dynamic reorganization, expression, and regulation transformed not only biological but also all sciences. While such understanding has helped solve many of life's mysteries, much more remains to be discovered, primarily in the realms of biodiversity evolution, genome regulation, cellular mechanisms, organism–environment interaction, the origin of life, cells, biochemical networks, nonrandom adaptive mutations, epigenetics, and repetitive DNA regulatory function. Resolving the numerous open questions in these fields will highlight the dynamics of biological systems, based on the evolving genome, environmental stresses, and adaptive changes, all affecting the dynamic genome read–write memory system underlying evolution. Deciphering these and other mysteries will contribute to the advancement of evolutionary theory and biodiversity evolution and will transform medicine and serve as a basis for improving domestication processes in agriculture, a central pillar of human civilization.

Prof. Nevo is a leading Israeli expert in evolution. He established and directed the Institute of Evolution at the University of Haifa, Israel (1972–2008) and the International Graduate Center of Evolution (2004–2008). Both the Institute of Evolution and the International Graduate Center of Evolution are truly unique. His scientific career involved interdisciplinary studies of biodiversity evolution, adaptation, and speciation across life, based on the science he advanced dealing with evolutionary functional ecological genetics and genomics across life from viruses and bacteria through fungi, plants, and animals to humans. Prof. Nevo has published more than 1500 scientific papers in top scientific journals, including 14 in *Science* and *Nature*, and approximately 70 in the *Proceedings of the National Academy USA*, and is author, coauthor, and/or editor of 36 books. In collaboration with many colleagues, he has contributed substantially to our understanding of correlates and predictors of genetic diversity in nature under diverse environmental stresses (chemical, climatic, thermal, biotic, abiotic, and atomic). His studies involve genes, genomes, repeatomes, phenomes, populations, species, and ecosystems of bacteria, fungi, plants, animals, and humans, focusing on the structure, function, and causation of genetic diversity in nature at the local, regional, and global scales. His major models of biodiversity evolution involve the evolution of blind subterranean mammals (1948-ongoing), wild cereals (1975-ongoing), marine organisms as indicators of pollution (1980s). Moreover, the evolution in action model presented by "Evolution Canyons," caused by interslope microclimatic divergence at four microsites in Israel which he dubbed as the "Israeli Galapagos" and extended to "Evolution Plateau" and "Evolution Slope," geologic-edaphic microsites in the Upper Galilee (1990-ongoing). Studies on the evolution of Dead Sea fungi (1997-ongoing) and of chimpanzees in the savannahs of Mali mirroring early human evolution (2001-ongoing) were developed with his doctoral and postdoctoral students. He has conducted genetic–ecological studies locally (in six natural "Evolution Canyon" laboratories in Israel), regionally (in Israel and the Fertile Crescent of the Near East as natural genetic laboratories), and globally (across all continents, treated as genetic laboratories). These interdisciplinary studies link genetics and ecology as ecological–genetics and ecological–genomics, bridging genotypes and phenotypes, integrating molecular and organismal biology, organism–environment relationships, and elucidating patterns and causation of genetic diversity in nature. Notably, these studies link ecological stresses with the level of genetic polymorphism in proteins and DNA across life (bacteria, fungi, plants, and animals) and the entire planet (all continents, except Antarctica). Prof. Nevo established the environmental theory of genetic diversity proposing that, in general, genetic polymorphism at all scales, namely, local, regional, and global, and across life, is positively correlated with and predictable by ecological stress. The Evolution Canyon model initiated by Prof. Nevo became a classical model of biodiversity evolution at the microscale. The Evolution Canyon was created by the sharp microclimatic interslope divergence that occurs across short distances in terms of the biotic representation (biomes) of two continents, Africa and Europe. The 250 papers and four books published on the Evolution Canyon microsite model addressed diverse fundamental problems of evolutionary biology. These include biodiversity evolution, genetic polymorphism, transposon, and retrotransposon dynamics and their effects on genome size, DNA repair, mutation, recombination, gene conversion rates, adaptation and speciation, methylation associated with stress, lateral transfer, splice variation, and genome-wide gene expression, as well as the twin evolutionary processes of adaptation and both full and incipient sympatric speciation across life from viruses and bacteria through fungi, plants, and animals, from invertebrates to mammals.

In terms of applied science, Prof. Nevo has advanced a novel genetic methodology to safeguard the quality of marine environments (as documented in 25 scientific papers). Similarly, he substantiated the idea that wild progenitors of cereals and other cultivars harbor rich genetic resources that should be conserved in situ and ex situ and used in crop improvement (as documented in 420 papers and a book). The wild relatives of crops represent the best hope for future genetic crop improvement, advancing the second genetic green revolution, and guaranteeing the increase and stabilization of world food production, a matter of cardinal importance, particularly with the world population growing at an everincreasing rate. In practice, he and his colleagues have mapped several candidate genes and

QTLs related to adaptation and domestication. Prof. Nevo and colleagues discovered 77 species of filamentous fungi in the Dead Sea, cloned several genes, sequenced the first eukaryote filamentous fungus from the Dead Sea, and introduced the HOG gene into yeast and *Arabidopsis* and showed that, in principle, genetic resources of the Dead Sea fungi could revolutionize saline agriculture (documented in dozens of papers and a book). Prof. Nevo also advanced the study of subterranean mammals across the globe, considering such animals as having participated in a singularly important global evolutionary experiment. Indeed, he and his colleagues have written more than 380 scientific papers and two books on subterranean mammals from all continents. In studying blind subterranean mole rats (species of the genus *Spalax*), Prof. Nevo and colleagues identified five species in Israel, four of which originated allopathically, and one species found in the upper Galilee that arose from sympatric speciation. Similarly, Prof. Nevo and his coworkers identified hundreds of hypoxia-tolerant genes linked to cancer resistance, stroke, and cardiovascular disease resistance that could revolutionize medicine, space flight, and ocean diving. They also transformed *Spalax* VEGF into ischemic mice and were thus able to save the leg of an experimental mouse after severing its main blood vessel by generating extensive capillarization and vascularization.

Prof. Nevo founded (1973) the Institute of Evolution at the University of Haifa and served as Institute Director until October 2008. The Institute of Evolution is a world center of excellence, conducting integrative research in biodiversity, molecular, genomic, and organismal evolutions, linking field, laboratory, and theoretical research programs across life, focusing on ecological stress and genetic evolution. In 2004 Prof. Nevo established the International Graduate Center of Evolution with 77 doctoral students from 13 countries, thus cultivating future world leaders in the fields of biodiversity and genetic diversity, adaptation, and speciation in nature. Finally, Prof. Nevo and colleagues have studied the effects of atomic radiation resulting from the Chernobyl disaster and found numerous molecular mutations associated with cancer in the offspring of those who cleaned the site and in those who were born after the event, indicating that the mutations had passed through the germline from parent to offspring.

Prof. Nevo is not only a great scientist but also a great man. He is honest, optimistic, a man of his word, a wonderful father and grandfather, and a trusted friend. Prof. Nevo is also a great teacher (72 students from his laboratory have received PhD degrees with more than 50% summa cum laude). He is, moreover, a great traveler (in the last 3 years alone, he visited Vietnam, the Republic of Georgia, and Tanzania). He is a well-rounded man with interests in different fields, not only in science. He is well versed in classical music, art, and philosophy. Finally, Prof. Nevo is a walking encyclopedia and Wikipedia. He is intimately familiar with classical and modern biology, evolution, the biodiversity of different groups of living organisms, genetics, molecular biology, climatology, ecology, and geology. As a recognized generator of diverse scientific ideas, including the impact of ecology on genetics and evolution, Prof. Nevo has received numerous awards and accolades. He is an international member of the National Academy of Sciences of the United States and in 2016 won the most prestigious prize in Israel, the Israeli Prize in Life Sciences.

We believe that the book will have widespread appeal for active scientists at all levels of training interested in the biological sciences from the cellular to the organismal, including naturalists, experimentalists, and theoreticians, and also undergraduate and graduate students from other branches of sciences, since evolution is the core of all science. Current dynamics of the biological sciences is dramatic, and, as noted in the title of the classic 1973 paper on evolutionary theory by Theodosius Dobzhansky, "Nothing in Biology Makes Sense Except in the Light of Evolution." This highlights the tremendous importance of studies from the Nevo laboratory for the future of mankind, given his contributions to medicine, agriculture, pharmacology, and other industries associated with life sciences and human civilization.

Solomon P. Wasser[1] and Milana Frenkel-Morgenstern[2]
[1]Haifa University, Israel
[2]Bar-Ilan University, Israel
2020

CHAPTER 1

How should we think about evolution in the age of genomics?

James A. Shapiro
Department of Biochemistry and Molecular Biology, University of Chicago, Gordon Center for Integrative Science, Chicago, IL, United States

Background: tribute to a unique evolutionary biologist

This 90th Jubilee Symposium and book, *New Horizons in Evolution*, pay tribute to Eviatar Nevo, a prolific pioneer best known today for his work at Evolution Canyon, where two distinct ecologies sit side-by-side [1]. Exploiting this special geography, Eibi and his many colleagues have been able to observe the effects of ecological differences on real-time processes of evolutionary change of numerous species from microbes to mammals at the population [2−4], organismal [5,6], karyotypic [7], and molecular levels [8].

Eibi's research highlights the complexity of evolutionary responses to ecological parameters using the most up-to-date tools available. In particular, he has documented the impacts of ecological stresses [9] on specific processes that bring about adaptive genome change: DNA repair [10], mutation [8], chromosome rearrangements [11], amplification, and movement of repetitive and mobile DNA elements [12−14].

Eibi's extremely broad documentation of how real-world ecological challenges integrate with genome change operations presents us with an opportunity to reconsider the basic principles of evolutionary biology in the age of genomics. Does our contemporary knowledge of how genomes change in the course of evolution confirm traditional principles established in the 19th and 20th centuries? Or does that body of information force us to adopt a new, more contemporary set of principles? This article will argue in favor of the latter position, based on contemporary empirical data cited in a pair of recent reviews [15,16] and also posted under various headings on my University of Chicago web page (online link 1).

New Horizons in Evolution
DOI: https://doi.org/10.1016/B978-0-323-90752-1.00010-9

Basic principles of evolutionary change necessary to encompass Eibi's work

Traditional evolutionary theory in the 19th and early 20th centuries focused on accidental changes in isolated genomes. Under the principle of "Descent with Modification," evolutionary biologists had to explain where and how hereditary changes took place. Each species was assumed to have its own genome, comprising the nuclear chromosomes. The nuclear genome was isolated from the genomes of other species by sexual incompatibility (the Bateson–Dobzhansky–Muller model [17]) and from life history events by the Weismann Barrier between the soma and the germline [18,19]. Within these isolated genomes, hereditary changes were assumed to occur by random accidents during germline reproduction. When DNA was identified as the molecular carrier of genetic information, the random accident theory was updated to unavoidable copying errors in the course of DNA replication [20,21].

The traditional perspective did not allow any possibilities for ecological inputs, like those Eibi's research has documented, into the process of hereditary variation. Today, of course, we can see that the "isolationist" view of hereditary determination was unjustifiably restrictive in several fundamental ways. The second half of the 20th century made us aware of many biochemical processes of hereditary change, ranging from the active movement of mobile DNA elements in the genomes of maize and all other organisms (online link 2) [22] to DNA transfer, repair, rearrangement, and mutator functions (online link 3) [23,24]. Similar to all physiological activities (and as Eibi's research demonstrates so well), these molecular "Natural Genetic Engineering" (NGE) processes of DNA change are subject to cellular regulation (online link 4) and operate in a manner that is sensitive to ecological inputs (online link 5). As we shall see subsequently, genomic analysis has documented important evolutionary adaptations that result from NGE action.

In addition to the regulated physiological processes that alter DNA molecules, we now have overwhelming evidence that genomes are not as isolated from each other or from the environment as traditional theories assumed [16]. There are multiple forms of "infectious heredity" understood in its broadest sense (online link 6) [25]. All organisms reproduce in the presence of abundant environmental DNA as well as a staggering density of viruses, vesicles, and other forms of enclosed DNA molecules. Multiple mechanisms exist for cells of different organisms to exchange

DNA molecules, and there is abundant genomic data that adaptive horizontal DNA transfers have occurred across virtually all taxonomic boundaries. Moreover, as we shall discuss shortly, there is no question that cell fusions can combine unrelated genomes in a single hereditary lineage to form modified or new kinds of organisms, including the first mitochondrion-bearing ancestor of all eukaryotes. The reproductive boundaries between species are not as absolute as once assumed, and we shall see that mating across species boundaries is a major stimulus to evolutionary innovation in sexually reproducing organisms.

The Weissman Barrier soma-germline separation is also not absolute, as some exponents of traditional theory have claimed. Such a barrier cannot exist in cells that proliferate by vegetative multiplication, such as all prokaryotes and lower unicellular eukaryotes, where there is a direct hereditary connection between "somatic" and "germline" cells. Even in unicellular eukaryotes that undergo sexual differentiation and reproduction, vegetative cells are the direct precursors of spores and gametes. The same cell lineage connection exists between soma and germline in plants. Since flowers containing plant sexual organs develop out of somatic tissues, the genomic consequences of life history events can be incorporated into pollen and ovules that merge to form the next generation. Only in animals, where germline cells separate from somatic tissues early in multicellular development, formation of a Weismann Barrier is a realistic possibility. Nonetheless, we now know about processes of macromolecular transport in animals which facilitate the transfer of somatically acquired genomic information to animal sperm cells [26].

From the foregoing summary, we can see that discoveries based on genomic data provide us a 21st century picture of ecologically sensitive evolutionary processes that coincides with what Eibi's amazingly productive research has revealed.

Cell fusions produced foundational evolutionary innovations

The deepest evolutionary divides among living organisms are the separation of all cells into three distinct lineages: *Bacteria*, *Archaea*, and *Eukarya* [27,28]. It is a salutary reminder of the power of genomic analysis and of the capacity for new data to transform our understanding of fundamental evolution principles to recognize that our knowledge of *Archaea* as a distinct cell type only dates from analysis of ribosomal RNA sequences in

1977, less than 50 years ago [29,30]. The same kind of early genomic evidence made it clear that the mitochondrion of the earliest known eukaryotic ancestor descended from an endosymbiotic gram-negative *Proteobacterium* [31–34]. Parallel sequence analysis confirmed the evolutionary origins of light-harvesting plastid organelles in a wide range of algae, plants, and other photosynthetic eukaryotes as descendants of endosymbiotic *Cyanobacteria* (online link 8).

Fossil and genomic data tell us that both *Bacteria* and *Archaea* are the most ancient cell lineages (dating from >3.4 GYA) and that the primordial symbiogenetic event in evolution of *Eukarya* occurred approximately 1.6–1.8 GYA, as the Earth's atmosphere was accumulating a significant concentration of molecular oxygen (O_2), following the evolution of oxygenic photosynthesis in *Cyanobacteria* [35–37]. The O_2 concentration is important because the best genomic evidence indicates that the host cell in the primordial symbiogenesis was an anaerobic archaeal cell encoding many proteins once considered to be exclusive to eukaryotes [35,38–41]. Since *Proteobacteria* are aerobic and contemporary mitochondria are the loci of oxidative energy-yielding metabolism in eukaryotic cells, it is evident that acquisition of an aerobic endosymbiont would provide a significant metabolic advantage to an anaerobic host cell in an environment with a growing atmospheric O_2 concentration.

During the course of transformation from an independent cell to a subcellular organelle in the proto-eukaryotic cell, the endosymbiont *Proteobacterium* underwent a series of major changes in genome content (online link 9). In all eukaryotes, mitochondrial DNA coding content is only a fraction of that in *Proteobacteria.* The largest mitochondrial genome encodes only 100 protein and RNA molecules compared with over 800 for the smallest *Proteobacterium* cell. A typical animal mitochondrion like ours encodes only 37 molecules [42]. DNA containing the vast majority of bacterial coding sequences required for mitochondrial maintenance and metabolism transferred to the nuclear genome of the evolving eukaryotic host cells. The resulting nuclear-encoded mitochondrial proteins are synthesized like other eukaryotic proteins and imported into the mitochondrion organelle by newly evolved protein transport systems. Since overall genome size, physical DNA structure, and coding content differs greatly between the mitochondria in the cells of various eukaryotic lineages, it is clear that mitochondrial genome evolution has involved an ongoing and complex series of taxonomically specific DNA restructuring processes [32].

Although sequence data indicate that the symbiogenetic event originating mitochondrial evolution appears to have been unique, multiple cell fusion events have transferred plastids encoding photosynthetic capabilities to diverse eukaryotic lineages (online link 10). The oldest one involved a common cyanobacterial progenitor of the different plastids in four distinct groups of organisms: green algae (*Chlorophyta*), red algae (*Rhodophyta*), blue-gray algae (*Glaucophyta*), and green plants (*Embryophyta*). Such a cyanobacterial fusion is called a "primary symbiogenesis," and there has been a second, much more recent primary symbiogenesis of a distinct species of *Cyanobacteria* creating a single photosynthetic amoeba, *Paulinella chromatophora*. Furthermore "secondary" symbiogenetic events have occurred when a photosynthetic eukaryote, generally one of the algae, has fused with a nonphotosynthetic eukaryote cell type to create a novel photosynthetic lineage [43]. If the product of a secondary photosynthetic fusion merges with another nonphotosynthetic lineage, that creates a "tertiary" symbiogenesis.

Many of the most important photosynthetic organisms on Earth, such as diatoms, have resulted from these secondary and tertiary symbiogeneses. Clearly photosynthetic cell fusions have occurred over a prolonged period of evolutionary time and are quite likely to be taking place now. As with mitochondria, there has been significant DNA transfer from plastids to the nuclear genome, and plastid-specific protein transport has evolved to incorporate into the plastids nuclear-encoded proteins needed for photosynthesis. Also similar to mitochondria, there are lineage-specific differences in plastid DNA content, physical DNA structures, and coding capacities. In cases of secondary and tertiary symbiogenesis, there are also significant rearrangements and losses of nuclear DNA from the eukaryotic endosymbionts.

In addition to these various cases of photosynthetic symbiogenesis, there are numerous other cases of cell fusions and endosymbiosis that have profound adaptive significance [44–46]. Bacteria can invade other bacteria as well as virtually all kinds of eukaryotic cells [47,48]. A smaller number of endosymbiotic Archaea have been documented, but more cases will doubtless appear as genomic analysis comes to bear on more types of organisms from other parts of the biosphere. Eukaryotic microbes can become endosymbionts of other eukaryotes, just as they do in secondary and tertiary photosynthetic fusions [49]. The adaptive significance of these endosymbiotic relationships involve various adaptive characteristics, such as synthesis of important nutrients (vitamins, amino acids, etc.), utilization of particular food sources (e.g., digestion of plant polymers), or protection against predators or infectious agents [45,50]. So-called "obligate" endosymbiosis occurs when the cell fusion becomes essential for reproduction of the host organism or the endosymbiont (often due to genome reduction, similar to what occurred in mitochondria and plastids) [51,52].

Microbiomes and holobionts

Besides cell fusions, microbes and multicellular organisms establish important symbiotic relationships simply by growing in close proximity or by the microbes colonizing the cells or interior cavities of major organ systems, like the intestine (online link 11). Each multicellular organism has its own "microbiome," the generic term for all the associated microorganisms [53]. The microbes interact biochemically with each other and with the multicellular host in ways that affect the overall phenotype.

Different organs or regions of a single host can have distinct microbiome compositions with unique phenotypic consequences. In plants, for example, the microbiomes on leaves and roots are dramatically different and play radically different roles in transport of nutrients and responses to biotic and abiotic stresses [54]. Of particular importance for all plants and animals is the role a healthy microbiome at each site plays in blocking infection by microbial pathogens.

The microbiome plays important roles in many adaptive phenotypes. We are becoming familiar with discussions of how the "human microbiome" (usually meaning the intestinal microbiome) affects our health and well-being, metabolism and digestion, pregnancy, immune responses, and even mood and states of mind. Microbiome species synthesize critical nutrients for the macroscopic holobiont host, ranging from amino acids in aphids [55] to a wide range of essential metabolites in primitive marine animals [56] to signaling molecules that affect functioning of our own metabolic, neural, and innate immune systems [57]. In *Drosophila*, the intestinal microbiome affects growth factor signaling and morphogenesis [58], volatile pheromone production and social attraction [59], and neuropeptide synthesis and locomotor behavior [60].

Clearly the adaptive properties of the host plus its microbiome result from expression of microbial and host genomes. Typically the protein coding capacity of the total microbiome genome is far more diverse than that of the host genome [61]. In our own case, the human gut microbiome is estimated to encode from 3.3 to 9.9 million distinct proteins or 150–450 times greater than the basic nucleus-encoded human proteome.

From an evolutionary point of view, the recognition of microbiome contributions to whole organism phenotypes poses a definitional challenge. What is the evolving entity? To deal with this question, the terms "holobiont" and "hologenome" were invented to describe the evolving entity and its genetic endowment [62–64]. A holobiont is composed of a multicellular plant or animal together with its associated microbiome, and this terminology has been widely adopted in the relevant literature. Holobiont heredity differs radically from Mendelian principles and has been described as far more similar to schemes proposed by Lamarck [65] and Darwin ("gemmules") [66]. Frequently microbiome components are transmitted horizontally to the oocyte (pre- or postfertilization) or developing embryo by maternal tissues [67,68], but horizontal transmission also occurs paternally [69,70]. It is safe to say that our knowledge of transgenerational microbiome maintenance is very partial and requires a great deal of further research [71].

Because of their composite natures, holobionts can rapidly evolve complex adaptive phenotypes by acquisition or loss of microbiome constituents, outcomes not achievable simply by changes to the host genome. In *Drosophila*, mosquitoes and other invertebrates acquisition of bacterial endosymbionts from the *Wolbachia* group affects various important characteristics, such as resistance to viruses and parasites, and also frequently generates mating incompatibility between colonized and *Wolbachia*-free hosts [50,72,73]. Since mating incompatibility between two populations is often the first step of divergence into separate species, *Wolbachia* entry into the microbiome has been characterized as stimulating "speciation by symbiosis" [74].

Interspecific hybridization

Cell fusions are an essential feature of sexual reproduction. Contrary to the idealized assumption of complete reproductive isolation between different species, there is abundant genomic evidence of mating between related but distinct microbial, plant, and animal species, including real-time observations (online link 12). Interspecific matings are ecologically sensitive because their frequency will increase when mating population sizes decline and conspecific mates become harder to find.

The consequences of interspecific mating are high levels of genome instability (such as chromosome rearrangements, activation of mobile DNA elements, whole-genome duplications—online link 13) and the formation of novel species with phenotypic traits that are more than simple mixtures of characters from the two parents. This kind of hybrid speciation was long ago characterized as "Cataclysmic Evolution" by the distinguished evolutionary biologist G. Ledyard Stebbins [75]. Typically the karyotype of the hybrid species contains a diploid number equal to the sum of the chromosomes in the two parental species. The increase in chromosome number results from the whole-genome duplications necessary for the initial hybrid to undergo successful meiosis.

Although hybrid speciation was long known to occur in plants and serve as the source of useful agricultural crop species, its importance in animals was not appreciated before genomics provided evidence for many hybrid species. Of particular interest are Darwin's finches in the Galapagos Islands, an important evolutionary model cited by Darwin [76], and freshwater cichlid fishes that have become models for rapid speciation and phenotypic diversification [77–79]. In the case of the Galapagos finches, it is worthwhile noting that interspecific hybridization has also been followed in real time by Rosemary and Peter Grant and colleagues, who have documented abrupt changes in beak morphology in hybrid birds rather than the gradual changes postulated by Darwin [80–82].

Protein evolution

Ever since the articulation of the "one gene-one protein" hypothesis in the middle of the 20th century [83], the formation of new protein sequences to execute novel functions has been seen as central to adaptive evolutionary change. With the identification of DNA as the genetic material and the elucidation of the coding relationships between genomic DNA and the sequence of amino acids in each protein [84,85], protein evolution was widely assumed to occur largely by a gradual succession of single amino acid substitutions due to random mutations in the underlying DNA code. However, DNA sequencing and genomic comparisons led to a number of unexpected insights which indicated that protein evolution involves far more active cellular DNA manipulation than initially believed.

Proteins as systems

The first pair of insights concerned the organization of protein molecules and of the DNA that encodes them. Most proteins consist of structurally and functionally independent "domains" joined together, often connected by short linker peptides [86,87]. Different proteins are generally similar to each other in one or more domains but not in others. Because each domain has specific functional characteristics, the overall activity of a given protein is determined by the integration of its various domains into a functional system. This means that proteins can evolve functionally by amplifying, acquiring and rearranging their domain contents to generate novel combinations [88–94].

Mechanistically amplifying and rearranging domains occurs by joining together distinct DNA coding regions not by mutational changes altering particular amino acids. DNA joining involves many distinct biochemical processes and proteins that have to be coordinated and synthesized at the same time. Frequently domain rearrangements involve mobile DNA NGE functions (online link 14).

While many domains are shared across broad phylogenetic distances, patterns of multidomain architectures are specific to each taxon [95,96]. Domain organization indicates that the primary object of evolutionary change is often the domain rather than the whole protein [96–100]. There are protein domain databases [101,102], and it is common practice in contemporary comparative genomics to describe a coding region and its cognate protein product by its domain content. One major question in protein evolution involves the sources of new domain architectures [103]. Domain loss and the appearance of novel kinds of domains are genomic signatures at the emergence of major new taxonomic groups [104].

Coding sequences in pieces

When DNA sequencing was applied to mammalian regions encoding well-known proteins, a surprising result emerged. The sequences for a single polypeptide chain were not continuous but consisted of a series of expressed DNA elements ("exons") encoding segments of the chain separated by intervening DNA elements ("introns") [105,106]. The intron–exon coding pattern is widespread among eukaryotes. In the process of protein synthesis, introns are "spliced" from the primary transcript to form the continuously coding mRNA that is translated on the ribosomes [107]. Although some introns are self-splicing ribozymes [108], splicing generally takes place on a complex "spliceosome" organelle [109].

There are multiple ways that the exon–intron–exon protein coding structure contributes to protein diversity and evolution. In coding region transcripts with multiple exons and introns, not all splicing events necessarily produce only a single combination of joined exons. "Alternative splicing" that creates different combinations of exons from a single pre-mRNA transcript enhances an organism's protein repertoire [110]. Regulation of alternative splicing means that different conditions can control the expression of distinct proteins from a single coding region [111–113]. In many organisms, there is even "trans-splicing," where exons from two different pre-mRNA transcripts can be joined together to produce hybrid proteins encoded by two different coding regions [114,115]. Since the sequences of exon–intron boundaries are important determinants of where and when splicing events occur, one way that novel protein architectures evolve is by changes in splicing patterns rather than by alteration in amino acid coding sequences [116,117].

By and large, there is a good (but not absolute) correspondence between exons and protein domains [118]. This means that protein evolution by domain rearrangement often involves mobilizing the corresponding exons, sometimes with associated introns, into new genomic sites. Since it is easier to mobilize a domain coding sequence that is isolated as one exon (or several exons in series), split coding regions facilitate functional protein evolution [119]. It has been documented that the requisite DNA restructuring events often involve the DNA rearrangement activities associated with mobile DNA elements (online link 14).

Another major question in protein evolution concerns the origins of new domains. It turns out that novel protein coding sequences have a variety of sources in both "coding" and "non-coding" DNA sequences (online link 15). The ability of supposedly "non-coding" genetic elements to contribute to protein coding came as a surprise to many evolutionary theorists, who called such elements "selfish or junk DNA" [120–122]. In 2001 it was first established that repetitive mobile DNA elements contribute directly to many protein coding sequences [123]. Since then, repetitive mobile DNA elements have proven to be a rich source for the origination of exons encoding novel domains, often by acquiring novel splicing signals so that previously intronic segments form new exons (online link 16) [124].

Combined with the above-cited capacity of mobile DNA elements to help mobilize exon rearrangements, it appears from their role as substrates for novel domain coding sequences that the mobile DNA component of various genomes serve as major facilitators of protein evolution. We will see below that mobile DNA elements play a parallel role in the evolution of regulatory and transcription networks for complex adaptations in advanced plant and animal species. These discoveries show that once poorly understood so-called "junk DNA" elements can play important roles in evolution. Recognizing the validity of that statement should serve as an object lesson about how dangerous it is to misinterpret unexpected observations (in this case, the abundance of repetitive DNA in genomes of advanced organisms) and base broad generalizations upon our ignorance rather than our understanding.

Horizontal DNA transfers

An important and unexpected aspect of rapid evolutionary adaptation first became evident when antibiotics were widely used to combat bacterial infections following World War II. Rather than acquire resistance by mutation, the mechanism well-documented by laboratory experiments, the vast majority of resistant bacteria isolated in clinical settings were found to contain genetic elements encoding high levels of resistance that were able to transmit that resistance to other, phylogenetically distant strains of bacteria, thereby helping to explain the rapid spread of antibiotic resistance (online link 17). These "resistance transfer factors" (R-factors) encoded various resistance mechanisms that included chemical inactivation of specific antibiotics, antibiotic removal from the host bacterium, and modification of the cellular targets of antibiotic action. Many R-factors combined coding information for multiple activities conferring resistance to several antibiotics at once [125].

Interbacterial transmission became known as a form of "infective heredity" [126–128] and provided a virtually instantaneous way of acquiring new adaptive traits. No extended process of developing a new character was necessary. Since the new genetic information came from another cell, infective heredity constituted a "horizontal transfer" of genetic information, quite distinct from the normal vertical transmission from ancestral cells. Over time, it became evident that many different kinds of adaptive traits in bacteria (and later in archaea) are subject to horizontal transfer, such as metabolic pathways, surface attachment structures, virulence factors, ability to establish symbiotic relationships, and synthesis of lethal compounds attacking unrelated bacteria [129–134]. Horizontal transfer became so universal a feature of bacterial genetics that some scientists proposed the concept of a shared pan-genome which individual strains of bacteria sampled freely according to the demands of the ecological niches they inhabited [135].

With the advent of widespread DNA sequence analysis, the protein-coding complement of many eukaryotic genomes was established. Comparisons of these coding sequences helped define the phylogenetic relationships of various species by their protein repertoires. While these relationships were largely consistent with shared ancestries and protein diversification across the generations (including the emergence of novel proteins and domains), there were also instances of protein-coding sequences appearing in lineages that were absent from the genomes of ancestral species but highly similar to those of unrelated organisms, often from a completely different domain of life. These "misplaced" coding sequences must have been acquired by horizontal DNA transfer either directly or indirectly from their original source. Multiple cellular mechanisms exist for horizontal DNA transfer (online link 18), but the genomic data provide no indication of how any particular transfer actually occurred.

Horizontal DNA transfers have been documented across virtually all taxonomic boundaries (online link 19). They involve many important adaptive traits. For example, both Bdelloid rotifers (a class of microscopic animal) and herbivorous nematode worms have acquired coding sequences from bacteria and fungi on multiple occasions that allow them to synthesize enzymes to break down otherwise indigestible plant polymers [136–139]. Since different rotifers have acquired hundreds of distinct bacterial and fungal sequences, it is clear that horizontal DNA acquisition by these tiny animals has comprised ongoing molecular incorporations [138]. The ability to short-circuit the process of protein adaptation and immediately extend the organism's range of food resources illustrates the kind of evolutionary advantage conferred by cellular capacities for DNA uptake and genome integration.

Horizontal transfers can involve DNA segments encoding whole proteins or only encoding one or more individual domains. In either case, the transfers involve the introduction of new domains into a distant lineage and set up conditions for the evolution of novel domain architectures in multiple proteins [140,141]. In this way, horizontal transfer can initiate the evolution of new protein families and the specialized adaptive traits they support. Examples include the proliferation of plant cell wall destabilizing proteins following transfer of expansin domains from plants to bacteria and fungi [142], interacting domains and regulatory networks in oomycetes (filamentous fungus-like water molds) [143], and the "effector" proteins with eukaryotic domains that *Legionella* bacteria inject into target cells to commandeer host functions during infection [144–147].

Major adaptive changes have been found to involve horizontal DNA transfers. One such change is the use of bacterial sequences to foster the emergence from thermophilic ancestors of new Archaeal lineages capable of growth under mesophilic conditions [148,149]. Another recently documented case is the ability of autotrophic and osmotrophic eukaryotic lineages to assimilate environmental nitrate as their sole source of metabolic nitrogen [150].

Although the precise mechanism of horizontal DNA transfer is indeterminate for any particular case, the kinds of interactions that occur throughout the biosphere provide us with multiple potential paths for this kind of genomic exchange [16]. One of the most important, the ubiquity of genomic information protected inside virus capsids, is our next topic.

The virosphere as an evolutionary R&D sector (online link 20)

Viruses are the most abundant biological entities on planet Earth [151]. They are found at astonishing concentrations in the soil, in bodies of water ($\sim 10^{10}$/L), and in the atmosphere (falling at $\sim 10^{9}/m^{2}$/day) [152,153]. In addition to viruses, there are a variety of virus-like particles (VLPs) that contain nucleic acids but not viral genomes [154]. Among the VLPs are dedicated DNA transport particles called "gene transfer agents" (GTAs) [155]. While these viral and virus-like agents inject DNA or RNA into cells that have surface receptors, which limits their range of target cells, some have been documented to participate in cross-species DNA transfers. VLP donors capable of transferring genetic information to mesophilic *Escherichia coli* and *Bacillus subtilis* bacteria have been reported to include microbial mats of hyperthermophilic bacteria from hot springs and marine bacteria from the oceans [156,157].

Viruses, and especially the bacteriophages which infect bacteria, are the most numerous reservoirs of protein coding sequences on planet

Earth. These coding sequences can be divided into two groups with quite distinct evolutionary potentials:

1. The first group includes sequences encoding proteins similar to those found in cells that participate in established metabolic routines, ranging from the different steps of alternative photosynthesis systems to various forms of intracellular energy metabolism, phosphate recycling, apoptosis (programmed cell death), cell surface structures, virulence, and infectivity [154,158]. These reservoirs of established cell physiology functions can be utilized to repair damaged cells or to extend the metabolic capabilities of an organism adapting to novel ecological circumstances [159–162].
2. The second group consists of uniquely viral protein coding information that cells do not possess and which, therefore, have the potential to initiate the evolution of entirely new types of cellular proteins [163]. Over 90% of all unique coding sequences in the genomic databases occur in viral genomes [164]. Since they are unique, they do not exist in cells. When one of these unique sequences is exapted by a cell (often for a function related to its source, like antivirus defense), a novel protein sequence and structure appears in the genome. This new motif is then available to combine with other protein domains or mutate to a new functionality and thus generate a capability which did not previously exist, either in the virosphere or cellular realm. This creative capability is part of what makes the virosphere an evolutionary R and D domain of life. In this vein, retroviruses have been cited as source of new genes in vertebrates [165].

Virus particle capsids naturally break down over time and liberate their nucleic acid cargos into the environment. The virosphere is thus a major contributor to the free DNA molecules present in all ecologies. Many types of cells have been found to be capable of taking up DNA from their surroundings and incorporating it into their genomes (online link 6): Archaea, "competent" bacteria with dedicated DNA import complexes, yeast, red algae, and invertebrate and vertebrate sperm. Under experimental conditions, DNA "transformation" has further been extended to cultured animal cells, suggesting that similar processes may occur under natural conditions. The role of viruses as sources of environmental DNA for uptake has recently been demonstrated in experiments with so-called "super-spreader" bacteriophages that transfer plasmid DNA from *E. coli* bacteria to unrelated phage-resistant *Streptococcus pneumonia* bacteria competent for DNA import [166]. From a public health point of view, bacteriophage particles have also been identified as environmental reservoirs for antibiotic resistance elements [167].

In addition to spreading cellular DNA in the environment, viruses also serve directly as vectors for intercellular DNA transfer [168]. When a viral particle incorporates and transfers a fragment of host genomic DNA from an infected virus-producing cell to a virus-sensitive cell of the same taxonomic group, the process is called "transduction" [169]. Typically the transduced fragment is free of viral DNA, does not replicate, and comes from many regions of the host cell genome. For any particular cellular locus this transfer process occurs at low frequency and is called "generalized transduction." But there are other cases where a fragment of cellular DNA becomes linked to virus DNA in such a way that it can replicate together with the viral genome and incorporate into large numbers of virus particles. When those particles infect virus-sensitive taxonomically related cells, they carry multiple copies of a particular donor DNA sequence at high frequency to the recipients, and the transfer process is designated "specialized transduction." Examples of such specialized transduction include bacteriophages carrying chromosomal loci [170,171] and animal tumor viruses carrying cellular oncogenes [172–176].

It is significant that Rous Sarcoma Virus (RSV) and many other animal tumor viruses are RNA-containing retroviruses that must reverse-transcribe their genomes into DNA prior to incorporation into the infected cell genome. This means that host loci transduced retrovirally are in RNA and consequently in spliced intron-free configuration. During the retroviral replication cycle, read-through transcription into adjacent chromosomal regions is one mechanism for incorporation of host cell sequences into the retrovirus genome [177]. This mechanism also applies to retrovirus-like elements that have lost the ability to produce infectious particles but can still proliferate through the genome as endogenous retrovirus (ERV) elements. One plant example is the *BS1* retrovirus-like element in maize which retrotransduces a spliced coding sequence (*pma*) for a proton-transporting ATPase [178,179]. This retrotransposition process has been documented in a wide range of eukaryotes from yeast to primates, with significant evolutionary impact on genome content [180–190].

Viral transduction of cellular DNA also occurs in the totally different recently recognized family of nucleocytoplasmic large DNA viruses (NCLDVs) (online link 21) [191]. The NCLDVs are characterized by their large DNA genomes, which range in size from about 100 kb to over

2.5 MB [192,193]. They include the *Poxviridae* (*variola* smallpox virus, *vaccinia* virus, cowpox virus, etc.) and can infect a wide range of eukaryotes, including amoebae, green algae, invertebrates, and vertebrates. An outstanding characteristic of NCLDV genomes is expansion by the acquisition (horizontal transfer) of DNA sequences from eukaryotic host cells or from bacterial endosymbionts of the host; one author has dubbed this acquisition potential a "genomic accordion" [194,195]. Thus the larger NCLDV genomes acquire a mixture of eukaryotic and bacterial DNA sequences that they carry as they infect new host cells.

DNA acquired by NCLDVs is often rich in sequences encoding RNAs and proteins involved in translation of mRNA into protein, which may be adaptive, but they also include sequences encoding functions of questionable utility for viral reproduction, such as bacterial restriction-modification systems and mobile DNA elements. The viruses reproducing in amoeba and other protists have a higher content of bacterial DNA, which reflects the ability of many bacteria to proliferate in these eukaryotic hosts. The multiple transfers between eukaryotes, viruses and bacteria have earned the infected amoebae and similar cells the appellation of "evolutionary melting pots" [196–198]. This name is even more appropriate when we consider that many of the same bacteria growing in amoebae are also able to infect the cells of plants and animals, where they typically act as pathogens.

Although the standard view of viruses is that they are basically destructive parasites lethal to their host cells, many viruses regularly integrate their genomes into the host cell genome and reproduce as part of the augmented biological combination. As we shall see, such virus integrations have profound evolutionary implications for both microbes and complex multicellular organisms. DNA bacteriophage genomes integrate into the bacterial genome to generate "prophages" at one or a few defined sites using site-specific recombinases [199,200] or at many sites throughout the host genome using transposase functions [201,202]. Bacterial strains carrying prophages are termed "lysogenic" because various treatments can induce the prophages to undergo viral reproduction and produce free phages that can go on to lyse sensitive bacteria [203].

When a prophage includes sequences encoding one or more proteins that affect host cell adaptation, the prophage-bearing strain is said to have undergone "lysogenic conversion" (online link 21). Lysogenic conversion was first noted in bacterial pathogens where the prophage encoded a disease-specific toxin (*Corynebacterium diphtheria*, *Clostridium botulinum*, *Vibrio cholerae*, and *Shigella dysenteriae*), but other bacterial pathogens proved to contain multiple prophages encoding a variety of virulence factors each contributing partially to overall pathogenicity [204,205].

Lysogenic conversion also affects bacterial functions not necessarily related to pathogenesis, such as biofilm formation and sporulation [206,207]. The association of converting bacteriophages and their host bacteria most likely did not first arise in the context of human disease but rather in other natural environments, where toxin production may protect the bacteria from grazing by lower eukaryote predators, which are sensitive to the same toxins as human cells [154,208,209]. In any case, it is clear that lysogenic conversion provides a kind of "quantum leap" in adaptive evolution that is similar to symbiont acquisition by holobionts.

The holobiont analogy may also apply to eukaryotic viruses [210]. Single- and double-stranded DNA and RNA eukaryotic virus genomes have also been found to integrate into host cell chromosomes [211–218]. In many cases, the integration mechanisms are not understood, but acquisition of viral sequences is often protective against superinfection by the same or related viruses (online link 20). One case where the integration details are known in considerable detail comprises retroviruses, including RSV and other tumor viruses discussed above. In addition to their oncogenic potential, retroviral insertions can have profound implications for mammalian adaptive evolution by formatting complex genomic networks.

When a reverse-transcribed and double-stranded cDNA copy of a retroviral genome integrates into an infected cell chromosome, the resulting provirus is called an endogenous retrovirus (ERV) and becomes a template both for virus production and also for retrotransposition of further ERV copies into new genomic locations [219,220]. Loss of sequences needed for infection but not for retrotransposition converts the mutant ERV into a retrotransposon, which cannot produce infectious virus but can still mobilize ERV sequences in the genome. Because of this DNA mobilization capacity, infection of a host germline by a retrovirus is typically followed by a burst of new ERV or derivative retrotransposon copies accumulating around the genome.

The distributed ERV copies provide a network of related sequence elements which can coordinate expression of unlinked genetic loci for complex adaptive functionalities. ERV-derived transcription networks have been well-documented in mammals for the following traits:

- pluripotency and stem cell proliferation in humans [221–224];
- interferon-inducible innate immunity in humans [225]; and
- placental development in mice and humans [226–229].

ERVs provide key transcription signals (promoters and enhancers) for these critical mammalian adaptations. In addition, each mammalian order exapts one or two ERV envelope proteins (Env), which fuse viral and target cell membranes during retrovirus infection, to serve another adaptive function as a cell fusion protein ("Syncytin") during development of the placental multinuclear syncytiotrophoblast structure [230].

What is remarkable about these different ERV holobiont contributions to mammalian adaptation is that they tend to involve recently acquired, species-specific retroviruses, rather than ancient retroviruses shared by many mammals. For example, over 300 transcriptional signals used for placental development in mice come from a superfamily of mouse-specific ERVs, most of which are not even found in rats [227]. Similarly each mammalian order has its own distinct set of Syncytin proteins [230,231]. This is not the pattern we would expect to find if these ERV-derived functions arose early in mammalian evolution and were retained as different orders of mammals radiated over time. It is almost as though networks for these particular traits evolved *de novo* in each species as its genome became populated with new ERV elements. As more species genomes are analyzed functionally, it will be important to determine whether they also show recently acquired ERVs providing transcriptional formatting for shared mammalian traits.

Extracellular vesicles—stress responses and crossing the Weismann Barrier in animals

The more we learn about cells and their genomes, the more we discover about intricate communication systems that transmit useful information affecting genome expression and cellular function. Since their discovery in cultures of maturing red blood cells attracted broad scientific attention [232], membrane-enclosed extracellular vesicles (EVs) have revealed a whole new realm of intercellular communication vectors for transport of DNA, RNA, and protein cargoes [233]. Online databases catalogue the compositions of these extracellular bodies: Vesiclepedia (www.microvesicles.org/), EVpedia (www.evpedia.info), and ExoCarta (www.exocarta.org). EVs have been found to transmit genomic DNA [234–238], functional mRNAs, all manner of noncoding regulatory RNAs, and an apparently limitless repertoire of proteins.

EVs appear to be ubiquitous in the biosphere (online link 7). They are produced by gram-negative and gram-positive bacteria [239], archaea

[240–242], many different eukaryotic microbes [243–246], plants [247], and animals (online link 21). In bacteria, where cause and effect are most easily demonstrated, EV production occurs nonrandomly and is regulated by stress [248], by environmental conditions [249], by a sensor kinase protein [250], by multicomponent regulatory systems [251,252], and by quorum signaling between cells [253,254]. Moreover, vesicles clearly operate functionally. Here is a list of a small fraction of the biological phenomena documented to involve EVs:

- bacterial carcinogenesis [255–257],
- bacterial pathogenesis of animals [258–260],
- bacterial pathogenesis of plants [261,262],
- bacterial signaling [263–266],
- biofilm formation [267–269],
- cancer proliferation and metastasis [270–272],
- defense against bacteriophage infection [273],
- fungal infections of humans [274],
- intercellular and interorgan metabolic signaling in normal growth and pregnancy [275],
- *M. xanthus* predation [276],
- plant defenses against fungal infection [277,278],
- signaling radiation damage in mouse cells [279],
- transcription regulation in cancer cells [280], and
- bioluminescent *Vibrio fischerii* symbionts triggering light organ biogenesis in their squid host [281,282].

Although they form by different mechanisms and have distinct molecular compositions in the various organisms, the basic physics of membrane-enclosed EVs fusing with target cell membranes imparts a potential for interaction across any cellular or taxonomic barrier. Vesicles transfer their macromolecular cargoes between cells of the same species, between cells of different species, across major taxonomic divides, and through the circulatory systems of multicellular organisms [283]. What the existence of EV transport means is that there are no insurmountable physical barriers to macromolecular communication between cells.

From an evolutionary perspective, perhaps the most significant consequence of EV macromolecular transport is sperm-mediated communication between somatic and germline cells in animals, that is, crossing the so-called Weismann barrier [18]. As part of normal development, sperm acquire a sequence of small RNA cargoes as they mature in the epididymis from somatic vesicles called "epididymosomes" [284]. Mature sperm transmit the last of these cargoes to the fertilized oocyte and thus provide epigenetic information that is essential for successful embryonic development [285].

In an unprogrammed manner, sperm cells also acquire somatic nucleic acids from circulating EVs from different tissues and transmit them to the developing embryo [26,286,287]. These somatic DNAs and RNAs can also be transmitted to the fertilized oocyte and inherited in the developing embryo and its later progeny by a form of episomal replication and non-Mendelian inheritance. RNA inheritance depends on sperm and embryonic reverse transcriptase activity [288]. Both sperm-carried somatic DNA and RNA molecules are expressed in progeny tissues and have been documented to affect the properties of the progeny animals according to environmental and other effects on their fathers [289–292].

In other words, by virtue of EVs and sperm cells, somatically acquired epigenetic information is transmissible through the germline to animal progeny. The results cited above show that the Weismann Barrier is not inviolable and that there can be a type of Lamarckian [65]/Darwinian [66] inheritance of acquired characteristics in animals. Clearly there is no germline/soma separation for microorganisms or plants, where flowers carrying the germline cells develop from somatic tissue. Thus we now know of cellular and subcellular (vesicular) mechanisms for transmitting somatically acquired traits in the course of evolutionary variation in all organisms. It remains to be determined empirically how important these modes of non-Mendelian inheritance are in shaping the outcomes of major evolutionary transitions.

Evolutionary changes in genome composition

Central to all contemporary thinking about evolutionary processes is our conception of the molecular and cellular basis for the transmission of hereditary traits. As we have just seen in the discussion of sperm-mediated transfer of epigenetic information to the next generation and its progeny, there is no unitary mechanism. It remains possible that, as with EVs, we may yet discover new forms of hereditary transmission. Moreover, Darwin formulated his ideas about evolution in the absence of any clear information about the mechanism of hereditary transmission [76].

Nonetheless, while recognizing the incompleteness of our knowledge, it has been true ever since the rediscovery of Mendelism at the start of the

20th century [293], that virtually all evolutionary thinking has focused on the genome. Now that we can analyze genomes by their complete DNA sequences, it is instructive to review how our knowledge and thinking have developed over the past 120 years.

In the early decades of the 20th century, the genome was thought to consist of a collection of factors that determined inherited characteristics and followed Mendel's Laws. In 1905 Johannsen coined the term "gene" to indicate any such factor independently of its structure or function [294,295]. Quickly geneticists discovered that genes did not always behave independently in inheritance and could be grouped into linkage groups, which were quickly found to represent the chromosomes visible in the eukaryotic cell nucleus [296–298]. (Bacteria did not fully acquire the right to claim genetics and genomes until after WWII, in large part because they lacked visible chromosomes.) Gradually in these early decades of the 20th century, Johannsen's intentionally neutral term took on a life of its own, and the hereditary apparatus came to be thought of as a "genome" (="gene body") composed of individual "gene" entities collected together in the chromosomes. Naturally two related questions arose: (1) What does a gene do to determine inherited traits? (2) What is a gene made of?

The middle of the 20th century provided intellectually satisfying answers to both questions. Analysis of biochemical traits in the fungus *Neurospora* indicated that different genes determined the function of specific enzymatic reactions [83,299], leading to the "one gene-one enzyme" hypothesis, which was quickly generalized to the "one gene-one protein" or "one gene-one polypeptide" hypothesis. Thus genes came to be considered the hereditary units that encoded the biochemically active proteins in all living cells. In less than a dozen years, the chemical nature of the hereditary material was identified as DNA [300,301], and the base-paired structure of DNA was determined to be ideal for carrying and replicating linear code determining the order of amino acids in protein chains [84,302,303]. The problems of gene function and hereditary transmission were considered solved. The double helix revealed the "Secret of Life."

Nonetheless, two lines of genetics research indicated that genomes were not as simple as assumed in 1953. One was the branch called cytogenetics, which centered on visualizing and studying chromosomes under the microscope. Cytogenetic analysis revealed that eukaryotic chromosomes are not uniform structures. They contain clearly visible physically and functionally distinct regions, such as centromeres and telomeres essential for chromosome duplication and transmission at cell division [304], plus other structures whose functions would not become evident until the era of molecular biology, such as the nucleolus organizing region (NOR) [305] that consisted of tandemly repeating DNA templates for ribosomal RNA as well as other still poorly characterized sequences. (Highly repetitive DNA regions are particularly difficult to sequence accurately.)

A second line of research that indicated greater genome complexity beyond protein coding sequences came out of studying how cells regulate specific protein synthesis [306]. These studies led to the recognition of DNA sequences that help determine what conditions lead to transcription of specific DNA regions into mRNA that can then be translated into proteins on the ribosomes. Initially these studies involved bacterial systems, where the noncoding regulatory sequences are relatively simple and mostly concern transcription, but extension to eukaryotic cells revealed many complexities in genome expression, such as widely spaced transcriptional control signals [307–309], chromatin formatting [310–312], and sequences that guide splicing of introns out of pre-mRNA primary transcripts to form mRNAs carrying continuous coding regions [107,313]. In other words, studies of genome maintenance and genome expression revealed a compositional and functional complexity unsuspected in the mid-20th century. Detailed molecular analysis of genome function has shown every genome to be an elaborately formatted database that contains a great deal of information in addition to protein-coding sequences.

A third line of research which indicated that genomes are more than just collections of genes for different proteins came from analysis of the physical properties of genomic DNA from humans and other complex eukaryotes. By analyzing how rapidly denatured single-stranded DNA could recover base-pairing to a double-stranded state, Britten and Kohne found that genomic DNA from higher organisms contained a very large fraction of repetitive DNA sequences, completely different from the largely unique sequences predicted by gene theories [314]. The presence of this repetitive DNA fraction took geneticists by surprise. They could not understand what functions DNA repeats could fulfill, and distinguished scholars labeled them "junk DNA" [120] or "selfish DNA" [122]. A radical philosophy of strict neo-Darwinism was even erected on the basis of "the selfish gene" hypothesis [121].

Eventually a more sophisticated understanding of genome function and evolution began to form in this century around the notion that genomes contain much more than genes. McClintock had discovered in the late 1940s that maize chromosomes contained "controlling elements" capable of moving through the genome and altering developmental patterns [315]. Britten and Davidson pointed out that repetitive DNA elements can serve to form regulatory networks analogous to computer circuits and documented some of those circuits in invertebrate genomes [316–318]. The need for repetitive elements to provide signals formatting the genome for its diverse functional roles became increasingly evident [319,320]. Moreover, the role of repetitive mobile DNA elements (e.g., McClintock's controlling elements) in generating evolutionary novelties gained broad recognition (online link 22).

One sign of the growing appreciation for the so-called "noncoding" portion of the genome was a graph showing the correlation of organismal complexity (measured by different cell types) with genome content [321]. While protein-coding DNA content peaked at about 35,000 proteins and did not get larger with genome size, noncoding DNA content continued to increase at greater organismal complexity without diminution. In other words, the data seemed to indicate that noncoding genome sequences became more important as organisms became more complex. We do not fully understand why this correlation seems to hold, but part of the answer may lie in our growing appreciation of the complex regulatory roles played by so-called "noncoding" RNA transcripts [322,323], both smaller regulatory molecules and "long noncoding" lncRNAs (online link 23) [324].

Evolutionary thinking in the years to come

As Eibi has taught us all, evolution is a complex ongoing process intimately connected with ecological dynamics. The genomics-based phenomena reviewed here complement his insights and lead us to three major conclusions:

- Cell, organism, and genome evolution are active biological processes that provide important adaptive benefits, especially triggered at times of ecological change. In other words, RW DNA databases are biologically superior to ROM databases [325].
- Evolutionary change typically results from biosphere interactions that involve communication between two or more genomes in cells,

viruses or vesicles. As Evolution Canyons all over the planet teach us, no genome is an island [16].

- Ecological change stimulates evolutionary innovation by fostering creative biosphere interactions and triggering natural genetic engineering functions. In other words, as Eibi has been documenting for decades, outside events trigger active internal change [9,326].

These three conclusions recognize the inherent complexity of evolutionary processes occurring at the ecology/biology interface. As we enter the third decade of the 21st century, genomics has documented many important evolutionary phenomena outside the reductionist thinking that characterized the Modern Synthesis and other oversimplifications of the last century. For example, conventional theory envisaged hereditary changes occurring within the isolated genome of each species [17], but we now know that virtually all genomes interact with external genomes in the course of evolutionary change [16].

It is time to acknowledge that our current genomics-based knowledge of evolutionary change is too diverse to formulate a single comprehensive theory encompassing roles for all the ecological and biological interactions documented to be involved in stimulating and executing evolutionary changes. Consequently we need to focus on experimental evolution to uncover the evolutionary principles we do not yet have. We need to abandon artificial simplified culture conditions [327] and adapt the path forward that Eibi has indicated. In other words, we need to devise 21st century versions of synthetic Evolution Canyons, controlled but complex microcosms where realistic simulations of ecological and biological interactions can trigger meaningful adaptive evolutionary changes. That is how we will learn new principles involving the dynamic connections of ecology to evolutionary innovation.

List of abbreviations used

ERV endogenous retrovirus
EV extracellular vesicle
GTA gene transfer agent
NCLDV nucleocytoplasmic large DNA virus
NGE natural genetic engineering
ROM read-only memory
RSV Rous Sarcoma Virus
RW read-write
VLP virus-like particle

References

[1] E. Nevo, "Evolution Canyon," a potential microscale monitor of global warming across life, Proc. Natl. Acad. Sci. U. S. A. 109 (8) (2012) 2960–2965. Available from: https://doi.org/10.1073/pnas.1120633109.

[2] S. Raz, et al., Scorpion biodiversity and interslope divergence at "evolution canyon", lower Nahal Oren microsite, Mt. Carmel, Israel, PLoS One 4 (4) (2009) e5214. Available from: https://doi.org/10.1371/journal.pone.0005214.

[3] E. Nevo, Ecological genomics of natural plant populations: the Israeli perspective, Methods Mol. Biol. 513 (2009) 321–344. Available from: https://doi.org/10.1007/978-1-59745-427-8_17.

[4] T.K. Ezov, et al., Molecular-genetic biodiversity in a natural population of the yeast *Saccharomyces cerevisiae* from "Evolution Canyon": microsatellite polymorphism, ploidy and controversial sexual status, Genetics 174 (3) (2006) 1455–1468. Available from: https://doi.org/10.1534/genetics.106.062745.

[5] S. Raz, J.H. Graham, A. Cohen, B.L. de Bivort, I. Grishkan, E. Nevo, Growth and asymmetry of soil microfungal colonies from "Evolution Canyon," Lower Nahal Oren, Mount Carmel, Israel, PLoS One 7 (4) (2012) e34689. Available from: https://doi.org/10.1371/journal.pone.0034689.

[6] E. Nevo, Y.B. Fu, T. Pavlicek, S. Khalifa, M. Tavasi, A. Beiles, Evolution of wild cereals during 28 years of global warming in Israel, Proc. Natl. Acad. Sci. U. S. A. 109 (9) (2012) 3412–3415. Available from: https://doi.org/10.1073/pnas.1121411109.

[7] E. Ivanitskaya, L. Rashkovetsky, E. Nevo, Chromosomes in a hybrid zone of Israeli mole rats (Spalax, Rodentia), Genetika 46 (10) (2010) 1301–1304.

[8] B.C. Lamb, S. Mandaokar, B. Bahsoun, I. Grishkan, E. Nevo, Differences in spontaneous mutation frequencies as a function of environmental stress in soil fungi at "Evolution Canyon," Israel, Proc. Natl. Acad. Sci. U. S. A. 105 (15) (2008) 5792–5796. Available from: https://doi.org/10.1073/pnas.0801995105.

[9] E. Nevo, Evolution under environmental stress at macro- and microscales, Genome Biol. Evol. 3 (2011) 1039–1052. Available from: https://doi.org/10.1093/gbe/evr052.

[10] A. Lupu, E. Nevo, I. Zamorzaeva, A. Korol, Ecological-genetic feedback in DNA repair in wild barley, Hordeum spontaneum, Genetica 127 (1-3) (2006) 121–132. Available from: https://doi.org/10.1007/s10709-005-2611-0.

[11] O. Raskina, J.C. Barber, E. Nevo, A. Belyayev, Repetitive DNA and chromosomal rearrangements: speciation-related events in plant genomes, Cytogenet. Genome Res. 120 (3–4) (2008) 351–357. Available from: https://doi.org/10.1159/000121084.

[12] R. Kalendar, J. Tanskanen, S. Immonen, E. Nevo, A.H. Schulman, Genome evolution of wild barley (*Hordeum spontaneum*) by BARE-1 retrotransposon dynamics in response to sharp microclimatic divergence, Proc. Natl. Acad. Sci. U. S. A. 97 (12) (2000) 6603–6607. Available from: https://doi.org/10.1073/pnas.110587497 [pii].

[13] A. Belyayev, R. Kalendar, L. Brodsky, E. Nevo, A.H. Schulman, O. Raskina, Transposable elements in a marginal plant population: temporal fluctuations provide new insights into genome evolution of wild diploid wheat, Mob. DNA 1 (1) (2010) 6.

[14] A. Beiles, S. Raz, Y. Ben-Abu, E. Nevo, Putative adaptive inter-slope divergence of transposon frequency in fruit flies (*Drosophila melanogaster*) at "Evolution Canyon", Mount Carmel, Israel, Biol. Direct 10 (2015) 58. Available from: https://doi.org/10.1186/s13062-015-0074-5.

[15] J.A. Shapiro, Living organisms author their read-write genomes in evolution, Biology (Basel) 6 (4) (2017). Available from: https://doi.org/10.3390/biology6040042.

[16] J.A. Shapiro, No genome is an island: toward a 21st century agenda for evolution, Ann. N. Y. Acad. Sci. 1447 (1) (2019) 21–52. Available from: https://doi.org/10.1111/nyas.14044.
[17] H.A. Orr, Dobzhansky, Bateson, and the genetics of speciation, Genetics 144 (4) (1996) 1331–1335.
[18] A. Weismann, The Germ-Plasm: A Theory of Heredity, Charles Scribner's Sons, New York, 1893.
[19] A. Weismann, Das Keimplasma: eine Theorie der Vererbung (The Germ Plasm: A Theory of Inheritance), Fischer, Jena, 1892.
[20] S. Brenner, L. Barnett, F.H.C. Crick, A. Orgel, The theory of mutagenesis, J. Mol. Biol. 3 (1961) 121–124.
[21] S. Brenner, Turing centenary: life's code script, Nature 482 (7386) (2012) 461. Available from: https://doi.org/10.1038/482461a.
[22] N.L. Craig, M. Chandler, M. Gellert, A.L. Lambowitz, P.A. Rice, S.B. Sandmeyer (Eds.), Mobile DNA III, American Society for Microbiology, Washington, DC, 2015.
[23] J.A. Shapiro, Evolution: A View from the 21st Century, FT Press Science, Upper Saddle River, NJ, 2011, p. 272.
[24] J.E. Haber, Genome Stability: DNA Repair and Recombination, first ed., Garland Scientific, New york, 2013.
[25] N. Gontier (Ed.), Reticulate Evolution: Symbiogenesis, Lateral Gene Transfer, Hybridization and Infectious Heredity (Interdisciplinary Evolution Research), Springer, Switzerland, 2015.
[26] C. Spadafora, The "evolutionary field" hypothesis. Non-Mendelian transgenerational inheritance mediates diversification and evolution,", Prog. Biophys. Mol. Biol. 134 (2018) 27–37. Available from: https://doi.org/10.1016/j.pbiomolbio.2017.12.001.
[27] C.R. Woese, On the evolution of cells, Proc. Natl. Acad. Sci. U. S. A. 99 (13) (2002) 8742–8747.
[28] M.W. Gray, Lynn Margulis and the endosymbiont hypothesis: 50 years later, Mol. Biol. Cell 28 (10) (2017) 1285–1287.
[29] C.R. Woese, G.E. Fox, The concept of cellular evolution, J. Mol. Evol. 10 (1) (1977) 1–6.
[30] C.R. Woese, G.E. Fox, Phylogenetic structure of the prokaryotic domain: the primary kingdoms, Proc. Natl. Acad. Sci. U. S. A. 74 (11) (1977) 5088–5090.
[31] C.R. Woese, Endosymbionts and mitochondrial origins, J. Mol. Evol. 10 (2) (1977) 93–96.
[32] M.W. Gray, Mitochondrial evolution, Cold Spring Harb. Perspect. Biol. 4 (9) (2012) a011403. Available from: https://doi.org/10.1101/cshperspect.a011403.
[33] J. Martijn, J. Vosseberg, L. Guy, P. Offre, T.J.G. Ettema, Deep mitochondrial origin outside the sampled alphaproteobacteria, Nature 557 (7703) (2018) 101–105. Available from: https://doi.org/10.1038/s41586-018-0059-5.
[34] M.P. Ferla, J.C. Thrash, S.J. Giovannoni, W.M. Patrick, New rRNA gene-based phylogenies of the Alphaproteobacteria provide perspective on major groups, mitochondrial ancestry and phylogenetic instability, PLoS One 8 (12) (2013) e83383.
[35] H.C. Betts, M.N. Puttick, J.W. Clark, T.A. Williams, P.C.J. Donoghue, D. Pisani, Integrated genomic and fossil evidence illuminates life's early evolution and eukaryote origin, Nat. Ecol. Evol. 2 (10) (2018) 1556–1562. Available from: https://doi.org/10.1038/s41559-018-0644-x.
[36] E.J. Javaux, The early eukaryotic fossil record, Adv. Exp. Med. Biol. 607 (2007) 1–19.
[37] L.M. Ward, J.L. Kirschvink, W.W. Fischer, Timescales of oxygenation following the evolution of oxygenic photosynthesis, Orig. Life Evol. Biosph. 46 (1) (2016) 51–65. Available from: https://doi.org/10.1007/s11084-015-9460-3.

[38] A. Spang, et al., Asgard archaea are the closest prokaryotic relatives of eukaryotes, PLoS Genet. 14 (3) (2018) e1007080.

[39] L. Eme, A. Spang, J. Lombard, C.W. Stairs, T.J.G. Ettema, Archaea and the origin of eukaryotes, Nat. Rev. Microbiol. 15 (12) (2017) 711–723. Available from: https://doi.org/10.1038/nrmicro.2017.133.

[40] A. Spang, et al., Proposal of the reverse flow model for the origin of the eukaryotic cell based on comparative analyses of Asgard archaeal metabolism, Nat. Microbiol. 4 (7) (2019) 1138–1148. Available from: https://doi.org/10.1038/s41564-019-0406-9.

[41] K. Zaremba-Niedzwiedzka, et al., Asgard archaea illuminate the origin of eukaryotic cellular complexity, Nature 541 (7637) (2017) 353–358. Available from: https://doi.org/10.1038/nature21031.

[42] D.C. Chan, Mitochondria: dynamic organelles in disease, aging, and development, Cell 125 (7) (2006) 1241–1252. Available from: https://doi.org/10.1016/j.cell.2006.06.010.

[43] P.J. Keeling, The number, speed, and impact of plastid endosymbioses in eukaryotic evolution, Annu. Rev. Plant Biol. 64 (2013) 583–607. Available from: https://doi.org/10.1146/annurev-arplant-050312-120144.

[44] P.J. Keeling, J.P. McCutcheon, W.F. Doolittle, Symbiosis becoming permanent: survival of the luckiest, Proc. Natl. Acad. Sci. U. S. A. 112 (33) (2015) 10101–10103. Available from: https://doi.org/10.1073/pnas.1513346112.

[45] J.P. McCutcheon, From microbiology to cell biology: when an intracellular bacterium becomes part of its host cell, Curr. Opin. Cell Biol. 41 (2016) 132–136.

[46] N.A. Moran, G.M. Bennett, The tiniest tiny genomes, Annu. Rev. Microbiol. 68 (2014) 195–215. Available from: https://doi.org/10.1146/annurev-micro-091213-112901.

[47] J.P. McCutcheon, C.D. von Dohlen, An interdependent metabolic patchwork in the nested symbiosis of mealybugs, Curr. Biol. 21 (16) (2011) 1366–1372. Available from: https://doi.org/10.1016/j.cub.2011.06.051.

[48] F. Husnik, J.P. McCutcheon, Repeated replacement of an intrabacterial symbiont in the tripartite nested mealybug symbiosis, Proc. Natl. Acad. Sci. U. S. A. 113 (37) (2016) E5416–E5424.

[49] N. Corradi, Microsporidia: eukaryotic intracellular parasites shaped by gene loss and horizontal gene transfers, Annu. Rev. Microbiol. 69 (2015) 167–183. Available from: https://doi.org/10.1146/annurev-micro-091014-104136.

[50] A.R.I. Lindsey, T. Bhattacharya, I.L.G. Newton, R.W. Hardy, Conflict in the intracellular lives of endosymbionts and viruses: a mechanistic look at Wolbachia-mediated pathogen-blocking, Viruses 10 (4) (2018).

[51] J.P. McCutcheon, N.A. Moran, Extreme genome reduction in symbiotic bacteria, Nat. Rev. Microbiol. 10 (1) (2011) 13–26. Available from: https://doi.org/10.1038/nrmicro2670.

[52] M.A. Campbell, P. Lukasik, C. Simon, J.P. McCutcheon, Idiosyncratic genome degradation in a bacterial endosymbiont of periodical cicadas, Curr. Biol. 27 (22) (2017) 3568–3575. Available from: https://doi.org/10.1016/j.cub.2017.10.008. e3.

[53] B. Koskella, L.J. Hall, C.J.E. Metcalf, The microbiome beyond the horizon of ecological and evolutionary theory, Nat. Ecol. Evol. 1 (11) (2017) 1606–1615. Available from: https://doi.org/10.1038/s41559-017-0340-2.

[54] P. Vandenkoornhuyse, A. Quaiser, M. Duhamel, A. Le Van, A. Dufresne, The importance of the microbiome of the plant holobiont, N. Phytol. 206 (4) (2015) 1196–1206. Available from: https://doi.org/10.1111/nph.13312.

[55] A.C. Wilson, et al., Genomic insight into the amino acid relations of the pea aphid, Acyrthosiphon pisum, with its symbiotic bacterium *Buchnera aphidicola*, Insect Mol. Biol. 19 (2) (2010) 249–258. Available from: https://doi.org/10.1111/j.1365-2583.2009.00942.x.

[56] N. Dubilier, C. Bergin, C. Lott, Symbiotic diversity in marine animals: the art of harnessing chemosynthesis, Nat. Rev. Microbiol. 6 (10) (2008) 725–740.
[57] J.K. Nicholson, et al., Host-gut microbiota metabolic interactions, Science 336 (6086) (2012) 1262–1267. Available from: https://doi.org/10.1126/science.1223813.
[58] S.C. Shin, et al., Drosophila microbiome modulates host developmental and metabolic homeostasis via insulin signaling, Science 334 (6056) (2011) 670–674. Available from: https://doi.org/10.1126/science.1212782.
[59] I. Venu, Z. Durisko, J. Xu, R. Dukas, Social attraction mediated by fruit flies' microbiome, J. Exp. Biol. 217 (Pt. 8) (2014) 1346–1352. Available from: https://doi.org/10.1242/jeb.099648.
[60] C.E. Schretter, et al., A gut microbial factor modulates locomotor behaviour in Drosophila, Nature 563 (2018) 402–406. Available from: https://doi.org/10.1038/s41586-018-0634-9.
[61] K.R. Theis, et al., Getting the hologenome concept right: an eco-evolutionary framework for hosts and their microbiomes, mSystems 1 (2) (2016). Available from: https://doi.org/10.1128/mSystems.00028-16.
[62] L. Margulis, Symbiogenesis and symbionticism, in: L. Margulis, R. Fester (Eds.), Symbiosis as a Source of Evolutionary Innovation: Speciation and Morphogenesis, MIT Press, Cambridge, MA, 1991, pp. 49–92.
[63] S.R. Bordenstein, K.R. Theis, Host biology in light of the microbiome: ten principles of holobionts and hologenomes, PLoS Biol. 13 (8) (2015) e1002226. Available from: https://doi.org/10.1371/journal.pbio.1002226.
[64] E. Rosenberg, I. Zilber-Rosenberg, Microbes drive evolution of animals and plants: the hologenome concept, MBio 7 (2) (2016) e01395. Available from: https://doi.org/10.1128/mBio.01395-15.
[65] J.-B. Lamarck, Philosophie Zoologique, Original Edition of 1809 with Introduction by Andre Pichot, Flammarion, Paris, 1994.
[66] C. Darwin, The Variation of Animals and Plants under Domestication, 2, Organe Judd, NewYork, 1868.
[67] L.J. Funkhouser, S.R. Bordenstein, Mom knows best: the universality of maternal microbial transmission, PLoS Biol. 11 (8) (2013) e1001631.
[68] S.F. Gilbert, A holobiont birth narrative: the epigenetic transmission of the human microbiome, Front. Genet. 5 (2014) 282. Available from: https://doi.org/10.3389/fgene.2014.00282.
[69] M. Bright, S. Bulgheresi, A complex journey: transmission of microbial symbionts, Nat. Rev. Microbiol. 8 (3) (2010) 218–230. Available from: https://doi.org/10.1038/nrmicro2262.
[70] D.M. Drown, P.C. Zee, Y. Brandvain, M.J. Wade, Evolution of transmission mode in obligate symbionts, Evol. Ecol. Res. 15 (1) (2013) 43–59.
[71] B.M. Fitzpatrick, Symbiote transmission and maintenance of extra-genomic associations, Front. Microbiol. 5 (2014) 46.
[72] J.H. Werren, L. Baldo, M.E. Clark, Wolbachia: master manipulators of invertebrate biology, Nat. Rev. Microbiol. 6 (10) (2008) 741–751.
[73] F. Landmann, The Wolbachia endosymbionts, Microbiol. Spectr. 7 (2) (2019). Available from: https://doi.org/10.1128/microbiolspec.BAI-0018-2019.
[74] J.D. Shropshire, S.R. Bordenstein, Speciation by symbiosis: the microbiome and behavior, MBio 7 (2) (2016). Available from: https://doi.org/10.1128/mBio.01785-15.
[75] G.L. J. Stebbins, Cataclysmic evolution, Sci. Am. 184 (4) (1951) 54–59.
[76] C. Darwin, On the Origin of Species by Means of Natural Selection, or the Preservation of Favoured Races in the Struggle for Life, John Russel, London, 1859.

[77] W. Salzburger, Understanding explosive diversification through cichlid fish genomics, Nat. Rev. Genet. 19 (11) (2018) 705–717. Available from: https://doi.org/10.1038/s41576-018-0043-9.
[78] J.I. Meier, D.A. Marques, C.E. Wagner, L. Excoffier, O. Seehausen, Genomics of parallel ecological speciation in Lake Victoria Cichlids, Mol. Biol. Evol. 35 (6) (2018) 1489–1506. Available from: https://doi.org/10.1093/molbev/msy051.
[79] J.I. Meier, D.A. Marques, S. Mwaiko, C.E. Wagner, L. Excoffier, O. Seehausen, Ancient hybridization fuels rapid cichlid fish adaptive radiations, Nat. Commun. 8 (2017) 14363.
[80] S. Lamichhaney, et al., Evolution of Darwin's finches and their beaks revealed by genome sequencing, Nature 518 (7539) (2015) 371–375. Available from: https://doi.org/10.1038/nature14181.
[81] S. Lamichhaney, F. Han, M.T. Webster, L. Andersson, B.R. Grant, P.R. Grant, Rapid hybrid speciation in Darwin's finches, Science (2017). Available from: https://doi.org/10.1126/science.aao4593.
[82] B.R. Grant, P.R. Grant, Watching speciation in action, Science 355 (6328) (2017) 910–911. Available from: https://doi.org/10.1126/science.aam6411.
[83] G.W. Beadle, The genes of men and molds, Sci. Am. 179 (3) (1948) 30–39.
[84] F. Crick, On protein synthesis, Symp. Soc. Exp. Biol. 12 (1958) 138–163.
[85] F.H. Crick, The genetic code, Sci. Am. 207 (1962) 66–74.
[86] R.F. Doolittle, The multiplicity of domains in proteins, Annu. Rev. Biochem. 64 (1995) 287–314 [Online]. Available: <http://www.ncbi.nlm.nih.gov/entrez/query.fcgi?cmd = Retrieve&db = PubMed&dopt = Citation&list_uids = 7574483>.
[87] C.P. Ponting, R.R. Russell, The natural history of protein domains, Annu. Rev. Biophys. Biomol. Struct. 31 (2002) 45–71. Available from: https://doi.org/10.1146/annurev.biophys.31.082901.134314 [pii].
[88] E. Bornberg-Bauer, F. Beaussart, S.K. Kummerfeld, S.A. Teichmann, J. Weiner 3rd, The evolution of domain arrangements in proteins and interaction networks, Cell Mol. Life Sci. 62 (4) (2005) 435–445.
[89] A.K. Bjorklund, D. Ekman, S. Light, J. Frey-Skott, A. Elofsson, Domain rearrangements in protein evolution, J. Mol. Biol. 353 (4) (2005) 911–923.
[90] A.K. Bjorklund, D. Ekman, A. Elofsson, Expansion of protein domain repeats, PLoS Comput. Biol. 2 (8) (2006) e114.
[91] J. Weiner 3rd, F. Beaussart, E. Bornberg-Bauer, Domain deletions and substitutions in the modular protein evolution, FEBS J. 273 (9) (2006) 2037–2047. Available from: https://doi.org/10.1111/j.1742-4658.2006.05220.x.
[92] M. Buljan, A. Frankish, A. Bateman, Quantifying the mechanisms of domain gain in animal proteins, Genome Biol. 11 (7) (2010) R74. Available from: https://doi.org/10.1186/gb-2010-11-7-r74.
[93] J.G. Lees, N.L. Dawson, I. Sillitoe, C.A. Orengo, Functional innovation from changes in protein domains and their combinations, Curr. Opin. Struct. Biol. 38 (2016) 44–52. Available from: https://doi.org/10.1016/j.sbi.2016.05.016.
[94] E.S. Lander, et al., Initial sequencing and analysis of the human genome, Nature 409 (6822) (2001) 860–921.
[95] M. Levitt, Nature of the protein universe, Proc. Natl. Acad. Sci. U. S. A. 106 (27) (2009) 11079–11084. Available from: https://doi.org/10.1073/pnas.0905029106.
[96] S. Yang, P.E. Bourne, The evolutionary history of protein domains viewed by species phylogeny, PLoS One 4 (12) (2009) e8378. Available from: https://doi.org/10.1371/journal.pone.0008378.
[97] E.J. Deeds, H. Hennessey, E.I. Shakhnovich, Prokaryotic phylogenies inferred from protein structural domains, Genome Res. 15 (3) (2005) 393–402. Available from: https://doi.org/10.1101/gr.3033805.

[98] M. Liu, H. Walch, S. Wu, A. Grigoriev, Significant expansion of exon-bordering protein domains during animal proteome evolution, Nucleic Acids Res. 33 (1) (2005) 95–105.
[99] S. Yang, R.F. Doolittle, P.E. Bourne, Phylogeny determined by protein domain content, Proc. Natl. Acad. Sci. U. S. A. 102 (2) (2005) 373–378. Available from: https://doi.org/10.1073/pnas.0408810102.
[100] M. Buljan, A. Bateman, The evolution of protein domain families, Biochem. Soc. Trans. 37 (Pt. 4) (2009) 751–755.
[101] H.D. Carroll, J.L. Spouge, M. Gonzalez, MultiDomainBenchmark: a multi-domain query and subject database suite, BMC Bioinforma. 20 (1) (2019) 77.
[102] M.W. Gonzalez, W.R. Pearson, RefProtDom: a protein database with improved domain boundaries and homology relationships, Bioinformatics 26 (18) (2010) 2361–2362. Available from: https://doi.org/10.1093/bioinformatics/btq426.
[103] M. Toll-Riera, M.M. Alba, Emergence of novel domains in proteins, BMC Evol. Biol. 13 (2013) 47. Available from: https://doi.org/10.1186/1471-2148-13-47.
[104] A. Nasir, K.M. Kim, G. Caetano-Anolles, Global patterns of protein domain gain and loss in superkingdoms, PLoS Comput. Biol. 10 (1) (2014) e1003452. Available from: https://doi.org/10.1371/journal.pcbi.1003452.
[105] W. Gilbert, Why genes in pieces? Nature 271 (1978) 501. Available from: https://doi.org/10.1038/271501a0 [Online].
[106] W. Gilbert, The exon theory of genes, Cold Spring Harb. Symp. Quant. Biol. 52 (1987) 901–905.
[107] P.A. Sharp, Split genes and RNA splicing, Cell 77 (1994) 805–815.
[108] T.R. Cech, Self-splicing of group I introns, Annu. Rev. Biochem. 59 (1990) 543–568. Available from: https://doi.org/10.1146/annurev.bi.59.070190.002551.
[109] Y. Shi, Mechanistic insights into precursor messenger RNA splicing by the spliceosome, Nat. Rev. Mol. Cell Biol. 18 (11) (2017) 655–670. Available from: https://doi.org/10.1038/nrm.2017.86.
[110] X. Yang, et al., Widespread expansion of protein interaction capabilities by alternative splicing, Cell 164 (4) (2016) 805–817. Available from: https://doi.org/10.1016/j.cell.2016.01.029.
[111] W.Y. Tarn, Cellular signals modulate alternative splicing, J. Biomed. Sci. 14 (4) (2007) 517–522. Available from: https://doi.org/10.1007/s11373-007-9161-7.
[112] J.P. Venables, J. Tazi, F. Juge, Regulated functional alternative splicing in Drosophila, Nucleic Acids Res. 40 (1) (2012) 1–10. Available from: https://doi.org/10.1093/nar/gkr648.
[113] A.M. Mastrangelo, D. Marone, G. Laido, A.M. De Leonardis, P. De Vita, Alternative splicing: enhancing ability to cope with stress via transcriptome plasticity, Plant Sci. 185-186 (2012) 40–49. Available from: https://doi.org/10.1016/j.plantsci.2011.09.006.
[114] T. Blumenthal, Trans-splicing and operons in *C. elegans*, WormBook (2012) 1–11. Available from: https://doi.org/10.1895/wormbook.1.5.2.
[115] Q. Lei, C. Li, Z. Zuo, C. Huang, H. Cheng, R. Zhou, Evolutionary insights into RNA trans-splicing in vertebrates, Genome Biol. Evol. 8 (3) (2016) 562–577. Available from: https://doi.org/10.1093/gbe/evw025.
[116] C. Zhang, H. Yang, H. Yang, Evolutionary character of alternative splicing in plants, Bioinform. Biol. Insights 9 (1) (2015) 47–52. Available from: https://doi.org/10.4137/bbi.s33716.
[117] S.J. Bush, L. Chen, J.M. Tovar-Corona, A.O. Urrutia, Alternative splicing and the evolution of phenotypic novelty, Philos. Trans. R. Soc. Lond. B Biol. Sci. 372 (1713) (2017). Available from: https://doi.org/10.1098/rstb.2015.0474.

[118] M. Liu, A. Grigoriev, Protein domains correlate strongly with exons in multiple eukaryotic genomes—evidence of exon shuffling? Trends Genet. 20 (9) (2004) 399–403.

[119] Y. Luo, et al., Splicing-related features of introns serve to propel evolution, PLoS One 8 (3) (2013) e58547. Available from: https://doi.org/10.1371/journal.pone.0058547.

[120] S. Ohno, So much "junk" DNA in our genome, Brookhaven Symp. Biol. 23 (1972) 366–370.

[121] R. Dawkins, The Selfish Gene, Oxford University Press, Oxford, 1976.

[122] L.E. Orgel, F.H. Crick, Selfish DNA: the ultimate parasite, Nature 284 (5757) (1980) 604–607.

[123] A. Nekrutenko, W.H. Li, Transposable elements are found in a large number of human protein-coding genes, Trends Genet. 17 (11) (2001) 619–621 [Online]. Available: <http://www.ncbi.nlm.nih.gov/entrez/query.fcgi?cmd = Retrieve&db = PubMed&dopt = Citation&list_uids = 11672845>.

[124] F. Abascal, M.L. Tress, A. Valencia, Alternative splicing and co-option of transposable elements: the case of TMPO/LAP2alpha and ZNF451 in mammals, Bioinformatics 31 (14) (2015) 2257–2261. Available from: https://doi.org/10.1093/bioinformatics/btv132.

[125] A.I. Bukhari, J.A. Shapiro, S.L. Adhya (Eds.), DNA Insertion Elements, Plasmids and Episomes, Cold Spring Harbor Press, Cold Spring Harbor, NY, 1977.

[126] J. Lederberg, L.L. Cavalli, E.M. Lederberg, Sex compatibility in *Escherichia coli*, Genetics 37 (6) (1952) 720–730.

[127] L.L. Cavalli, J. Lederberg, E.M. Lederberg, An infective factor controlling sex compatibility in Bacterium coli, J. Gen. Microbiol. 8 (1) (1953) 89–103. Available from: https://doi.org/10.1099/00221287-8-1-89.

[128] T. Watanabe, Infective heredity of multiple drug resistance in bacteria, Bacteriol. Rev. 27 (1963) 87–115.

[129] T.J. Treangen, E.P. Rocha, Horizontal transfer, not duplication, drives the expansion of protein families in prokaryotes, PLoS Genet. 7 (1) (2011) e1001284. Available from: https://doi.org/10.1371/journal.pgen.1001284.

[130] A. Kanhere, M. Vingron, Horizontal gene transfers in prokaryotes show differential preferences for metabolic and translational genes, BMC Evol. Biol. 9 (2009) 9.

[131] Y.I. Wolf, K.S. Makarova, N. Yutin, E.V. Koonin, Updated clusters of orthologous genes for Archaea: a complex ancestor of the Archaea and the byways of horizontal gene transfer, Biol. Direct 7 (2012) 46. Available from: https://doi.org/10.1186/1745-6150-7-46.

[132] D.M. Faguy, W.F. Doolittle, Horizontal transfer of catalase-peroxidase genes between archaea and pathogenic bacteria, Trends Genet. 16 (5) (2000) 196–197 [Online]. Available: <http://www.ncbi.nlm.nih.gov/entrez/query.fcgi?cmd = Retrieve&db = PubMed&dopt = Citation&list_uids = 10782109>.

[133] N.C. Kyrpides, G.J. Olsen, Archaeal and bacterial hyperthermophiles: horizontal gene exchange or common ancestry? Trends Genet. 15 (8) (1999) 298–299.

[134] E.V. Koonin, Horizontal gene transfer: essentiality and evolvability in prokaryotes, and roles in evolutionary transitions, F1000Res 5 (2016). Available from: https://doi.org/10.12688/f1000research.8737.1.

[135] S. Sonea, M. Panisset, A New Bacteriology, Jones and Batlett, Boston, MA, 1983.

[136] E.G. Danchin, et al., Multiple lateral gene transfers and duplications have promoted plant parasitism ability in nematodes, Proc. Natl. Acad. Sci. U. S. A. 107 (41) (2010) 17651–17656. Available from: https://doi.org/10.1073/pnas.1008486107.

[137] A. Haegeman, J.T. Jones, E.G. Danchin, Horizontal gene transfer in nematodes: a catalyst for plant parasitism? Mol. Plant Microbe Interact. 24 (8) (2011) 879–887. Available from: https://doi.org/10.1094/mpmi-03-11-0055.

[138] I. Eyres, et al., Horizontal gene transfer in bdelloid rotifers is ancient, ongoing and more frequent in species from desiccating habitats, BMC Biol. 13 (1) (2015) 90. Available from: https://doi.org/10.1186/s12915-015-0202-9.
[139] L. Szydlowski, C. Boschetti, A. Crisp, E.G. Barbosa, A. Tunnacliffe, Multiple horizontally acquired genes from fungal and prokaryotic donors encode cellulolytic enzymes in the bdelloid rotifer Adineta ricciae, Gene 566 (2) (2015) 125–137. Available from: https://doi.org/10.1016/j.gene.2015.04.007.
[140] C.X. Chan, A.E. Darling, R.G. Beiko, M.A. Ragan, Are protein domains modules of lateral genetic transfer? PLoS One 4 (2) (2009) e4524.
[141] C.X. Chan, R.G. Beiko, A.E. Darling, M.A. Ragan, Lateral transfer of genes and gene fragments in prokaryotes, Genome Biol. Evol. 1 (2009) 429–438. Available from: https://doi.org/10.1093/gbe/evp044.
[142] N. Nikolaidis, N. Doran, D.J. Cosgrove, Plant expansins in bacteria and fungi: evolution by horizontal gene transfer and independent domain fusion, Mol. Biol. Evol. 31 (2) (2014) 376–386. Available from: https://doi.org/10.1093/molbev/mst206.
[143] P.F. Morris, L.R. Schlosser, K.D. Onasch, T. Wittenschlaeger, R. Austin, N. Provart, Multiple horizontal gene transfer events and domain fusions have created novel regulatory and metabolic networks in the oomycete genome, PLoS One 4 (7) (2009) e6133. Available from: https://doi.org/10.1371/journal.pone.0006133.
[144] K.S. de Felipe, et al., Legionella eukaryotic-like type IV substrates interfere with organelle trafficking, PLoS Pathog. 4 (8) (2008) e1000117. Available from: https://doi.org/10.1371/journal.ppat.1000117.
[145] K.S. de Felipe, et al., Evidence for acquisition of Legionella type IV secretion substrates via interdomain horizontal gene transfer, J. Bacteriol. 187 (22) (2005) 7716–7726.
[146] I.S. Franco, H.A. Shuman, X. Charpentier, The perplexing functions and surprising origins of *Legionella pneumophila* type IV secretion effectors, Cell Microbiol. 11 (10) (2009) 1435–1443. Available from: https://doi.org/10.1111/j.1462-5822.2009.01351.x.
[147] C.T. Price, S. Al-Khodor, T. Al-Quadan, Y. Abu Kwaik, Indispensable role for the eukaryotic-like ankyrin domains of the ankyrin B effector of *Legionella pneumophila* within macrophages and amoebae, Infect. Immun. 78 (5) (2010) 2079–2088. Available from: https://doi.org/10.1128/iai.01450-09.
[148] P. Deschamps, Y. Zivanovic, D. Moreira, F. Rodriguez-Valera, P. Lopez-Garcia, Pangenome evidence for extensive interdomain horizontal transfer affecting lineage core and shell genes in uncultured planktonic thaumarchaeota and euryarchaeota, Genome Biol. Evol. 6 (7) (2014) 1549–1563. Available from: https://doi.org/10.1093/gbe/evu127.
[149] P. Lopez-Garcia, Y. Zivanovic, P. Deschamps, D. Moreira, Bacterial gene import and mesophilic adaptation in archaea, Nat. Rev. Microbiol. 13 (7) (2015) 447–456. Available from: https://doi.org/10.1038/nrmicro3485.
[150] E. Ocana-Pallares, S.R. Najle, C. Scazzocchio, I. Ruiz-Trillo, Reticulate evolution in eukaryotes: origin and evolution of the nitrate assimilation pathway, PLoS Genet. 15 (2) (2019) e1007986. Available from: https://doi.org/10.1371/journal.pgen.1007986.
[151] T. Olszak, A. Latka, B. Roszniowski, M.A. Valvano, Z. Drulis-Kawa, Phage life cycles behind bacterial biodiversity, Curr. Med. Chem. 24 (36) (2017) 3987–4001. Available from: https://doi.org/10.2174/0929867324666170413100136.
[152] I. Reche, G. D'Orta, N. Mladenov, D.M. Winget, C.A. Suttle, Deposition rates of viruses and bacteria above the atmospheric boundary layer, ISME J. 12 (4) (2018) 1154–1162.
[153] M. Breitbart, C. Bonnain, K. Malki, N.A. Sawaya, Phage puppet masters of the marine microbial realm, Nat. Microbiol. 3 (7) (2018) 754–766. Available from: https://doi.org/10.1038/s41564-018-0166-y.

[154] F. Rohwer, R.V. Thurber, Viruses manipulate the marine environment, Nature 459 (7244) (2009) 207–212. Available from: https://doi.org/10.1038/nature08060.
[155] A.S. Lang, A.B. Westbye, J.T. Beatty, The distribution, evolution, and roles of gene transfer agents in prokaryotic genetic exchange, Annu. Rev. Virol. 4 (1) (2017) 87–104. Available from: https://doi.org/10.1146/annurev-virology-101416-041624.
[156] H.X. Chiura, Generalized gene transfer by virus-like particles from marine bacteria, Aquat. Microb. Ecol. 13 (1997) 75–83 [Online]. Available: <https://www.int-res.com/abstracts/ame/v13/n1/p75-83/>.
[157] H.X. Chiura, Broad host range xenotrophic gene transfer by virus-like particles from a hot spring, Microbes Env. 17 (2002) 53–58. Available from: https://doi.org/10.1264/jsme2.2002.53s.
[158] I. Sharon, et al., Comparative metagenomics of microbial traits within oceanic viral communities, ISME J. 5 (7) (2011) 1178–1190. Available from: https://doi.org/10.1038/ismej.2011.2.
[159] C.A. Suttle, Environmental microbiology: viral diversity on the global stage, Nat. Microbiol. 1 (11) (2016) 16205. Available from: https://doi.org/10.1038/nmicrobiol.2016.205.
[160] D. Paez-Espino, et al., Uncovering Earth's virome, Nature 536 (7617) (2016) 425–430. Available from: https://doi.org/10.1038/nature19094.
[161] S. Roux, et al., Ecogenomics and potential biogeochemical impacts of globally abundant ocean viruses, Nature 537 (7622) (2016) 689–693. Available from: https://doi.org/10.1038/nature19366.
[162] B.C.M. Ramisetty, P.A. Sudhakari, Bacterial 'grounded' prophages: hotspots for genetic renovation and innovation, Front. Genet. 10 (2019) 65.
[163] E.V. Koonin, M. Krupovic, The depths of virus exaptation, Curr. Opin. Virol. 31 (2018) 1–8. Available from: https://doi.org/10.1016/j.coviro.2018.07.011.
[164] E.V. Koonin, V.V. Dolja, A virocentric perspective on the evolution of life, Curr. Opin. Virol. 3 (5) (2013) 546–557. Available from: https://doi.org/10.1016/j.coviro.2013.06.008.
[165] M. Naville, et al., Not so bad after all: retroviruses and long terminal repeat retrotransposons as a source of new genes in vertebrates, Clin. Microbiol. Infect. 22 (4) (2016) 312–323. Available from: https://doi.org/10.1016/j.cmi.2016.02.001.
[166] E.C. Keen, et al., Novel "Superspreader" bacteriophages promote horizontal gene transfer by transformation, MBio 8 (1) (2017). Available from: https://doi.org/10.1128/mBio.02115-16.
[167] O. Larranaga, et al., Phage particles harboring antibiotic resistance genes in fresh-cut vegetables and agricultural soil, Env. Int. 115 (2018) 133–141. Available from: https://doi.org/10.1016/j.envint.2018.03.019.
[168] N.D. Zinder, J. Lederberg, Genetic exchange in Salmonella, J. Bacteriol. 64 (5) (1952) 679–699.
[169] N.D. Zinder, Transduction in bacteria, Sci. Am. 199 (5) (1958) 38–43.
[170] M.L. Morse, E.M. Lederberg, J. Lederberg, Transduction in *Escherichia coli* K-12, Genetics 41 (1) (1956) 142–156.
[171] K. Ippen, J.A. Shapiro, J.R. Beckwith, Transposition of the lac region to the gal region of the *Escherichia coli* chromosome: isolation of lambda-lac transducing bacteriophages, J. Bacteriol. 108 (1) (1971) 5–9 [Online]. Available: <http://www.ncbi.nlm.nih.gov/entrez/query.fcgi?cmd = Retrieve&db = PubMed&dopt = Citation&list_uids = 4941573>.
[172] H.E. Varmus, P.K. Vogt, J.M. Bishop, Integration of deoxyribonucleic acid specific for Rous sarcoma virus after infection of permissive and nonpermissive hosts, Proc. Natl. Acad. Sci. U. S. A. 70 (11) (1973) 3067–3071.

[173] S. Saule, M. Roussel, C. Lagrou, D. Stehelin, Characterization of the oncogene (erb) of avian erythroblastosis virus and its cellular progenitor, J. Virol. 38 (2) (1981) 409–419.
[174] T.J. Gonda, D.K. Sheiness, J.M. Bishop, Transcripts from the cellular homologs of retroviral oncogenes: distribution among chicken tissues, Mol. Cell Biol. 2 (6) (1982) 617–624.
[175] K.H. Klempnauer, J.M. Bishop, Transduction of c-myb into avian myeloblastosis virus: locating points of recombination within the cellular gene, J. Virol. 48 (3) (1983) 565–572.
[176] G. Symonds, E. Stubblefield, M. Guyaux, J.M. Bishop, Cellular oncogenes (c-erb-A and c-erb-B) located on different chicken chromosomes can be transduced into the same retroviral genome, Mol. Cell Biol. 4 (8) (1984) 1627–1630.
[177] A. Swain, J.M. Coffin, Mechanism of transduction by retroviruses, Science 255 (5046) (1992) 841–845. PMID: 1371365.
[178] M.G. Palmgren, Capturing of host DNA by a plant retroelement: Bs1 encodes plasma membrane H(+)-ATPase domains, Plant Mol. Biol. 25 (2) (1994) 137–140.
[179] T.E. Bureau, S.E. White, S.R. Wessler, Transduction of a cellular gene by a plant retroelement, Cell 77 (4) (1994) 479–480 [Online]. Available: <http://www.ncbi.nlm.nih.gov/entrez/query.fcgi?cmd = Retrieve&db = PubMed&dopt = Citation&list_uids = 8187172>.
[180] B. Song, et al., Comparative genomics reveals two major bouts of gene retroposition coinciding with crucial periods of symbiodinium evolution, Genome Biol. Evol. 9 (8) (2017) 2037–2047.
[181] S. Tan, et al., LTR-mediated retroposition as a mechanism of RNA-based duplication in metazoans, Genome Res. 26 (12) (2016) 1663–1675. Available from: https://doi.org/10.1101/gr.204925.116.
[182] Y. Hahn, Evidence for the dissemination of cryptic non-coding RNAs transcribed from intronic and intergenic segments by retroposition, Bioinformatics 29 (13) (2013) 1593–1599. Available from: https://doi.org/10.1093/bioinformatics/btt258.
[183] A. Bohne, A. Darras, H. D'Cotta, J.F. Baroiller, D. Galiana-Arnoux, J.N. Volff, The vertebrate makorin ubiquitin ligase gene family has been shaped by large-scale duplication and retroposition from an ancestral gonad-specific, maternal-effect gene, BMC Genomics 11 (2010) 721. Available from: https://doi.org/10.1186/1471-2164-11-721.
[184] M. Osterholz, L. Walter, C. Roos, Retropositional events consolidate the branching order among New World monkey genera, Mol. Phylogenet. Evol. 50 (3) (2009) 507–513. Available from: https://doi.org/10.1016/j.ympev.2008.12.014.
[185] J.N. Volff, J. Brosius, Modern genomes with retro-look: retrotransposed elements, retroposition and the origin of new genes, Genome Dyn. 3 (2007) 175–190.
[186] W. Wang, et al., High rate of chimeric gene origination by retroposition in plant genomes, Plant Cell 18 (8) (2006) 1791–1802.
[187] A. Pavlicek, A.J. Gentles, J. Paces, V. Paces, J. Jurka, Retroposition of processed pseudogenes: the impact of RNA stability and translational control, Trends Genet. 22 (2) (2006) 69–73. Available from: https://doi.org/10.1016/j.tig.2005.11.005.
[188] A.C. Marques, I. Dupanloup, N. Vinckenbosch, A. Reymond, H. Kaessmann, Emergence of young human genes after a burst of retroposition in primates, PLoS Biol. 3 (11) (2005) e357.
[189] J. Schacherer, Y. Tourrette, J.L. Souciet, S. Potier, J. De Montigny, Recovery of a function involving gene duplication by retroposition in *Saccharomyces cerevisiae*, Genome Res. 14 (7) (2004) 1291–1297. Available from: https://doi.org/10.1101/gr.2363004.
[190] J. Brosius, The contribution of RNAs and retroposition to evolutionary novelties, Genetica 118 (2003) 99–116.

[191] N. Yutin, Y.I. Wolf, D. Raoult, E.V. Koonin, Eukaryotic large nucleo-cytoplasmic DNA viruses: clusters of orthologous genes and reconstruction of viral genome evolution, Virol. J. 6 (2009) 223.

[192] N. Philippe, et al., Pandoraviruses: amoeba viruses with genomes up to 2.5 Mb reaching that of parasitic eukaryotes, Science 341 (6143) (2013) 281–286. Available from: https://doi.org/10.1126/science.1239181.

[193] M. Legendre, et al., Diversity and evolution of the emerging Pandoraviridae family, Nat. Commun. 9 (1) (2018) 2285.

[194] J. Filee, Route of NCLDV evolution: the genomic accordion, Curr. Opin. Virol. 3 (5) (2013) 595–599. Available from: https://doi.org/10.1016/j.coviro.2013.07.003.

[195] J. Filee, Genomic comparison of closely related Giant Viruses supports an accordion-like model of evolution, Front. Microbiol. 6 (2015) 593. Available from: https://doi.org/10.3389/fmicb.2015.00593.

[196] C. Moliner, P.E. Fournier, D. Raoult, Genome analysis of microorganisms living in amoebae reveals a melting pot of evolution, FEMS Microbiol. Rev. 34 (3) (2010) 281–294. Available from: https://doi.org/10.1111/j.1574-6976.2010.00209.x.

[197] M. Boyer, et al., Giant Marseillevirus highlights the role of amoebae as a melting pot in emergence of chimeric microorganisms, Proc. Natl. Acad. Sci. U. S. A. 106 (51) (2009) 21848–21853.

[198] Z. Wang, M. Wu, Comparative genomic analysis of Acanthamoeba endosymbionts highlights the role of amoebae as a "melting pot" shaping the Rickettsiales evolution, Genome Biol. Evol. 9 (11) (2017) 3214–3224.

[199] A.M. Segall, N.L. Craig, New wrinkles and folds in site-specific recombination, Mol. Cell 19 (4) (2005) 433–435. Available from: https://doi.org/10.1016/j.molcel.2005.08.003.

[200] N.D. Grindley, K.L. Whiteson, P.A. Rice, Mechanisms of site-specific recombination, Annu. Rev. Biochem. 75 (2006) 567–605.

[201] B. Hallet, D.J. Sherratt, Transposition and site-specific recombination: adapting DNA cut-and-paste mechanisms to a variety of genetic rearrangements, FEMS Microbiol. Rev. 21 (2) (2010) 157–178. doi: S0168-6445(97)00055-7 [pii].

[202] A. Toussaint, P.A. Rice, Transposable phages, DNA reorganization and transfer, Curr. Opin. Microbiol. 38 (2017) 88–94. Available from: https://doi.org/10.1016/j.mib.2017.04.009.

[203] A. Lwoff, Lysogeny, Bacteriol. Rev. 17 (4) (1953) 269–337.

[204] H. Brussow, C. Canchaya, W.D. Hardt, Phages and the evolution of bacterial pathogens: from genomic rearrangements to lysogenic conversion, Microbiol. Mol. Biol. Rev. 68 (3) (2004) 560–602.

[205] L.C. Fortier, O. Sekulovic, Importance of prophages to evolution and virulence of bacterial pathogens, Virulence 4 (5) (2013) 354–365.

[206] T.H. Silver-Mysliwiec, M.G. Bramucci, Bacteriophage-enhanced sporulation: comparison of spore-converting bacteriophages PMB12 and SP10, J. Bacteriol. 172 (4) (1990) 1948–1953.

[207] S.A. Rice, et al., The biofilm life cycle and virulence of Pseudomonas aeruginosa are dependent on a filamentous prophage, ISME J. 3 (3) (2009) 271–282.

[208] J.A. Grasis, et al., Species-specific viromes in the ancestral holobiont Hydra, PLoS One 9 (10) (2014) e109952. Available from: https://doi.org/10.1371/journal.pone.0109952.

[209] C.B. Silveira, F.L. Rohwer, Piggyback-the-Winner in host-associated microbial communities, NPJ Biofilms Microb. 2 (2016) 16010.

[210] F.P. Ryan, Viral symbiosis and the holobiontic nature of the human genome, APMIS 124 (1-2) (2016) 11–19. Available from: https://doi.org/10.1111/apm.12488.

[211] V.A. Belyi, A.J. Levine, A.M. Skalka, Unexpected inheritance: multiple integrations of ancient bornavirus and ebolavirus/marburgvirus sequences in vertebrate genomes, PLoS Pathog. 6 (7) (2010) e1001030. Available from: https://doi.org/10.1371/journal.ppat.1001030.

[212] M. Krupovic, P. Forterre, Single-stranded DNA viruses employ a variety of mechanisms for integration into host genomes, Ann. N. Y. Acad. Sci. 1341 (2015) 41–53. Available from: https://doi.org/10.1111/nyas.12675.

[213] S.N. Pantry, P.G. Medveczky, Latency, integration, and reactivation of human Herpesvirus-6, Viruses 9 (7) (2017).

[214] J.L. Hurwitz, B.G. Jones, E. Charpentier, D.L. Woodland, Hypothesis: RNA and DNA viral sequence integration into the mammalian host genome supports long-term B cell and T cell adaptive immunity, Viral Immunol. 30 (9) (2017) 628–632. Available from: https://doi.org/10.1089/vim.2017.0099.

[215] N. Wallaschek, et al., The telomeric repeats of human herpesvirus 6A (HHV-6A) are required for efficient virus integration, PLoS Pathog. 12 (5) (2016) e1005666.

[216] M.G. Fischer, T. Hackl, Host genome integration and giant virus-induced reactivation of the virophage mavirus, Nature 540 (7632) (2016) 288–291. Available from: https://doi.org/10.1038/nature20593.

[217] U. Palatini, et al., Comparative genomics shows that viral integrations are abundant and express piRNAs in the arboviral vectors *Aedes aegypti* and *Aedes albopictus*, BMC Gen. 18 (1) (2017) 512.

[218] Y. Suzuki, et al., Uncovering the repertoire of endogenous flaviviral elements in *Aedes* mosquito genomes, J. Virol. 91 (15) (2017).

[219] J.M. Coffin, S.H. Hughes, H.E. Varmus, Retroviruses, Cold Spring Harbor Laboratory Press, Cold Spring Harbor, NY, 1997.

[220] D.L. Mager, J.P. Stoye, Mammalian endogenous retroviruses, Microbiol. Spectr. 3 (1) (2015). Available from: https://doi.org/10.1128/microbiolspec.MDNA3-0009-2014.

[221] J. Goke, et al., Dynamic transcription of distinct classes of endogenous retroviral elements marks specific populations of early human embryonic cells, Cell Stem Cell 16 (2) (2015) 135–141. Available from: https://doi.org/10.1016/j.stem.2015.01.005.

[222] E.J. Grow, et al., Intrinsic retroviral reactivation in human preimplantation embryos and pluripotent cells, Nature 522 (7555) (2015) 221–225. Available from: https://doi.org/10.1038/nature14308.

[223] J. Ito, et al., Systematic identification and characterization of regulatory elements derived from human endogenous retroviruses, PLoS Genet. 13 (7) (2017) e1006883.

[224] G. Glinsky, et al., Single cell expression analysis of primate-specific retroviruses-derived HPAT lincRNAs in viable human blastocysts identifies embryonic cells co-expressing genetic markers of multiple lineages, Heliyon 4 (6) (2018) e00667.

[225] E.B. Chuong, N.C. Elde, C. Feschotte, Regulatory evolution of innate immunity through co-option of endogenous retroviruses, Science 351 (6277) (2016) 1083–1087. Available from: https://doi.org/10.1126/science.aad5497.

[226] E.B. Chuong, M.A. Rumi, M.J. Soares, J.C. Baker, Endogenous retroviruses function as species-specific enhancer elements in the placenta, Nat. Genet. 45 (3) (2013) 325–329. Available from: https://doi.org/10.1038/ng.2553.

[227] E.B. Chuong, Retroviruses facilitate the rapid evolution of the mammalian placenta, Bioessays 35 (10) (2013) 853–861. Available from: https://doi.org/10.1002/bies.201300059.

[228] C.E. Dunn-Fletcher, et al., Anthropoid primate-specific retroviral element THE1B controls expression of CRH in placenta and alters gestation length, PLoS Biol. 16 (9) (2018) e2006337.

[229] E.B. Chuong, The placenta goes viral: retroviruses control gene expression in pregnancy, PLoS Biol. 16 (10) (2018) e3000028.
[230] C. Lavialle, et al., Paleovirology of 'syncytins', retroviral env genes exapted for a role in placentation, Philos. Trans. R. Soc. Lond. B Biol. Sci. 368 (1626) (2013) 20120507. Available from: https://doi.org/10.1098/rstb.2012.0507.
[231] J. Denner, Expression and function of endogenous retroviruses in the placenta, APMIS 124 (1-2) (2016) 31–43. Available from: https://doi.org/10.1111/apm.12474.
[232] C.V. Harding, J.E. Heuser, P.D. Stahl, Exosomes: looking back three decades and into the future, J. Cell Biol. 200 (4) (2013) 367–371.
[233] M. Yanez-Mo, et al., Biological properties of extracellular vesicles and their physiological functions, J. Extracell. Vesicles 4 (2015) 27066.
[234] M. Gaudin, et al., Hyperthermophilic archaea produce membrane vesicles that can transfer DNA, Env. Microbiol. Rep. 5 (1) (2013) 109–116. Available from: https://doi.org/10.1111/j.1758-2229.2012.00348.x.
[235] S. Fischer, et al., Indication of horizontal DNA gene transfer by extracellular vesicles, PLoS One 11 (9) (2016) e0163665. Available from: https://doi.org/10.1371/journal.pone.0163665.
[236] J. Cai, G. Wu, P.A. Jose, C. Zeng, Functional transferred DNA within extracellular vesicles, Exp. Cell Res. 349 (1) (2016) 179–183. Available from: https://doi.org/10.1016/j.yexcr.2016.10.012.
[237] F. Tran, J.Q. Boedicker, Genetic cargo and bacterial species set the rate of vesicle-mediated horizontal gene transfer, Sci. Rep. 7 (1) (2017) 8813.
[238] N.J. Bitto, et al., Bacterial membrane vesicles transport their DNA cargo into host cells, Sci. Rep. 7 (1) (2017) 7072.
[239] M. Toyofuku, N. Nomura, L. Eberl, Types and origins of bacterial membrane vesicles, Nat. Rev. Microbiol. 17 (1) (2019) 13–24. Available from: https://doi.org/10.1038/s41579-018-0112-2.
[240] K.S. Makarova, N. Yutin, S.D. Bell, E.V. Koonin, Evolution of diverse cell division and vesicle formation systems in Archaea, Nat. Rev. Microbiol. 8 (10) (2010) 731–741. Available from: https://doi.org/10.1038/nrmicro2406.
[241] D.H. Choi, et al., Extracellular vesicles of the hyperthermophilic archaeon "*Thermococcus onnurineus*" NA1T, Appl. Env. Microbiol. 81 (14) (2015) 4591–4599.
[242] E. Marguet, et al., Membrane vesicles, nanopods and/or nanotubes produced by hyperthermophilic archaea of the genus Thermococcus, Biochem. Soc. Trans. 41 (1) (2013) 436–442. Available from: https://doi.org/10.1042/bst20120293.
[243] B.L. Deatherage, B.T. Cookson, Membrane vesicle release in bacteria, eukaryotes, and archaea: a conserved yet underappreciated aspect of microbial life, Infect. Immun. 80 (6) (2012) 1948–1957.
[244] S. Rayner, S. Bruhn, H. Vallhov, A. Andersson, R.B. Billmyre, A. Scheynius, Identification of small RNAs in extracellular vesicles from the commensal yeast *Malassezia sympodialis*, Sci. Rep. 7 (2017) 39742.
[245] X. Sisquella, et al., Malaria parasite DNA-harbouring vesicles activate cytosolic immune sensors, Nat. Commun. 8 (1) (2017) 1985.
[246] D.S. Goncalves, et al., Extracellular vesicles and vesicle-free secretome of the protozoa *Acanthamoeba castellanii* under homeostasis and nutritional stress and their damaging potential to host cells, Virulence 9 (1) (2018) 818–836.
[247] S. Robatzek, Vesicle trafficking in plant immune responses, Cell Microbiol. 9 (1) (2007) 1–8. Available from: https://doi.org/10.1111/j.1462-5822.2006.00829.x.
[248] I.A. Macdonald, M.J. Kuehn, Stress-induced outer membrane vesicle production by *Pseudomonas aeruginosa*, J. Bacteriol. 195 (13) (2013) 2971–2981.
[249] N. Orench-Rivera, M.J. Kuehn, Environmentally controlled bacterial vesicle-mediated export, Cell Microbiol. 18 (11) (2016) 1525–1536.

[250] S. Shibata, K.L. Visick, Sensor kinase RscS induces the production of antigenically distinct outer membrane vesicles that depend on the symbiosis polysaccharide locus in *Vibrio fischeri*, J. Bacteriol. 194 (1) (2012) 185–194.
[251] U. Resch, et al., A two-component regulatory system impacts extracellular membrane-derived vesicle production in group A Streptococcus, MBio 7 (6) (2016).
[252] K.J. McMahon, M.E. Castelli, E. Garcia Vescovi, M.F. Feldman, Biogenesis of outer membrane vesicles in Serratia marcescens is thermoregulated and can be induced by activation of the Rcs phosphorelay system, J. Bacteriol. 194 (12) (2012) 3241–3249.
[253] C. Florez, J.E. Raab, A.C. Cooke, J.W. Schertzer, Membrane distribution of the pseudomonas quinolone signal modulates outer membrane vesicle production in *Pseudomonas aeruginosa*, MBio 8 (4) (2017).
[254] Y. Kang, E. Goo, J. Kim, I. Hwang, Critical role of quorum sensing-dependent glutamate metabolism in homeostatic osmolality and outer membrane vesiculation in *Burkholderia glumae*, Sci. Rep. 7 (2017) 44195.
[255] I. Polakovicova, S. Jerez, I.A. Wichmann, A. Sandoval-Borquez, N. Carrasco-Veliz, A.H. Corvalan, Role of microRNAs and exosomes in *Helicobacter pylori* and Epstein–Barr virus associated gastric cancers, Front. Microbiol. 9 (2018) 636.
[256] N.S. Barteneva, et al., Extracellular vesicles in gastrointestinal cancer in conjunction with microbiota: on the border of kingdoms, Biochim. Biophys. Acta Rev. Cancer 1868 (2) (2017) 372–393. Available from: https://doi.org/10.1016/j.bbcan.2017.06.005.
[257] K. Chitcholtan, M.B. Hampton, J.I. Keenan, Outer membrane vesicles enhance the carcinogenic potential of *Helicobacter pylori*, Carcinogenesis 29 (12) (2008) 2400–2405. Available from: https://doi.org/10.1093/carcin/bgn218.
[258] Y. Liu, K.A.Y. Defourny, E.J. Smid, T. Abee, Gram-positive bacterial extracellular vesicles and their impact on health and disease, Front. Microbiol. 9 (2018) 1502.
[259] F. Askarian, et al., *Staphylococcus aureus* membrane-derived vesicles promote bacterial virulence and confer protective immunity in murine infection models, Front. Microbiol. 9 (2018) 262.
[260] Y.J. Yu, X.H. Wang, G.C. Fan, Versatile effects of bacterium-released membrane vesicles on mammalian cells and infectious/inflammatory diseases, Acta Pharmacol. Sin. 39 (4) (2018) 514–533.
[261] M. Ionescu, P.A. Zaini, C. Baccari, S. Tran, A.M. da Silva, S.E. Lindow, Xylella fastidiosa outer membrane vesicles modulate plant colonization by blocking attachment to surfaces, Proc. Natl. Acad. Sci. U. S. A. 111 (37) (2014) E3910–E3918.
[262] L. Katsir, O. Bahar, Bacterial outer membrane vesicles at the plant-pathogen interface, PLoS Pathog. 13 (6) (2017) e1006306.
[263] L. Mashburn-Warren, et al., Interaction of quorum signals with outer membrane lipids: insights into prokaryotic membrane vesicle formation, Mol. Microbiol. 69 (2) (2008) 491–502.
[264] L.M. Mashburn, M. Whiteley, Membrane vesicles traffic signals and facilitate group activities in a prokaryote, Nature 437 (7057) (2005) 422–425. Available from: https://doi.org/10.1038/nature03925.
[265] M. Toyofuku, et al., Membrane vesicle-mediated bacterial communication, ISME J. 11 (6) (2017) 1504–1509.
[266] S. Brameyer, L. Plener, A. Muller, A. Klingl, G. Wanner, K. Jung, Outer membrane vesicles facilitate trafficking of the hydrophobic signaling molecule CAI-1 between *Vibrio harveyi* cells, J. Bacteriol. (2018). Mar 19.
[267] R. Grande, et al., *Helicobacter pylori* ATCC 43629/NCTC 11639 outer membrane vesicles (OMVs) from biofilm and planktonic phase associated with extracellular DNA (eDNA), Front. Microbiol. 6 (2015) 1369. Available from: https://doi.org/10.3389/fmicb.2015.01369.

[268] S. Liao, et al., Streptococcus mutans extracellular DNA is upregulated during growth in biofilms, actively released via membrane vesicles, and influenced by components of the protein secretion machinery, J. Bacteriol. 196 (13) (2014) 2355–2366. Available from: https://doi.org/10.1128/jb.01493-14.
[269] Y. Zhu, S.G. Dashper, Y.Y. Chen, S. Crawford, N. Slakeski, E.C. Reynolds, *Porphyromonas gingivalis* and *Treponema denticola* synergistic polymicrobial biofilm development, PLoS One 8 (8) (2013) e71727.
[270] C. Ciardiello, et al., Focus on extracellular vesicles: new frontiers of cell-to-cell communication in cancer, Int. J. Mol. Sci. 17 (2) (2016) 175.
[271] V.R. Minciacchi, M.R. Freeman, D. Di Vizio, Extracellular vesicles in cancer: exosomes, microvesicles and the emerging role of large oncosomes, Semin. Cell Dev. Biol. 40 (2015) 41–51.
[272] P. Fonseca, I. Vardaki, A. Occhionero, T. Panaretakis, Metabolic and signaling functions of cancer cell-derived extracellular vesicles, Int. Rev. Cell Mol. Biol. 326 (2016) 175–199. Available from: https://doi.org/10.1016/bs.ircmb.2016.04.004.
[273] T. Reyes-Robles, et al., *Vibrio cholerae* outer membrane vesicles inhibit bacteriophage infection, J. Bacteriol. 200 (15) (2018). Available from: https://doi.org/10.1128/jb.00792-17.
[274] L.S. Joffe, L. Nimrichter, M.L. Rodrigues, M. Del Poeta, Potential roles of fungal extracellular vesicles during infection, mSphere 1 (4) (2016).
[275] I. Samuelson, A.J. Vidal-Puig, Fed-EXosome: extracellular vesicles and cell-cell communication in metabolic regulation, Essays Biochem. 62 (2) (2018) 165–175. Available from: https://doi.org/10.1042/ebc20170087.
[276] A.G. Evans, et al., Predatory activity of *Myxococcus xanthus* outer-membrane vesicles and properties of their hydrolase cargo, Microbiology 158 (Pt. 11) (2012) 2742–2752. Available from: https://doi.org/10.1099/mic.0.060343-0.
[277] Q. Cai, et al., Plants send small RNAs in extracellular vesicles to fungal pathogen to silence virulence genes, Science 360 (6393) (2018) 1126–1129. Available from: https://doi.org/10.1126/science.aar4142.
[278] M. Regente, M. Pinedo, H. San Clemente, T. Balliau, E. Jamet, L. de la Canal, Plant extracellular vesicles are incorporated by a fungal pathogen and inhibit its growth,", J. Exp. Bot. 68 (20) (2017) 5485–5495. Available from: https://doi.org/10.1093/jxb/erx355.
[279] T. Szatmari, et al., Extracellular vesicles mediate radiation-induced systemic bystander signals in the bone marrow and spleen, Front. Immunol. 8 (2017) 347.
[280] S. Vdovikova, S. Gilfillan, S. Wang, M. Dongre, S.N. Wai, A. Hurtado, Modulation of gene transcription and epigenetics of colon carcinoma cells by bacterial membrane vesicles, Sci. Rep. 8 (1) (2018) 7434.
[281] M.S. Aschtgen, K. Wetzel, W. Goldman, M. McFall-Ngai, E. Ruby, *Vibrio fischeri*-derived outer membrane vesicles trigger host development, Cell Microbiol. 18 (4) (2016) 488–499.
[282] M.S. Aschtgen, J.B. Lynch, E. Koch, J. Schwartzman, M. McFall-Ngai, E. Ruby, Rotation of *Vibrio fischeri* flagella produces outer membrane vesicles that induce host development, J. Bacteriol. 198 (16) (2016) 2156–2165.
[283] G. van Niel, G. D'Angelo, G. Raposo, Shedding light on the cell biology of extracellular vesicles, Nat. Rev. Mol. Cell Biol. 19 (4) (2018) 213–228. Available from: https://doi.org/10.1038/nrm.2017.125.
[284] U. Sharma, et al., Small RNAs are trafficked from the epididymis to developing mammalian sperm, Dev. Cell 46 (4) (2018) 481–494. e6.
[285] C.C. Conine, F. Sun, L. Song, J.A. Rivera-Perez, O.J. Rando, Small RNAs gained during epididymal transit of sperm are essential for embryonic development in mice, Dev. Cell 46 (4) (2018) 470–480. e3.

[286] C. Spadafora, Sperm-mediated transgenerational inheritance, Front. Microbiol. 8 (2017) 2401.

[287] C. Spadafora, Soma to germline inheritance of extrachromosomal genetic information via a LINE-1 reverse transcriptase-based mechanism, Bioessays 38 (8) (2016) 726–733. Available from: https://doi.org/10.1002/bies.201500197.

[288] I. Sciamanna, C. De Luca, C. Spadafora, The reverse transcriptase encoded by LINE-1 retrotransposons in the genesis, progression, and therapy of cancer, Front. Chem. 4 (2016) 6. Available from: https://doi.org/10.3389/fchem.2016.00006.

[289] M. Rassoulzadegan, V. Grandjean, P. Gounon, F. Cuzin, Inheritance of an epigenetic change in the mouse: a new role for RNA, Biochem. Soc. Trans. 35 (Pt. 3) (2007) 623–625. Available from: https://doi.org/10.1042/bst0350623.

[290] M. Rassoulzadegan, V. Grandjean, P. Gounon, S. Vincent, I. Gillot, F. Cuzin, RNA-mediated non-mendelian inheritance of an epigenetic change in the mouse, Nature 441 (7092) (2006) 469–474. Available from: https://doi.org/10.1038/nature04674.

[291] M. Rassoulzadegan, F. Cuzin, Epigenetic heredity: RNA-mediated modes of phenotypic variation, Ann. N. Y. Acad. Sci. 1341 (2015) 172–175. Available from: https://doi.org/10.1111/nyas.12694.

[292] S.A. Eaton, N. Jayasooriah, M.E. Buckland, D.I. Martin, J.E. Cropley, C.M. Suter, Roll over Weismann: extracellular vesicles in the transgenerational transmission of environmental effects, Epigenomics 7 (7) (2015) 1165–1171. Available from: https://doi.org/10.2217/epi.15.58.

[293] T.H. Morgan, A.H. Sturtevant, H.J. Muller, C.B. Bridges, The Mechanism of Mendelian Heredity, Henry Holt, New York, 1915.

[294] W.L. Johannsen, Arvelighedslærens Elementer (The Elements of Heredity). Copenhagen, 1905.

[295] W. Johannsen, The genotype conception of heredity, Am. Nat. 45 (531) (1911) 129–159. Available from: https://doi.org/10.1086/279202.

[296] A.H. Sturtevant, Linkage variation and chromosome maps, Proc. Natl. Acad. Sci. U. S. A. 7 (7) (1921) 181–183.

[297] A.H. Sturtevant, C.B. Bridges, T.H. Morgan, The spatial relations of genes, Proc. Natl. Acad. Sci. U. S. A. 5 (5) (1919) 168–173.

[298] H.B. Creighton, B. McClintock, The correlation of cytological and genetical crossing-over in *Zea mays*. A corroboration, Proc. Natl. Acad. Sci. U. S. A. 21 (3) (1935) 148–150.

[299] G.W. Beadle, E.L. Tatum, Genetic control of biochemical reactions in neurospora, Proc. Natl. Acad. Sci. U. S. A. 27 (11) (1941) 499–506.

[300] O.T. Avery, C.M. MacLeod, M. McCarty, Studies on the chemical nature of the substance inducing transformation of Pneumococcal types: induction of transformation by a desoxyribonucleic acid fraction isolated prom Pneumococcus Type III, J. Exp. Med. 79 (1944) 137–158.

[301] A.D. Hershey, M. Chase, Independent functions of viral protein and nucleic acid in growth of bacteriophage, J. Gen. Physiol. 36 (1) (1952) 39–56.

[302] J.D. Watson, F.H. Crick, Molecular structure of nucleic acids; a structure for deoxyribose nucleic acid, Nature 171 (4356) (1953) 737–738.

[303] J.D. Watson, F.H. Crick, Genetical implications of the structure of deoxyribonucleic acid, Nature 171 (1953) 964–967.

[304] E.B. Wilson, revised and enlarged The Cell in Development and Inheritance, third ed., MacMillan, New York, 1928.

[305] B. McClintock, The relation of a particular chromosomal element to the development of the nucleoli in *Zea mays*, Zeitschrift fur Zellforschung und mikroskopische Anatomie 21 (2) (1934) 294–328. Available from: https://doi.org/10.1007/BF00374060.

[306] F. Jacob, J. Monod, Genetic regulatory mechanisms in the synthesis of proteins, J. Mol. Biol. 3 (1961) 318–356.
[307] S. Schoenfelder, et al., Preferential associations between co-regulated genes reveal a transcriptional interactome in erythroid cells (in eng\)Nat. Genet. 42 (1) (2010) 53–61. Available from: https://doi.org/10.1038/ng.496.
[308] C.H. Eskiw, N.F. Cope, I. Clay, S. Schoenfelder, T. Nagano, P. Fraser, Transcription factories and nuclear organization of the genome, Cold Spring Harb. Symp. Quant. Biol. 75 (2010) 501–506. Available from: https://doi.org/10.1101/sqb.2010.75.046.
[309] S. Schoenfelder, et al., The pluripotent regulatory circuitry connecting promoters to their long-range interacting elements, Genome Res. 25 (4) (2015) 582–597. Available from: https://doi.org/10.1101/gr.185272.114.
[310] J.R. Dixon, D.U. Gorkin, B. Ren, Chromatin domains: the unit of chromosome organization, Mol. Cell 62 (5) (2016) 668–680. Available from: https://doi.org/10.1016/j.molcel.2016.05.018.
[311] K. Lee, G.A. Blobel, Chromatin architecture underpinning transcription elongation, Nucleus 7 (4) (2016) 1–8. Available from: https://doi.org/10.1080/19491034.2016.1200770.
[312] R.C. Allshire, H.D. Madhani, Ten principles of heterochromatin formation and function, Nat. Rev. Mol. Cell Biol. 19 (4) (2018) 229–244. Available from: https://doi.org/10.1038/nrm.2017.119.
[313] S.H. Schwartz, J. Silva, D. Burstein, T. Pupko, E. Eyras, G. Ast, Large-scale comparative analysis of splicing signals and their corresponding splicing factors in eukaryotes, Genome Res. 18 (1) (2008) 88–103.
[314] R. Britten, D.E. Kohne, Repeated sequences in DNA. Hundreds of thousands of copies of DNA sequences have been incorporated into the genomes of higher organisms, Science 161 (1968) 529–540.
[315] B. McClintock, Controlling elements and the gene, Cold Spring Harb. Symp. Quant. Biol. 21 (1952) 197–216.
[316] R.J. Britten, E.H. Davidson, Repetitive and non-repetitive DNA sequences and a speculation on the origins of evolutionary novelty, Q. Rev. Biol. 46 (2) (1971) 111–138.
[317] E.H. Davidson, D.H. Erwin, Gene regulatory networks and the evolution of animal body plans, Science 311 (5762) (2006) 796–800.
[318] E.H. Davidson, The Regulatory Genome, Academic Press, San Diego, CA, 2006.
[319] J.A. Shapiro, R. v Sternberg, Why repetitive DNA is essential to genome function, Biol. Revs. (Camb.) 80 (2005) 227–250.
[320] J.A. Shapiro, A 21st century view of evolution: genome system architecture, repetitive DNA, and natural genetic engineering, Gene 345 (1) (2005) 91–100 [Online]. Available: <http://www.ncbi.nlm.nih.gov/entrez/query.fcgi?cmd = Retrieve&db = PubMed&dopt = Citation&list_uids = 15716117>.
[321] G. Liu, J.S. Mattick, R.J. Taft, A meta-analysis of the genomic and transcriptomic composition of complex life, Cell Cycle 12 (13) (2013) 2061–2072. Available from: https://doi.org/10.4161/cc.25134.
[322] A.K. Eggleston, A. Eccleston, B. Marte, C. Lupp, Regulatory RNA, Nature 482 (7385) (2012) 321. Available from: https://doi.org/10.1038/482321a.
[323] K.V. Morris, J.S. Mattick, The rise of regulatory RNA, Nat. Rev. Genet. 15 (6) (2014) 423–437. Available from: https://doi.org/10.1038/nrg3722.
[324] J. Brate, M. Adamski, R.S. Neumann, K. Shalchian-Tabrizi, M. Adamska, Regulatory RNA at the root of animals: dynamic expression of developmental lincRNAs in the calcisponge Sycon ciliatum, Proc. Biol. Sci. 282 (1821) (2015). Available from: https://doi.org/10.1098/rspb.2015.1746.
[325] J.A. Shapiro, How life changes itself: the Read-Write (RW) genome, Phys. Life Rev. 10 (3) (2013) 287–323. Available from: https://doi.org/10.1016/j.plrev.2013.07.001.

[326] E. Nevo, Molecular evolution and ecological stress at global, regional and local scales: the Israeli perspective, J. Exp. Zool. 282 (1998) 95–119.

[327] T.J. Kawecki, R.E. Lenski, D. Ebert, B. Hollis, I. Olivieri, M.C. Whitlock, Experimental evolution, Trends Ecol. Evol. 27 (10) (2012) 547–560. Available from: https://doi.org/10.1016/j.tree.2012.06.001.

Declarations

The author is solely responsible for the contents of this article, has no competing interests, received no funding for this article, consents to publication in *New Horizons in Evolution*, and makes relevant references available on his university web page http://shapiro.bsd.uchicago.edu/.

Specific online links to the author's web site (each of these links leads to further, more detailed sets of references, most of which have a clickable link to the article's PubMed entry):

Online Link 1. http://shapiro.bsd.uchicago.edu/evolution21.shtml (2011), http://shapiro.bsd.uchicago.edu/Genome%20Writing%20by%20Natural%20 Genetic%20Engineering.html (2017), and http://shapiro.bsd.uchicago.edu/% 20No_Genome_Is_An_Island_Extra_References.html (2019).

Online Link 2. http://shapiro.bsd.uchicago.edu/DNA_Transposons.shtml (2011); http://shapiro.bsd.uchicago.edu/Endogenous_Retroviruses.html (2011); http://shapiro.bsd.uchicago.edu/LTR_Retrotransposons.shtml (2011); http://shapiro.bsd.uchicago.edu/Distributed_genome_network_innovation_ attributed_to_mobile_DNA_elements.html (2017).

Online Link 3. DNA transfer, repair, rearrangement and mutator functions: http://shapiro.bsd.uchicago.edu/ExtraRefs.MolecularMechanisms NaturalGeneticEngineering.shtml (2011); http://shapiro.bsd.uchicago.edu/ ExtraRefs.NaturalGeneticEngineeringPartNormalLifeCycle.shtml (2011); http://shapiro.bsd.uchicago.edu/ExtraRefs.ProofreadingDNAR eplication.shtml (2011); http://shapiro.bsd.uchicago.edu/Exonuclease_ proofreading.html (2011); http://shapiro.bsd.uchicago.edu/ExtraRefs. DNADamageRepairAndMutagenesis.shtml (2011); http://shapiro.bsd.uchicago.edu/Translesion_mutator_polymerases.html (2011); http://shapiro.bsd. uchicago.edu/Chromatin_in_DNA_proofreading_and_repair.html (2011).

Online Link 4. http://shapiro.bsd.uchicago.edu/ExtraRefs.Cellular RegulationNaturalGeneticEngineering.shtml (2011).

Online Link 5. http://shapiro.bsd.uchicago.edu/StimuliDocumented ActivateNGE.html (2011); http://shapiro.bsd.uchicago.edu/Ecological_ Factors_that_Induce_Mutagenic_DNA_Repair_or_Modulate_NGE_Responses.html (2017).

Online Link 6. http://shapiro.bsd.uchicago.edu/Modes_of_Horizontal_DNA_Transfer.html (2017).

Online Link 7. http://shapiro.bsd.uchicago.edu/No_Genome_is_an_Island_Extra_References_9.html (2019).

Online Link 8. http://shapiro.bsd.uchicago.edu/No_Genome_is_an_Island_Extra_References_5.html (2019).

Online Link 9. http://shapiro.bsd.uchicago.edu/No_Genome_is_an_Island_Extra_References_4.html (2019).

Online Link 10. http://shapiro.bsd.uchicago.edu/Photosynthetic_eukaryotic_lineages_resulting_from_symbiogenesis.html (2017).

Online Link 11. http://shapiro.bsd.uchicago.edu/No_Genome_is_an_Island_Extra_References_7.html (2019).

Online Link 12. http://shapiro.bsd.uchicago.edu/Hybrid_dysgenesis_interspecific_hybridization.html (2011); http://shapiro.bsd.uchicago.edu/Selected_Examples_of_Speciation_and_Adaptive_Radiation_Involving_Interspecific_Hybridization_and_Changes_in%20Chromosome_Number.html (2017).

Online Link 13. http://shapiro.bsd.uchicago.edu/Genomic_consequences_of_experimental_interspecific_hybridization_in_plants_and_animals.html (2017).

Online Link 14. http://shapiro.bsd.uchicago.edu/Exon_Rearrangements_and_Retroposition_by_Mobile_DNA_element_Activity.html (2017).

Online Link 15. http://shapiro.bsd.uchicago.edu/Instances_of_Orphan_Coding_Sequences_%28%E2%80%9CNew_genes%E2%80%9D_and_New_exons%29_Discovered_in_Sequenced_Genomes.html (2011).

Online Link 16. http://shapiro.bsd.uchicago.edu/Origination_of_Novel_Exons_from_Mobile_DNA_Elements.html (2017).

Online Link 17. http://shapiro.bsd.uchicago.edu/ExtraRefs.AntibioticResistanceAndHorizontalTransfer.shtml (2011).

Online Link 18. http://shapiro.bsd.uchicago.edu/Modes_of_Horizontal_DNA_Transfer.html (2017).

Online Link 19. Horizontal DNA transfers across taxonomic boundaries; http://shapiro.bsd.uchicago.edu/Interkingdom_and_Eukaryotic_Horizontal_Transfer.html (2011); http://shapiro.bsd.uchicago.edu/Examples_of_inter-phylum_adaptive_horizontal_DNA_transfers_based_on_genomic_data.html (2017).

Online Link 20. http://shapiro.bsd.uchicago.edu/No_Genome_is_an_Island_Extra_References_8.html (2019).

Online Link 21. http://shapiro.bsd.uchicago.edu/No_Genome_is_an_Island_Extra_References_9.html (2019).

Online Link 22. http://shapiro.bsd.uchicago.edu/Distributed_genome_network_innovation_attributed_to_mobile_DNA_elements.html (2017); http://shapiro.bsd.uchicago.edu/Reevaluation_of_the_%E2%80%9CJunk_DNA%E2%80%9D_Concept_for_Repetitive_Mobile_Genetic_Elements_in_Genomes_of_Advanced_Organisms.html (2017).

Online Link 23. http://shapiro.bsd.uchicago.edu/Regulatory_Functions_Reported_for_Long_Non-coding_lncRNA_molecules.html (2017).

CHAPTER 2

Experience and the genome: the role of epigenetics

Moshe Szyf
Department of Pharmacology and Therapeutics, Faculty of Medicine, McGill University, Montreal, QC, Canada

Introduction

Epigenetics emerged as a field attempting to explain how one genome drives numerous cell-, time-, and context-specific genomic programs in multicellular organisms. The term epigenetics was first coined by Waddington to provide a framework for the changes occurring in DNA during cellular differentiation [1,2]. Several interrelated mechanisms emerged to confer upon DNA the capacity to express numerous different stable phenotypes. Modification of DNA by enzyme-catalyzed addition of a methyl group from the methyl donor *S*-adenosyl methionine (SAM) to generate either N(6)-methyladenine or 5-methylcytosine [3] (Fig. 2.1) appears early in evolution and is present in bacteria. The first discovered role of DNA methylation in bacteria was in restriction modification [4]. DNA methylation serves as a protection against invasion by bacteriophages by differentiating "self" from invader. Sequence-specific methylating enzymes, DNA methyltransferases methylate the host DNA and protect it from a methylation resistant restriction enzyme. Invading unmethylated DNA is cleaved by the restriction enzyme [4]. This early evolutionary role of DNA methylation instructs us on the first principles of DNA methylation. Differential DNA methylation provides diverse identities to identical DNA sequences by interfering with the binding of proteins to DNA. Thus DNA methylation dramatically expands the functional capacity of DNA and provides it with potential plasticity. DNA methylation is involved in bacteria in multiple genomic functions; control of origins of replication, gene expression and mismatch repair (for a review see [5]). In bacteria, DNA methylation can also function in creating phenotypic variants of identical genetic populations which can adapt bacteria to changing environmental conditions [5]. Thus the

New Horizons in Evolution
DOI: https://doi.org/10.1016/B978-0-323-90752-1.00001-8

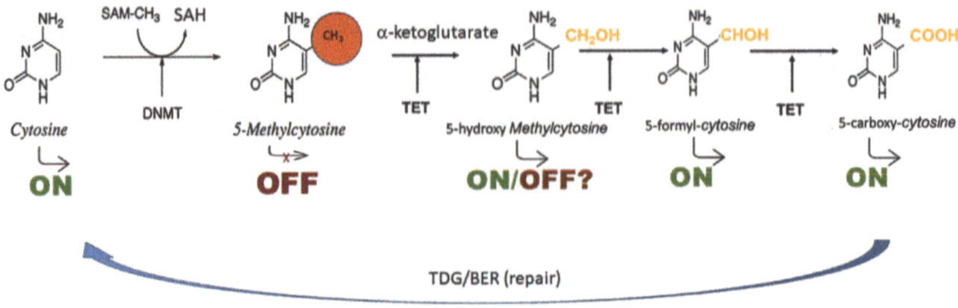

Figure 2.1 ***Cytosine modifications in vertebrate DNA.*** Cytosine is methylated at the 5′ position by a reaction catalyzed by DNA methyltransferases (DNMT). The methyl donor is *S*-adenosyl methionine (SAM). The products of the reaction are 5-methylcytosine and *S*-adenosyl homocysteine. 5-Methylcytosine is further oxidized to 5-hydroxymethylcytosine then to 5-formylcytosine and finally to 5-carboxyl cytosine in reactions catalyzed by Tet enzymes using a-ketoglutarate as a cofactor. 5-Formylcytosine and 5-carboxyl cytosine could be removed by a base excision repair (BER) catalyzed by thymidine deglycosylase (TDG) followed by repair with BER enzymes to generate unmethylated cytosine. When cytosine modifications are found in strategic positions in genes in promoters and enhancers, they alter the state of expression of a gene (horizontal arrow) which is indicated by ON or OFF signs. The different modifications of cytosine serve as different epigenetic signals.

involvement of DNA methylation in reversible adaptation to changing environments appears early in evolution and is present even in unicellular organisms.

The methyl moiety in 5-methylcytosine can be further modified by oxidation to 5-hydroxymethylcytosine. 5-hydroxymethylcytosine was first discovered in Escherichia *coli* phages T2, T4, and T6 [6]. However, in contrast to the situation in vertebrates, the 5-hydroxymethylcytosine in coliphages is synthesized prior to incorporation into DNA by a phage encoded enzyme using formaldehyde and tetrahydrofolate [7]. This phage catalyzed modification serves for protection of the phage DNA, since the 5-hydroxymethylcytosine is further modified by addition of UDP-glucose which blocks the activity of host restriction enzymes.

In vertebrates, the main studied DNA methylation is 5-methylcytosine (Fig. 2.1). 5-Methylcytosine was discovered by Hotchkiss in 1948 [8] and its potential biological role has been intriguing since. 6-Methyladenine has been detected in vertebrate DNA as well but it has received attention only recently and recent data suggest that it might be correlated with gene expression [9], however the presence of 6-methyladenine in vertebrate DNA has been disputed [10,11].

DNA methylation conferring cell type identity on DNA

DNA methylation in vertebrates is prevalent in the dinucleotide sequence CG, between 60% and 80% of CG dinucleotides are methylated in vertebrate tissues [12]. The use of a pair of restriction enzymes HpaII which is sensitive to methylation at CG and MspI which is insensitive to this methylation at CG dinucleotides (in the context of the sequence CCGG) [13] led to the discovery that the methylation of CGs is differentially distributed in tissues and that methylation at promoters of genes is negatively correlated with expression [14,15]. This discovery provided the first hint that in vertebrates, DNA methylation at CG dinucleotides is involved in tissue specific gene expression. For a pattern of methylation to be preserved across cell divisions, there should be a mechanism for inheritance. The CG dinucleotide is a palindrome and thus during DNA replication a nascent unmethylated cytosine in the dinucleotide sequence 5′CG3′ is positioned across a 3′GC5′ in the parental strand whose state of methylation serves as a template for DNA methyltransferase 1 (DNMT1) to copy to the nascent strand [16]. DNMT1 activity is enhanced by a hemimethylated substrate that is generated during replication of a methylated CG [16]. This provides a simple mechanism for preservation and clonal inheritance of DNA methylation patterns, which was demonstrated by showing that the methylation pattern of in vitro methylated ectopic DNA stably transfected into mouse cells in culture is preserved across many passages [17]. More recent data suggest that the maintenance of DNA methylation during replication involves additional proteins such as URHF1 which recognizes hemimethylated DNA and recruits DNMT1 to the replication fork [18].

Although CG methylation is prevalent in vertebrate genomes, methylation can happen in all other dinucleotide contexts CA, CT and CC. There is no known simple mechanism for inheritance of nonCG methylation since there is no methylated C in the parental strand to guide methylation of the nascent strand; for example, across a CA there is a TG sequence. NonCG methylation is therefore hard to preserve in replicating cells but they do accumulate in postmitotic neurons in the brain [19]. New DNA methylation patterns are established by de novo methylation enzymes that do not require template methylation, DNMT3a and DNMT3b [20–22]. De novo methyltransferases are particularly abundant in embryonal cells and the brain. In embryonal tissues, they are involved in building new DNA methylation patterns. In the brain, de novo methylation enzymes are possibly required for establishing new DNA methylation profiles during learning and other brain functions [19]. De novo methyltransferases are required for enhanced plasticity and diversity of DNA methylation patterns. They could potentially expand the DNA methylation repertoire in response to experience beyond the developmentally regulated conserved tissue specific patterns of methylation. NonCG methylation is probably catalyzed by de novo methyltransferase DNMT3a [23].

The 5 methyl moiety in cytosine is further modified by oxidation catalyzed by the dioxygenases Ten-Eleven Translocation-1, 2, and 3 (Tet1, Tet2, and Tet3) which convert 5-methylcytosine to 5-methylhydroxycytosine, 5-formylcytosine and 5-carboxylcytosine [24–26] (Fig. 2.1). These newly discovered modifications of methyl cytosine further expand the plastic capacity of the genome by offering further nuanced differentiation of genomic identities. Oxidation of 5-methylcytosine can also lead to demethylation either through passive loss of the oxidized methyl moiety during replication since it does not serve as a template for DNMT1, which requires a hemimethylated substrate [27], or by triggering a DNA repair activity involving DNA glycosylase (TDG) enzymes, which could lead to DNA demethylation even in nondividing cells such as brain neurons [19]. This is a possible mechanism for demethylation of genes during development (Fig. 2.1). It is unclear which mechanisms determine whether an oxidized methyl moiety is maintained as an epigenetic signal or whether it is removed in a demethylation reaction [28]. In any case, a significant number of oxidized 5-methylcytosine modifications are stably maintained in the genome [29,30].

DNA methylation and gene function

The early observations that DNA methylation patterns showed cell type specificity [31] and the inverse correlation between methylation and gene expression in promoters led to the hypothesis that DNA methylation controlled gene function and was potentially involved in cellular differentiation [32]. However, correlation does not prove causation; DNA methylation might be a consequence rather than a cause of gene expression. Three lines of evidence provide support to the hypothesis that DNA methylation in promoters silences gene expression and that an unmethylated promoter is required for gene activity. First, methylation of reporter genes in vitro silenced their expression when they were introduced into mammalian cell lines by DNA mediated gene transfer [33,34]. Second, inhibitors of DNA methylating enzymes (DNMTs) such as 5-azacytdine or its deoxy derivative turned on expression of methylated genes [35,36]. Third, depletion of *dnmt1* mRNA by either antisense inhibition [37] or genetic deletion [38] altered patterns of gene expression and turned on genes silenced by DNA methylation. *dnmt1* depletion in fibroblasts can trigger their differentiation to myocytes [37] while genetic depletion of *dnmt1* [38] leads to disruption of development and embryonal lethality, supporting the involvement of DNA methylation enzymes in defining cellular identity and development.

Although the combination of these lines of evidence is consistent with the idea that DNA methylation in strategic positions of genes such as promoters and enhancers silence gene expression and that removal of DNA methylation is necessary for induction of gene expression, each of these lines of evidence is confounded. Silencing ectopic DNA by methylation might be limited to foreign DNA and might not represent genomic regulation. 5-Azacytidine might induce gene expression by mechanisms other than inhibition of DNA methylation and changes in gene expression in response to *dnmt1* depletion might result from inhibition of DNA methylation independent gene suppression activities of DNMT1 enzyme rather than by inhibition of DNA methylation [39–41]. What is missing is evidence that a change in the state of methylation of a gene in its native position in the genome can alter its state of expression in both directions. Recently CRISPR/deltaCas9 gene editing tools were adapted to epigenetic editing by targeting either the de novo methylating enzyme DNMT3a or the 5-methylcytosine dioxygenase Tet to specific gene regulatory regions in living cells. These studies showed that targeted methylation by deltaCas9-DNMT3a or targeted "demethylation" by deltaCas9-Tet caused change in methylation and expression of the targeted genes [42]. However, it is unclear whether the targeted changes in expression were caused by the targeted changes in DNA methylation, by oxidation, or by other DNA methylation independent activities of the DNMT3a [43] and Tet enzymes [44,45].

Overlapping independently derived genome wide profiles of DNA methylation and gene expression using next generation sequencing methods show only partial inverse correlation between promoter DNA methylation and mRNA levels [46,47], or between promoter methylation and binding of RNApolII phosphorylated at serine 5 (RNAPolII-PS5), a form of RNApolII that is found during transcription onset on 5′ regions of genes [48]. Chromatin immunoprecipitation (ChIP) with antibodies directed against RNApolII-PS5 followed by bisulfite sequencing of ChIPped DNA was used to examine the state of methylation of DNA that was physically engaged in transcription onset. The state of DNA methylation of promoter sequences that are transcriptionally active and are physically bound to RNApolII-PS5 is unmethylated without exception [48]. These data suggest that transcription turn-on is incompatible with promoter DNA methylation. The partial overlap between overall expression and overall promoter methylation results from heterogeneity of expression and methylation across cells in a tissue, which is a mixture of cells where the promoter is unmethylated and highly expressed and others where the promoter is methylated and silenced. Overlapping transcription and methylation across the population of cells will show in this case both methylation and high expression while examining cells which express the gene will reveal unmethylated promoters only. Heterogeneity of methylation within a tissue further increases the plasticity of expression output as not only the level of methylation but also the extent of cell to cell heterogeneity defines the range, scope and variability of expression [48]. These data show that DNA methylation is inconsistent with expression, however it remains to be determined whether methylation defines expression or whether expression defines DNA methylation.

Several studies have shown that during activation of gene expression, demethylation follows rather than precedes transcription turn-on [49,50]. Transcription factor binding was proposed to define demethylated sites by either recruiting DNA demethylating activity or through masking of DNA methylation activity [32]. Data suggest that demethylated positions in enhancers correlate with transcription factor binding sites and that transcription factor binding is required for demethylation [51–56]. These data seem to support the idea that DNA methylation is a consequence rather than cause of gene activation. This does not exclude a possible important role for DNA demethylation in defining a variety of gene programs during differentiation. DNA demethylation could serve as a stable genomic memory of gene activation that is initiated prior and independently of DNA demethylation but nevertheless preserves the long-term active state of the gene [55]. It has been proposed that pioneering transcription factors program DNA demethylation during cellular differentiation, which in turn impacts on long-term accessibility of the promoters/enhancers to transcription factors and transcription machinery [52].

The genome wide analysis of unmethylated promoters reveals that although expressed genes are without exception unmethylated, numerous promoters are unmethylated and silenced [48]. It should be noted that gene activation requires the presence of transcription factors. Specific and cell type specific transcription factors are required to activate a gene. Thus in absence of transcription factors even unmethylated promoters will remain silenced. Demethylation prepares a gene for activation and anticipates future expression of the gene when the appropriate transcription factors will be available. Because of the hierarchical organization of gene function, demethylation and activation of a single transcription factor could result in activation of several unmethylated but silenced downstream targets [57]. Similarly demethylation and activation of a microRNA gene will result in downstream silencing of downstream targets. Thus DNA methylation programs are not a mere mirror image of transcriptional state of a cell. DNA methylation lays down the long-term potential of gene expression programming at future time points and contexts; anticipating future responses to physiological and developmental challenges. For example, demethylation of a glucocorticoid responsive gene might not affect its steady state expression but might anticipate the response to a glucocorticoid surge in the future triggered by a stressful response that would release high levels of glucocorticoid stress hormone [55]. Steady state mRNA expression does not reveal therefore the full potential implications of differential gene methylation. DNA methylation lays down a complex web of transcriptional programming that evolves with time and context. Therefore differences in DNA methylation manifest themselves differently at different times and contexts.

Mechanisms of silencing of expression by DNA methylation

The lesson learned from bacterial methylation is that DNA methylation alters interaction of proteins and their sequence-specific targets in DNA [4]. DNA methylation alters the chemical identity of the cytosine base in DNA and thus could potentially affect the interactions with sequence-specific transcription factors leading to silencing of gene expression. Several transcription factors were shown to be blocked by DNA methylation since 1990, when the first transcription factor activating enhancer binding protein 2 alpha (AP2) inhibited by methylation was described [58]. Inhibition of binding of AP2 by methylation of its target sequences resulted in inhibition of expression of *Proenkephalin* promoter expression vector transfected into rat C6 glioma cells [58]. A second mechanism involves attracting methylated DNA binding proteins family member such as the RETT syndrome protein methyl CpG binding protein 2 (MeCP2) [59]. MeCP2 and other members of this family such as methyl-CpG-binding domain protein 2 (MBD2) [60] attract chromatin repressor histone deacetylase complexes to methylated regions resulting in a silencing chromatin configuration and inhibition of transcription [61,62]. This second mechanism illustrates the interrelationship between chromatin modification and DNA methylation which was confirmed by different lines of study. For example, acetylation of histone 3 and lysine 9 (H3K9) is associated with active genes and hypomethylated promoters (for review see [63]). Inhibition of histone deacetylases (HDAC) with HDAC inhibitors (HDACi) such as TSA [64] butyric acid [65] or valproic acid [66] triggers demethylation and induction of gene activity. There are direct interactions between histone methylation and DNA methylating enzymes. For example, H3K4 tri-methylation, a marker of active promoters inhibits binding of DNMT3a and DNMT3b, preventing hypermethylation of active promoters. Histone H3K36 methylation at gene bodies of transcribed genes recruits DNA methylating enzymes DNMT3a and 3b (for a review see [67]). It is yet unclear how gene body methylation is involved in gene activity but it possibly silences spurious promoters in gene bodies [68].

Oxidized modifications of the methyl moiety in cytosine are most probably involved in epigenetic regulation of gene expression and it stands to reason that the different oxidation forms bear different epigenetic meaning. Although genome wide profiling has suggested that 5-hydroxymethylcytosine in enhancers is associated with cellular differentiation and gene activation [69], other studies have suggested a repressive function [70] (for a review see [28]). 5-Formylcytosine and 5-carboxycytosine which were believed to be intermediates in a DNA demethylation chain are also stably present in the genome, tend to concentrate in enhancers and are potentially involved in gene activation during cellular differentiation and development [71,72]. The role and mechanisms of these additional modifications of cytosine in gene expression and cellular differentiation will be surely elucidated in coming years.

Methylation of adenine N(6)-methyladenine which is ubiquitous in bacteria and lower eukaryotes was recently discovered in plants [73,74], drosophila [75], and humans [9] and is thus more evolutionary widespread than cytosine CG methylation which is absent in drosophila [76]. N-6 methyladenine is catalyzed by the methyltransferase N6AMT1 and demethylation of N6-methyladenine is catalyzed by the demethylase ALKBH1 [9]. The role of N(6)-methyladenine in gene expression is unknown as of yet. N(6)-methyladenine is localized at exon boundaries of actively transcribed genes [9] suggesting a role in gene activation while it was also shown to colocalize with H3K9me3 a silencing histone mark in human glioblastoma suggesting a silencing role [77]. While methylation at enhancers is associated with gene silencing, methylation in bodies of genes is associated with gene activity [78]. In drosophila brains, accumulation of N(6)-methyladenine by depleting the demethylase DMAD leads to transcriptional repression in coordination with polycomb proteins [79]. Although, the role of this newly discovered modification is uncertain, these early data point to a role in gene regulation and cellular differentiation like CG cytosine methylation. However, as discussed above recent data question the existence of N(6)-methyladenine in mammalian DNA [10,11].

In summary, a variety of DNA modifications that are involved in expanding the interpretation of genomic information and function have evolved. A diversity of potential combinations of these modifications expands exponentially genomic information enabling identical DNA sequences to develop a diversity of genome functionalities in time and space, which is critical for the development of a multicellular organism. DNA methylation-based modifications interact with histone modifications in a bilateral relationship which further enhances the plasticity of the genome.

Epigenetic programming by exposure and experience

A large body of data discussed above has established the role of DNA methylation in cellular differentiation and development. DNA methylation is a mechanism for conferring different cellular identities on identical DNA by altering the interactions between proteins and DNA. Experience, environment, and exposure are a driving force in evolution through the process of natural selection. Can DNA methylation play a role in creating differential experiential identities to DNA? This could be an important process in environmental adaptation and in generating phenotypic interindividual variation without a change in DNA sequence.

DNA methylation is programmed during development. It is therefore expected that chemical exposures during gestation that affect DNA methylation enzymes would alter DNA methylation processes, resulting in disruption of embryonal development and teratogenicity. However, the question arises whether mechanisms exist that respond to environmental cues early in life in an organized and predictive manner that result in long-term phenotypic changes rather than teratogenicity. The first evidence that maternal diet can affect offspring phenotype came from studies by Randy Jirtle that showed how the methyl content of maternal diet affects methylation during gestation and alters coat color driven by the state of methylation of a long terminal repeat in the mouse *agouti* gene [80]. In addition, maternal diet affected metabolic phenotypes in the offspring later in life [80,81].

Are these epigenetic alterations in response to early life exposures limited to chemical cues such as nutritional intake? Early life social environment has a large impact on behavior and physical health later in life [82]; are there biological mechanism that mediate phenotypic variation in response to behavioral and social signals? Can DNA methylation play a role in conferring "experiential identity" similar to its role in "cellular identity"?

Epigenetic programming by maternal care

First evidence for a plausibility of such a mechanism came from studies in rats examining the impact of differences in maternal care on offspring stress responsivity. There is a natural distribution of levels of maternal behaviors in rats which include Licking and grooming (LG), arch back nursing and nipple switching. Early observations showed that offspring of High LG mothers exhibited in adulthood a more controlled stress behavior than offspring of Low LG mothers [83]. The glucocorticoid stress hormone response to stress is controlled by negative feedback loop in the hippocampus through the glucocorticoid receptor; High LG offspring expressed higher levels of the glucocorticoid receptor than the Low LG offspring [83]. Differences in maternal behavior are nongenetically transmitted as shown by cross fostering experiments when offspring of High LG mothers were cross-fostered to Low LG mothers and vice versa [84]. The caring mother and not the biological mother defined the long-term stress-responsivity phenotype of the offspring.

How does maternal behavior define the life-long stress-response behavior of the offspring? Studies showed that maternal care programmed DNA methylation and histone acetylation of the offspring glucocorticoid receptor (*nr3c1*) exon 7 [85]. High LG offspring had lower DNA methylation and higher histone acetylation and higher *nr3c1* expression than the Low LG offspring. These differences in epigenetic programming by maternal care were established in the postnatal first week and remained stable for life. Cross-fostering reversed the epigenetic programming demonstrating that it was not genetically predetermined, supporting an epigenetic mechanism for maternal behavior programming [85]. What is the possible mechanism that translates maternal behavior into epigenetic programming in the brain? Maternal behavior triggers a serotoninergic signaling pathway in the hippocampus which activates a transcription factor (nerve growth factor induced A) NGFIA that binds the *nr3c1* exon 7 region resulting in recruitment of CREB binding protein and MBD2 to the region leading to increased acetylation and reduced methylation [86,87]. Differences in behavior between the mothers result in differential activation of this pathway and differences in epigenetic programming. This provides a working hypothesis describing a biochemical conduit between social exposure and DNA modification. We hypothesize that epigenetic responses to experience operate through highly conserved signaling pathways that sense the exposure and elicit specific signaling pathways that mediate downstream epigenetic reprogramming. The Low LG phenotype is transmitted across generations not through the germ line but through maternal behavior. Low LG mothers sire low LG offspring which will become Low LG mothers, thus stabilizing the phenotype across generations without germ line transmission. Maternal behavior programs offspring maternal behavior through differential methylation of the *estrogen receptor-alpha1b* promoter in the medial preoptic area of female offspring [88]. This provides an example of a stable epigenetic phenotype that is transferred across multiple generation by epigenetic programming which is both triggered by maternal behavior and programs maternal behavior in the offspring. A cycle of behavior and epigenetic programming stabilizes a phenotype across generations in absence of a sequence change and independent of germline transmission.

Epigenetic programming by maternal behavior is reversible

Epigenetic programs are laid down by bidirectional enzymatic activities and are therefore potentially reversible in contrast to germ line genetic mutations that are naturally selected. This points to the potential for reversing an epigenetic programmed phenotype by modulation of epigenetic enzymatic activities. When the HDAC inhibitor trichostatin A (TSA) was injected to the ventricle of a Low LG offspring adult rat, DNA methylation decreased, histone acetylation increased and the behavior of the treated rats reversed to resemble High LG rats [85]. In contrast, the behavior of High LG rats that were injected with methionine, the precursor of the methyl donor SAM, resembled Low LG rats [89]. These data showed the potential reversibility of epigenetic brain programs. This has implications on the prospect of developing natural products that could modulate the epigenome as treatments for behavioral and neuropsychiatric disorders. In difference from other drugs that treat mental disease symptomatically, epigenetic reprogramming might correct the underlying molecular causes of the disease [90,91].

Early life adversity triggers epigenetic reprogramming

The hypothesis that early life adversity results in epigenetic reprogramming of genes involved in controlling stress responses and other critical brain function was supported by other rodent studies. Early life exposure to an abusive mouse caretaker alters methylation *of brain-derived nerve growth factor* (*bdnf*) in the prefrontal cortex, [92] while early life stress causes demethylation of a gene which plays an important role in the stress response the *arginine vasopressin* (*Avp)* gene promoter [93].

Alterations in DNA methylation and chromatin modification in response to early life stress are broad and affect multiple gene networks

Early life stress is associated with many physiological and behavioral outcomes, it therefore stands to reason that if epigenetic programming is involved in this process it should not be limited to a candidate gene and must involve multiple gene networks and multiple tissues. Differences in maternal care cause changes in expression of hundreds of genes [94]. Analysis of differences in DNA methylation and histone H3K9 acetylation between High and Low LG adult offspring revealed differential methylation of broad genomic regions on both sides of the *nr3c1* locus. Remarkably the entire family of more than 70 protocadherin genes, which were formerly implicated in neuronal development and synaptic transmission [95] were differentially methylated and acetylated in response to differences in maternal care [96]. These data suggest that the epigenetic response to early life social environment is broad and involves several functional gene pathways.

The response to early life social environment is evolutionary conserved

Data from both nonhuman primates and humans show conservation of the epigenetic responses to early life stress both at the candidate gene and genome wide levels. A postmortem study examined methylation at the *Nr3c1* gene in a homologous promoter to the rat exon 1_7, exon 1f in brains from people that committed suicide who were either abused or not during early childhood as well as controls. Remarkably the group who were abused in childhood had higher methylation of the *Nr3c1* exon 1f promoter than people who were not abused in childhood, whether they committed suicide or not [97]. Although the human studies were correlational and were performed on a small number of people, the fact that the same gene is modified in both rodents and humans in response to early life adversity strongly supports the hypothesis that the changes in DNA methylation are caused by early life abuse in humans as well. The DNA methylation of protocadherin gene family is also affected in humans by early life adversity similar to rats [96].

The evolutionary conservation of epigenetic programming of stress response by early life adversity at the same genes and gene families points to its potential importance.

The response to early life adversity is system wide

Early life adversity anticipates challenges later in life that will affect multiple systems in addition to the brain, including the immune and metabolic systems. It is therefore plausible that the epigenetic response to behavioral signals evolved to include multiple physiological systems which are critical for survival in adverse environments. Nonhuman primates that are separated from their mother after birth and reared in a nursery have been examined as a model for studying the role of early life maternal attachment. Adult primates who were deprived of a mother early in life exhibit multiple phenotypic differences [98,99]. Genome wide analysis of promoter methylation showed a distinct signature of maternal deprivation in both the prefrontal cortex and in T cells of adult monkeys [100]. Differences in DNA methylation between nursery and maternally reared monkeys affect broad genomic regions, while some changes in DNA methylation overlap between prefrontal cortex and T cells, other changes remain unique for each tissue [100].

These data are consistent with the idea that the epigenetic response to early life adversity is genome and system wide. The response in T cells is most probably not just a "surrogate" of changes in the brain but possibly reflects the specific role of the immune system in the response to early life adversity.

Differences in maternal social status trigger changes in DNA methylation at birth. Placentae from high ranking monkeys are differently methylated than low ranking monkeys. Thus changes in social status have an impact on placenta methylome and appear at birth before the newborns had any social exposure [101]. These data support the idea of a system wide response to early life social stress. The epigenetic changes in the placenta might reflect the important role of the placenta in nourishment and development of the fetus. Placentae might be an important source of early life adversity biomarkers that predict developmental health trajectories and risk for developing different metabolic, immune and mental health disorders later in life [102]. Interestingly placental DNA methylation predicts behavioral phenotypes in adulthood in a mouse model [102].

Early life adversity affects dynamic developmental trajectories of DNA methylation

DNA methylation changes in response to early life adversity are not static and evolve during development. DNA methylation profiles of T cells in rhesus monkeys evolve during infancy, weaning and early adulthood [103]. Early life stress alters the developmental trajectory of DNA methylation during development. Thus, changes in DNA methylation during initial exposure to early life adversity do not necessarily remain into adulthood, however early life stress alters the normal trajectory of evolution of DNA methylation profiles leading to differences in DNA methylation in adulthood [103]. The differences in DNA methylation in infancy and adulthood are not the same. This could explain how early life adversity would be phenotypically manifested only in adulthood and how adult diseases might be caused by responses to early life events. Importantly the trajectories of DNA methylation and the effects of early life stress on the progression of DNA methylation are sex dependent [103]. This sex specificity is consistent with the different roles of the sexes that emerged in evolution and the idea that the epigenetic reprogramming in response to early life adversity plays an adaptive role preparing the animal for the lifelong environment.

Quebec Ice storm of 1998: a quasiexperimental design for studying early life adversity in humans

Although there is significant evidence for epigenetic programming by early life adversity in animals, it is very difficult to demonstrate similar mechanisms in humans. A first problem is that brain which is the most relevant tissue for social appraisal and behavior is inaccessible in living humans. Postmortem data give us a snapshot of the epigenetic state at the time of death, but do not capture the dynamic changes during life, nor is it clear whether the methylome at death reflects accurately the methylome during the life course. Since brain is inaccessible, peripheral tissues could be used. However, they do not represent the methylome of specific brain regions. Nevertheless, as discussed above there is evidence that changes in response to early child adversity are not limited to the brain and that there is an evolutionary and physiological explanation for system-wide responses to early child adversity particularly in the immune system. Many studies have shown associations between changes in stress related genes, DNA methylation in adults in blood and early life adversity as well as between genome wide DNA methylation and early life adversity. However, these studies are confounded by other variables including a multitude of exposures as well as genetic heterogeneity, which could not be completely excluded in a human study without randomization of the groups. For example, a recent large study of DNA methylation epigenome wide association in blood with early childhood trauma found no significant association after correction for confounding factors such as cigarette smoking [104]. It should be noted that a significant number of studies in humans have shown association between DNA methylation and early life adversity at candidate genes such *NR3C1* [105,106] exon 1f, the proximal regulator of glucocorticoid receptor *FKBP5* [107–111], *BDNF* [112], *OXTR* [112–114], *and the serotonin transporter SLC6A4* [115]. In any case, the discrepancy in results points to the challenges involved in human epigenetic studies of the effect of experience on the epigenome.

Natural disasters offer unique opportunities to examine the effect of randomly triggered stressors on the methylome since natural disasters appear randomly and target people in their path with no specific selection for genetics or past experiences. The 1998 ice storm in Quebec which left more than 6,000,000 people without electricity offered such an opportunity. King et al. recruited mothers who were pregnant during or became pregnant within three months of the storm, determined their objective stress during this period as well as their cognitive appraisal of the stress and followed up the newborn for more than 15 years [116]. The researchers found that the objective stress or maternal cognitive appraisal of the stress correlated with phenotypic variations among the children in IQ, language [117], autism [118], BMI and obesity, [119], insulin secretion [120], immunity [121], and asthma [122]. This is consistent with the idea that the response to early life experience is not limited to the brain but involves the other two critical systems for survival, metabolism and immunity, supporting a system wide coordinated response to early life stress as discussed above. DNA methylation analysis of children T cells DNA at the age of 13 correlating genome wide DNA methylation profiles and objective maternal stress identified a large number of CG sites whose methylation level correlated with objective stress [123]. Functional analysis of the pathways affected revealed gene pathways representing the three phenotypic dimensions that were discussed above: metabolism and obesity, immunity which was as expected predominant, and behavior [123]. The data provide evidence for the lasting impact of early life stress on the methylome. The data also provide an opportunity to examine whether these methylation differences mediate the phenotypic changes using statistical mediation analyses. These analyses showed that DNA methylation mediates the effect of maternal cognitive appraisal of the disaster on children insulin secretion [120], children BMI, and central adiposity [124], as well as the effect of maternal stress on childhood asthma [122] and cytokine production [125]. These data further support a system-wide response in the methylome to early life stress that targets metabolism, behavior, and immunity and that these are registered in the immune system. The immune system might be a major player in the system-wide adaptation to early life stress.

How can early life stress produce a system-wide epigenetic response that lasts into adulthood?

Stress is perceived by the brain; therefore it is expected that DNA methylation differences in newborns exposed to an early stressful environment will be detected only in the brain. How can early life stress produce a system-wide epigenetic response? One possible mechanism could involve the glucocorticoid stress hormone which has receptors in all tissues including sperm. An infant exposed to a stressful experience will respond by release of glucocorticoid hormone which will act on receptors in different tissues; binding of glucocorticoid (GC)-bound glucocorticoid receptor Nr3c1 to targets in different tissues will result in epigenetic reprogramming (Fig. 2.2). Since glucocorticoids are well established to act on metabolic [126–129], immune [130] and brain pathways [123–125], this could provide an explanation for the methylome and phenotypic effects seen in children exposed to early life stress. If this hypothesis is true, the glucocorticoid receptor *nr3c1* should be important for the evolution of DNA methylation during development. In accordance with this hypothesis, a mouse gene knockout of one copy of *nr3c1* in the fetuses but not the mothers resulted in sex dependent changes in DNA methylation across many loci in the fetal placenta [102]. DNA methylation changed in both males and females, however in the opposite directions. Differences in methylation between male and females who had one copy of *nr3c1* were dramatically larger than in wild type; a sex by genotype effect on DNA methylation. This study supports the idea that *nr3c1* plays a role in shaping the DNA methylation profile in peripheral tissues and positions

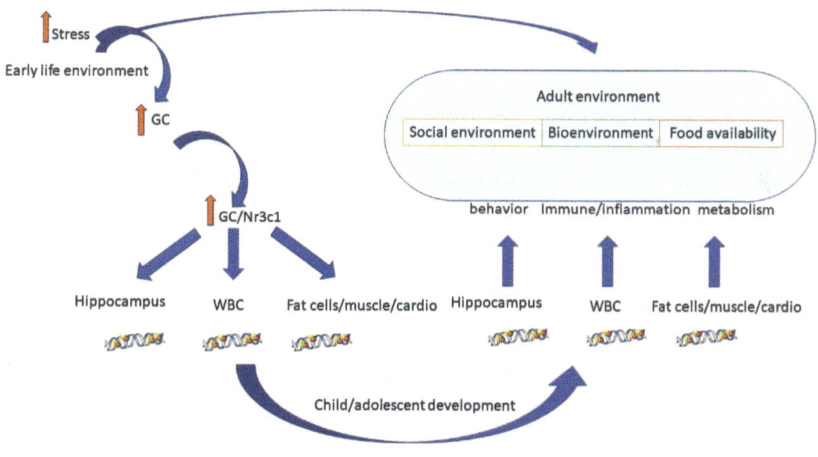

Figure 2.2 ***DNA methylation mediating system wide life-long adaptation to early life signals.*** Increased stress in postnatal early life environment anticipates harsh environment during adulthood. Stress leads to release of glucocorticoids (GC) in the infant which acts on glucocorticoid receptors (Nr3c1) in different tissues including the brain (hippocampus), white blood cells, and cardiovascular muscle and fat tissues. Binding of the glucocorticoid receptor at different positions in the genome leads to rearrangement of the methylome. During childhood and adolescence, the initial reprogramming affects the evolution of the DNA methylation landscape in these tissues. The adult DNA methylation landscape is programmed for expressing an adaptive behavioral, immune, and metabolic phenotype to cope with the anticipated adult environment.

the glucocorticoid hormone as a possible mediator of the system wide effect of maternal stress on offspring DNA methylation [102].

Other hormones and possibly microRNA which are systemically distributed may act similarly to modulate DNA methylation in target genes in multiple tissues.

Summary and prospective: DNA methylation mediating life-long adaptation to early life signals

Although natural selection during evolution adapts organisms to their environment, the environment undergoes dynamic changes at shorter time scales. Social environments are particularly dynamic. It stands to reason that species evolved mechanisms for phenotypic adaptation at a generation or subgeneration time scale. Social environments do not operate in isolation but are tightly linked to physical- and bio-environments. Thus phenotypic adjustment to social environments will affect behavior, metabolic and immune-inflammatory pathways. Maternal social signals early in life in mammals anticipate the nature of the environment that the offspring is about to encounter in his life. Social stress, which requires behavioral adjustments is also associated with lack of food requiring adjustment of metabolic phenotypes and potential challenges from predators necessitating an adjustment of immune and inflammatory pathways. Thus, the newborn must launch an integrated response to the signals conveyed by the mother (Fig. 2.2). Life-long phenotypic changes require long-term changes in gene function. Changes in gene function could be achieved by changes in gene sequences but these are rare and random events and they get established only through natural selection at a multigenerational time scale.

Epigenetic programming could alter gene function without changing the gene sequence. Multiple epigenetic mechanisms exist that are interrelated. This review focused on DNA modification which is part of the chemical structure of the genome but is laid down by an enzymatic machinery different than the machinery that replicates the DNA sequence. Thus two levels of information exist on the same DNA molecule. The sequence is restricted by Watson and Crick rules while DNA modification is plastic and highly attuned to cellular signaling pathways. The biochemistry of DNA modification provides multiple potential combinations that vastly increase the plasticity of the genome. Different combinations of CG sites from a genomic repertoire of 25×10^6 could be methylated in different cells, different combinations of regulatory elements from a vast repertoire could become methylated, Cs in sequence contexts other than CG could become methylated and preserved particularly in nondividing cells [19], the basic methyl moiety could be further oxidized to hydroxyl, formyl and carboxyl forms [22–24] and methylation of adenine has been recently shown to have functional significance [9]. Tissues and organs are composed of millions of cells which could also be differentially modified; within the same tissue some cells could be methylated and silenced while others are unmethylated and expressed, thus altering the overall output of the tissue [48]. Moreover, hierarchical organization of genes provides opportunity for silencing or activating sets of related downstream genes by altering methylation of just one upstream regulator [57]. Since DNA modification interacts with histone modifications, the potential plasticity is increased dramatically [63,67]. In the past five decades, epigenetic mechanisms and DNA methylation were studied in the context of embryonal development as a mechanism for conferring multiple cellular identities to identical DNA sequences in a multicellular organism.

In the past 15 years, it became clear that epigenetic mechanisms also confer different experiential identities. Thus epigenetic processes provide a mechanism for genomic adaptation to changing environments at a single generation and short time scales. Studies examining maternal care in rodents [85] and nonhuman primates [100] revealed that experiences early in life trigger changes in methylation. Further studies showed that even when the early life experience is social, the DNA methylation and phenotypic consequences are not limited to the brain and involve the immune system as well [100]. Particularly telling are the results from the Quebec ice storm of 1998 [123]. These studies, taking advantage of the randomness of natural disaster provide quasiexperimental evidence for the lifelong consequences of early life stress in humans, supporting the hypothesis that early social stress affects the three cardinal systems neural, metabolic, and immune and that DNA methylation is mediating the effects of early life stress on these lifelong phenotypes [120,124,125].

Epigenetic mediated phenotypic adaptation must be highly organized and predictably responsive to the experiential signals in order to be effective. There should be a molecular link between experiences and targets in the genome in the brain. If indeed metabolic and immune systems are involved, there should be a mechanism for targeting epigenetic changes across these systems as well. Studies have provided some early working hypotheses. For example, maternal care triggers signaling pathways in the hippocampus that lead to targeting of epigenetic proteins to the *nr3c1* gene which is responsible for feedback inhibition of stress responses [86,87]. The stress hormone is well positioned to coordinate an organized epigenetic response to early life stress across neuronal, metabolic, and immune systems as the physiological role of glucocorticoids is to regulate brain [126–128], metabolic [129–132] and immune-inflammatory functions [133] (Fig. 2.2). The glucocorticoid receptor is known to recruit epigenetic modulating proteins to genes and alter chromatin structure and DNA methylation [134]. Ectopic perinatal administration of glucocorticoids results in numerous epigenetic changes across the body [135], exposure of human hippocampal neuron progenitors to glucocorticoids results in long lasting changes in DNA methylation [136] and genetic depletion of *Nr3c1* in the fetus results in numerous alterations in DNA methylation in the placenta [102].

Adaptations in early life that will affect life-long phenotypes should be anticipatory; changes triggered early in life should guide phenotypic alterations that manifest themselves at different times and contexts later in life in adulthood. The biochemical, cellular and developmental properties of the epigenetic machinery position them to play this role. DNA methylation changes early in life could prime phenotypic changes that will manifest themselves later in life. For example, demethylation of glucocorticoid responsive elements will be expressed phenotypically later in life when there is a surge in glucocorticoids in response to stress [55,137], demethylation of a promoter will be expressed only when a transcription factor required for activating the gene is expressed at a different time [57]. DNA methylation profiles evolve during development; early life stress alters the trajectory of evolution of DNA methylation during development into adulthood and thus an early life event will determine changes in DNA methylation that will occur later in development [103] (Fig. 2.2). Interestingly placental DNA methylation predicts behavioral phenotypes in adulthood in a mouse model [102].

The mechanisms that are involved in regulation of expression and the phenotype by DNA modification are yet to be clarified, particularly the new oxidized modifications of the methyl moiety in cytosine as well as the intricate relationship between chromatin modification and DNA modifications. Nevertheless, the data to date support the hypothesis that DNA modification serves as a mechanism for genomic adaptation to anticipated lifelong environments. Although this review examined the impact of early life stress, the working model proposed here offers a working hypothesis for other early life exposures such as nutrition as well as infection, inflammation, and microbiota and other mediating hormones.

The advantage of epigenetic adaptation over genetic adaptation is that it is reversible. Since experiences are dynamic and changeable, this mechanism allows for reversal when conditions change as is exemplified in the cross fostering studies in rats [85]. The balance between stability of epigenetic changes and potential reversibility needs to be further clarified. Could epigenetics operate at an evolutionary time scale whereby phenotypes are altered stably across many generations? Do epigenetic alterations play a role in speciation? For example, a recent study indicated that domestication and dog-breed formation of dogs is associated with broad changes in DNA methylation in the brain suggesting a role in speciation of dogs from wolves [138]. Are epigenetic adaptations genetically fixed when the phenotype is stabilized? Since epigenetic reprogramming is guided by environment in a predictable and organized process as discussed above, is there a possibility for similar processes operating in evolution?

Author contributions

MS conceived and wrote the article.

Funding

The publication of this article was funded by the Canadian Institute for Health Research to MS PJT-159583.

References

[1] C.H. Waddington, Canalization of development and genetic assimilation of acquired characters, Nature 183 (4676) (1959) 1654–1655 [Online]. Available from: <http://www.ncbi.nlm.nih.gov/entrez/query.fcgi?cmd = Retrieve&db = PubMed&dopt = Citation&list_uids = 13666847>.

[2] E. Jablonka, M.J. Lamb, The changing concept of epigenetics, Ann. N. Y. Acad. Sci. 981 (2002) 82–96 [Online]. Available from: <http://www.ncbi.nlm.nih.gov/entrez/query.fcgi?cmd = Retrieve&db = PubMed&dopt = Citation&list_uids = 12547675>.

[3] J.C. Wu, D.V. Santi, On the mechanism and inhibition of DNA cytosine methyltransferases, Prog. Clin. Biol. Res. 198 (1985) 119–129 [Online]. Available from: <http://www.ncbi.nlm.nih.gov/htbin-post/Entrez/query?db = m&form = 6&dopt = r&uid = 0004070306>.

[4] W. Arber, S. Linn, DNA modification and restriction, Annu. Rev. Biochem. 38 (1969) 467–500. Available from: https://doi.org/10.1146/annurev.bi.38.070169.002343.

[5] M.A. Sanchez-Romero, J. Casadesus, The bacterial epigenome, Nat. Rev. Microbiol. (2019). Available from: https://doi.org/10.1038/s41579-019-0286-2.
[6] G.R. Wyatt, S.S. Cohen, The bases of the nucleic acids of some bacterial and animal viruses: the occurrence of 5-hydroxymethylcytosine, Biochem. J. 55 (5) (1953) 774–782 [Online]. Available from: <http://www.ncbi.nlm.nih.gov/pubmed/13115372> or <http://www.ncbi.nlm.nih.gov/pmc/articles/PMC1269533/pdf/biochemj01090-0067.pdf>.
[7] J.G. Flaks, J. Lichtenstein, S.S. Cohen, Virus-induced acquisition of metabolic function. II. Studies on the origin of the deoxycytidylate hydroxymethylase of bacteriophage-infected *E. coli*, J. Biol. Chem. 234 (6) (1959) 1507–1511 [Online]. Available from: <http://www.ncbi.nlm.nih.gov/pubmed/13654406>, <http://www.jbc.org/content/234/6/1507.full.pdf>.
[8] R.D. Hotchkiss, The quantitative separation of purines, pyrimidines, and nucleosides by paper chromatography, J. Biol. Chem. 175 (1) (1948) 315–332 [Online]. Available from: <http://www.ncbi.nlm.nih.gov/entrez/query.fcgi?cmd = Retrieve&db = PubMed&dopt = Citation&list_uids = 18873306>.
[9] C.L. Xiao, et al., N(6)-methyladenine DNA modification in the human genome, Mol. Cell 71 (2) (2018) 306–318. Available from: https://doi.org/10.1016/j.molcel.2018.06.015. e7.
[10] K. Douvlataniotis, M. Bensberg, A. Lentini, B. Gylemo, C.E. Nestor, No evidence for DNA N (6)-methyladenine in mammals, Sci. Adv. 6 (12) (2020) eaay3335. Available from: https://doi.org/10.1126/sciadv.aay3335.
[11] X. Liu, et al., N(6)-methyladenine is incorporated into mammalian genome by DNA polymerase, Cell Res. 31 (1) (2021) 94–97. Available from: https://doi.org/10.1038/s41422-020-0317-6.
[12] Y. Gruenbaum, R. Stein, H. Cedar, A. Razin, Methylation of CpG sequences in eukaryotic DNA, FEBS Lett. 124 (1) (1981) 67–71 [Online]. Available from: <http://www.ncbi.nlm.nih.gov/htbin-post/Entrez/query?db = m&form = 6&dopt = r&uid = 0007215556>.
[13] C. Waalwijk, R.A. Flavell, MspI, an isoschizomer of hpaII which cleaves both unmethylated and methylated hpaII sites, Nucleic Acids Res. 5 (9) (1978) 3231–3236 [Online]. Available from: <http://www.ncbi.nlm.nih.gov/htbin-post/Entrez/query?db = m&form = 6&dopt = r&uid = 704354>.
[14] L.H. van der Ploeg, R.A. Flavell, DNA methylation in the human gamma delta beta-globin locus in erythroid and nonerythroid tissues, Cell 19 (4) (1980) 947–958 [Online]. Available from: <http://www.ncbi.nlm.nih.gov/entrez/query.fcgi?cmd = Retrieve&db = PubMed&dopt = Citation&list_uids = 6247075>.
[15] C. Waalwijk, R.A. Flavell, DNA methylation at a CCGG sequence in the large intron of the rabbit beta-globin gene: tissue-specific variations, Nucleic Acids Res. 5 (12) (1978) 4631–4634 [Online]. Available from: <http://www.ncbi.nlm.nih.gov/htbin-post/Entrez/query?db = m&form = 6&dopt = r&uid = 745990>.
[16] S. Urieli-Shoval, Y. Gruenbaum, A. Razin, Sequence and substrate specificity of isolated DNA methylases from *Escherichia coli* C, J. Bacteriol. 153 (1) (1983) 274–280 [Online]. Available from: <http://www.ncbi.nlm.nih.gov/htbin-post/Entrez/query?db = m&form = 6&dopt = r&uid = 0006336735>.
[17] R. Stein, Y. Gruenbaum, Y. Pollack, A. Razin, H. Cedar, Clonal inheritance of the pattern of DNA methylation in mouse cells, Proc. Natl. Acad. Sci. U. S. A. 79 (1) (1982) 61–65 [Online]. Available from: <http://www.ncbi.nlm.nih.gov/htbin-post/Entrez/query?db = m&form = 6&dopt = r&uid = 0006459581>.
[18] P. Bashtrykov, G. Jankevicius, R.Z. Jurkowska, S. Ragozin, A. Jeltsch, The Uhrf1 protein stimulates the activity and specificity of the maintenance DNA

methyltransferase Dnmt1 by an allosteric mechanism, J. Biol. Chem. 289 (7) (2014) 4106–4115. Available from: https://doi.org/10.1074/jbc.M113.528893.

[19] R. Lister, et al., Global epigenomic reconfiguration during mammalian brain development, Science 341 (6146) (2013) 1237905. Available from: https://doi.org/10.1126/science.1237905.

[20] C.L. Hsieh, In vivo activity of murine de novo methyltransferases, Dnmt3a and Dnmt3b, Mol. Cell Biol. 19 (12) (1999) 8211–8218 [Online]. Available from: <http://www.ncbi.nlm.nih.gov/htbin-post/Entrez/query?db = m&form = 6&dopt = r&uid = 0010567546> or <http://mcb.asm.org/cgi/content/full/19/12/8211> or <http://mcb.asm.org/cgi/content/abstract/19/12/8211>.

[21] M. Okano, D.W. Bell, D.A. Haber, E. Li, DNA methyltransferases Dnmt3a and Dnmt3b are essential for de novo methylation and mammalian development, Cell 99 (3) (1999) 247–257 [Online]. Available from: <http://www.ncbi.nlm.nih.gov/htbin-post/Entrez/query?db = m&form = 6&dopt = r&uid = 0010555141>.

[22] M. Okano, S. Xie, E. Li, Cloning and characterization of a family of novel mammalian DNA (cytosine-5) methyltransferases, Nat. Genet. 19 (3) (1998) 219–220 [Online]. Available from: <http://www.ncbi.nlm.nih.gov/entrez/query.fcgi?cmd = Retrieve&db = PubMed&dopt = Citation&list_uids = 9662389>.

[23] B.H. Ramsahoye, D. Biniszkiewicz, F. Lyko, V. Clark, A.P. Bird, R. Jaenisch, Non-CpG methylation is prevalent in embryonic stem cells and may be mediated by DNA methyltransferase 3a, Proc. Natl. Acad. Sci. U. S. A. 97 (10) (2000) 5237–5242 [Online]. Available from: <http://www.ncbi.nlm.nih.gov/htbin-post/Entrez/query?db = m&form = 6&dopt = r&uid = 0010805783>.

[24] S. Kriaucionis, N. Heintz, The nuclear DNA base 5-hydroxymethylcytosine is present in Purkinje neurons and the brain, Science 324 (5929) (2009) 929–930. Available from: https://doi.org/10.1126/science.1169786. 1169786 [pii].

[25] M. Tahiliani, et al., Conversion of 5-methylcytosine to 5-hydroxymethylcytosine in mammalian DNA by MLL partner TET1, Science 324 (5929) (2009) 930–935. Available from: https://doi.org/10.1126/science.1170116.

[26] A. Maiti, A.C. Drohat, Thymine DNA glycosylase can rapidly excise 5-formylcytosine and 5-carboxylcytosine: potential implications for active demethylation of CpG sites, J. Biol. Chem. 286 (41) (2011) 35334–35338. Available from: https://doi.org/10.1074/jbc.C111.284620.

[27] Y. Gruenbaum, H. Cedar, A. Razin, Substrate and sequence specificity of a eukaryotic DNA methylase, Nature 295 (5850) (1982) 620–622 [Online]. Available from: <http://www.ncbi.nlm.nih.gov/htbin-post/Entrez/query?db = m&form = 6&dopt = r&uid = 0007057921>.

[28] M. Szyf, The elusive role of 5'-hydroxymethylcytosine, Epigenomics (2016). Available from: https://doi.org/10.2217/epi-2016-0076.

[29] M. Bachman, et al., 5-Formylcytosine can be a stable DNA modification in mammals, Nat. Chem. Biol. 11 (8) (2015) 555–557. Available from: https://doi.org/10.1038/nchembio.1848.

[30] M. Bachman, S. Uribe-Lewis, X. Yang, M. Williams, A. Murrell, S. Balasubramanian, 5-Hydroxymethylcytosine is a predominantly stable DNA modification, Nat. Chem. 6 (12) (2014) 1049–1055. Available from: https://doi.org/10.1038/nchem.2064.

[31] A. Razin, M. Szyf, DNA methylation patterns. Formation and function, Biochim. Biophys. Acta 782 (4) (1984) 331–342 [Online]. Available from: <http://www.ncbi.nlm.nih.gov/htbin-post/Entrez/query?db=m&form=6&dopt=r&uid=0006383476>.

[32] A. Razin, A.D. Riggs, DNA methylation and gene function, Science 210 (4470) (1980) 604–610 [Online]. Available from: <http://www.ncbi.nlm.nih.gov/htbin-post/Entrez/query?db = m&form = 6&dopt = r&uid = 0006254144>.

[33] L. Vardimon, I. Kuhlmann, W. Doerfler, H. Cedar, Methylation of adenovirus genes in transformed cells and in vitro: influence on the regulation of gene expression? Eur. J. Cell Biol. 25 (1) (1981) 13–15 [Online]. Available from: <http://www.ncbi.nlm.nih.gov/htbin-post/Entrez/query?db = m&form = 6&dopt = r&uid = 0006895196>.

[34] R. Stein, A. Razin, H. Cedar, In vitro methylation of the hamster adenine phosphoribosyltransferase gene inhibits its expression in mouse L cells, Proc. Natl. Acad. Sci. U. S. A. 79 (11) (1982) 3418–3422 [Online]. Available from: <http://www.ncbi.nlm.nih.gov/htbin-post/Entrez/query?db = m&form = 6&dopt = r&uid = 0006954487>.

[35] P.A. Jones, Effects of 5-azacytidine and its 2'-deoxyderivative on cell differentiation and DNA methylation, Pharmacol. Ther. 28 (1) (1985) 17–27 [Online]. Available from: <http://www.ncbi.nlm.nih.gov/htbin-post/Entrez/query?db = m&form = 6&dopt = r&uid = 0002414786>.

[36] P.A. Jones, Altering gene expression with 5-azacytidine, Cell 40 (3) (1985) 485–486 [Online]. Available from: <http://www.ncbi.nlm.nih.gov/htbin-post/Entrez/query?db = m&form = 6&dopt = r&uid = 0002578884>.

[37] M. Szyf, J. Rouleau, J. Theberge, V. Bozovic, Induction of myogenic differentiation by an expression vector encoding the DNA methyltransferase cDNA sequence in the antisense orientation, J. Biol. Chem. 267 (18) (1992) 12831–12836 [Online]. Available from: <http://www.ncbi.nlm.nih.gov/htbin-post/Entrez/query?db = m&form = 6&dopt = r&uid = 0001618783>.

[38] E. Li, T.H. Bestor, R. Jaenisch, Targeted mutation of the DNA methyltransferase gene results in embryonic lethality, Cell 69 (6) (1992) 915–926 [Online]. Available from: <http://www.ncbi.nlm.nih.gov/htbin-post/Entrez/query?db = m&form = 6&dopt = r&uid = 0001606615>.

[39] S. Milutinovic, S.E. Brown, Q. Zhuang, M. Szyf, DNA methyltransferase 1 knock down induces gene expression by a mechanism independent of DNA methylation and histone deacetylation, J. Biol. Chem. 279 (27) (2004) 27915–27927. Available from: https://doi.org/10.1074/jbc.M312823200.

[40] J. Espada, et al., Regulation of SNAIL1 and E-cadherin function by DNMT1 in a DNA methylation-independent context, Nucleic Acids Res. 39 (21) (2011) 9194–9205. Available from: https://doi.org/10.1093/nar/gkr658.

[41] X. Qiu, et al., Upregulation of DNMT1 mediated by HBx suppresses RASSF1A expression independent of DNA methylation, Oncol. Rep. 31 (1) (2014) 202–208. Available from: https://doi.org/10.3892/or.2013.2848.

[42] X.S. Liu, et al., Editing DNA methylation in the mammalian genome, Cell 167 (1) (2016) 233–247. Available from: https://doi.org/10.1016/j.cell.2016.08.056. e17.

[43] K.E. Bachman, M.R. Rountree, S.B. Baylin, Dnmt3a and Dnmt3b are transcriptional repressors that exhibit unique localization properties to heterochromatin, J. Biol. Chem. 276 (34) (2001) 32282–32287. Available from: https://doi.org/10.1074/jbc.M104661200.

[44] J. Zhong, et al., TET1 modulates H4K16 acetylation by controlling auto-acetylation of hMOF to affect gene regulation and DNA repair function, Nucleic Acids Res. 45 (2) (2017) 672–684. Available from: https://doi.org/10.1093/nar/gkw919.

[45] G.A. Kaas, et al., TET1 controls CNS 5-methylcytosine hydroxylation, active DNA demethylation, gene transcription, and memory formation, Neuron 79 (6) (2013) 1086–1093. Available from: https://doi.org/10.1016/j.neuron.2013.08.032.

[46] J.T. Bell, et al., DNA methylation patterns associate with genetic and gene expression variation in HapMap cell lines, Genome Biol. 12 (1) (2011) R10. Available from: https://doi.org/10.1186/gb-2011-12-1-r10.

[47] V.K. Rakyan, et al., An integrated resource for genome-wide identification and analysis of human tissue-specific differentially methylated regions (tDMRs), Genome Res. 18 (9) (2008) 1518–1529. Available from: https://doi.org/10.1101/gr.077479.108.

[48] R. Massart, M. Suderman, V. Mongrain, M. Szyf, DNA methylation and transcription onset in the brain, Epigenomics (2017). Available from: https://doi.org/10.2217/epi-2016-0184.
[49] A.C. D'Alessio, I.C. Weaver, M. Szyf, Acetylation-induced transcription is required for active DNA demethylation in methylation-silenced genes, Mol. Cell Biol. 27 (21) (2007) 7462–7474 [Online]. Available from: <http://www.ncbi.nlm.nih.gov/entrez/query.fcgi?cmd = Retrieve&db = PubMed&dopt = Citation&list_uids = 17709385>.
[50] A. Pacis, et al., Gene activation precedes DNA demethylation in response to infection in human dendritic cells, Proc. Natl. Acad. Sci. U. S. A. 116 (14) (2019) 6938–6943. Available from: https://doi.org/10.1073/pnas.1814700116.
[51] L. Boureau, A. Constantinof, V.G. Moisiadis, S.G. Matthews, M. Szyf, The DNA methylation landscape of enhancers in the guinea pig hippocampus, Epigenomics 10 (4) (2018) 349–365. Available from: https://doi.org/10.2217/epi-2017-0064.
[52] A. Kirillov, B. Kistler, R. Mostoslavsky, H. Cedar, T. Wirth, Y. Bergman, A role for nuclear NF-kappaB in B-cell-specific demethylation of the Igkappa locus, Nat. Genet. 13 (4) (1996) 435–441 [Online]. Available from: <http://www.ncbi.nlm.nih.gov/htbin-post/Entrez/query?db = m&form = 6&dopt = r&uid = 0008696338>.
[53] B. Kistler, B. Baumann, Y. Bergman, T. Wirth, RelB is a key player for both kappa B-dependent transcription and demethylation in B cells, Immunobiology 198 (1-3) (1997) 24–34 [Online]. Available from: <http://www.ncbi.nlm.nih.gov/entrez/query.fcgi?cmd = Retrieve&db = PubMed&dopt = Citation&list_uids = 9442375>.
[54] H. Kusaba, et al., Association of 5' CpG demethylation and altered chromatin structure in the promoter region with transcriptional activation of the multidrug resistance 1 gene in human cancer cells, Eur. J. Biochem. 262 (3) (1999) 924–932 [Online]. Available from: <http://www.ncbi.nlm.nih.gov/entrez/query.fcgi?cmd = Retrieve&db = PubMed&dopt = Citation&list_uids = 10411657>.
[55] H. Thomassin, M. Flavin, M.L. Espinas, T. Grange, Glucocorticoid-induced DNA demethylation and gene memory during development, EMBO J. 20 (8) (2001) 1974–1983. Available from: https://doi.org/10.1093/emboj/20.8.1974.
[56] A. Feldmann, R. Ivanek, R. Murr, D. Gaidatzis, L. Burger, D. Schubeler, Transcription factor occupancy can mediate active turnover of DNA methylation at regulatory regions, PLoS Genet. 9 (12) (2013) e1003994. Available from: https://doi.org/10.1371/journal.pgen.1003994.
[57] D. Cheishvili, et al., DNA methylation controls unmethylated transcription start sites in the genome in trans, Epigenomics (2017). Available from: https://doi.org/10.2217/epi-2016-0141.
[58] M. Comb, H.M. Goodman, CpG methylation inhibits proenkephalin gene expression and binding of the transcription factor AP-2, Nucleic Acids Res. 18 (13) (1990) 3975–3982 [Online]. Available from: <http://www.ncbi.nlm.nih.gov/htbin-post/Entrez/query?db = m&form = 6&dopt = r&uid = 0001695733>.
[59] R.R. Meehan, J.D. Lewis, A.P. Bird, Characterization of MeCP2, a vertebrate DNA binding protein with affinity for methylated DNA, Nucleic Acids Res. 20 (19) (1992) 5085–5092 [Online]. Available from: <http://www.ncbi.nlm.nih.gov/htbin-post/Entrez/query?db = m&form = 6&dopt = r&uid = 0001408825>.
[60] B. Hendrich, A. Bird, Identification and characterization of a family of mammalian methyl-CpG binding proteins, Mol. Cell Biol. 18 (11) (1998) 6538–6547 [Online]. Available from: <http://www.ncbi.nlm.nih.gov/htbin-post/Entrez/query?db = m&form = 6&dopt = r&uid = 0009774669>, <http://mcb.asm.org/cgi/content/full/18/11/6538>.
[61] X. Nan, et al., Transcriptional repression by the methyl-CpG-binding protein MeCP2 involves a histone deacetylase complex [see comments], Nature 393 (6683)

(1998) 386–389 [Online]. Available from: <http://www.ncbi.nlm.nih.gov/htbin-post/Entrez/query?db = m&form = 6&dopt = r&uid = 0009620804>.

[62] H.H. Ng, et al., MBD2 is a transcriptional repressor belonging to the MeCP1 histone deacetylase complex, Nat. Genet. 23 (1) (1999) 58–61. Available from: https://doi.org/10.1038/12659.

[63] A.C. D'Alessio, M. Szyf, Epigenetic tete-a-tete: the bilateral relationship between chromatin modifications and DNA methylation, Biochem. Cell Biol. 84 (4) (2006) 463–476 [Online]. Available from: <http://www.ncbi.nlm.nih.gov/entrez/query.fcgi?cmd = Retrieve&db = PubMed&dopt = Citation&list_uids = 16936820>.

[64] N. Cervoni, S. Sang-beom, D. Chakravarti, M. Szyf, The oncoprotein Set/TAF-1β, an inhibitor of histone acetyltransferase, inhibits active demethylation of DNA, integrating DNA methylation and transcriptional silencing, J. Biol. Chem. 277 (28) (2002) 25026–25031.

[65] M. Szyf, L. Eliasson, V. Mann, G. Klein, A. Razin, Cellular and viral DNA hypomethylation associated with induction of Epstein-Barr virus lytic cycle, Proc. Natl. Acad. Sci. U. S. A. 82 (23) (1985) 8090–8094 [Online]. Available from: <http://www.ncbi.nlm.nih.gov/htbin-post/Entrez/query?db = m&form = 6&dopt = r&uid = 0002999791>.

[66] N. Detich, V. Bovenzi, M. Szyf, Valproate induces replication-independent active DNA demethylation, J. Biol. Chem. 278 (30) (2003) 27586–27592 [Online]. Available from: <http://www.ncbi.nlm.nih.gov/entrez/query.fcgi?cmd = Retrieve&db = PubMed&dopt = Citation&list_uids = 12748177>.

[67] M.V.C. Greenberg, D. Bourc'his, The diverse roles of DNA methylation in mammalian development and disease, Nat. Rev. Mol. Cell Biol. 20 (10) (2019) 590–607. Available from: https://doi.org/10.1038/s41580-019-0159-6.

[68] F. Neri, et al., Intragenic DNA methylation prevents spurious transcription initiation, Nature 543 (7643) (2017) 72–77. Available from: https://doi.org/10.1038/nature21373.

[69] C.M. Greco, et al., DNA hydroxymethylation controls cardiomyocyte gene expression in development and hypertrophy, Nat. Commun. 7 (2016) 12418. Available from: https://doi.org/10.1038/ncomms12418.

[70] I. Choi, R. Kim, H.W. Lim, K.H. Kaestner, K.J. Won, 5-Hydroxymethylcytosine represses the activity of enhancers in embryonic stem cells: a new epigenetic signature for gene regulation, BMC Genomics 15 (2014) 670. Available from: https://doi.org/10.1186/1471-2164-15-670.

[71] M. Iurlaro, et al., In vivo genome-wide profiling reveals a tissue-specific role for 5-formylcytosine, Genome Biol. 17 (1) (2016) 141. Available from: https://doi.org/10.1186/s13059-016-1001-5.

[72] S.K. Sayeed, J. Zhao, B.K. Sathyanarayana, J.P. Golla, C. Vinson, C/EBPbeta (CEBPB) protein binding to the C/EBP|CRE DNA 8-mer TTGC|GTCA is inhibited by 5hmC and enhanced by 5mC, 5fC, and 5caC in the CG dinucleotide, Biochim. Biophys. Acta 1849 (6) (2015) 583–589. Available from: https://doi.org/10.1016/j.bbagrm.2015.03.002.

[73] Z. Liang, et al., DNA N(6)-adenine methylation in *Arabidopsis thaliana*, Dev. Cell 45 (3) (2018) 406–416. Available from: https://doi.org/10.1016/j.devcel.2018.03.012. e3.

[74] L. Kong, L. Zhang, i6mA-DNCP: computational identification of DNA N(6)-methyladenine sites in the rice genome using optimized dinucleotide-based features, Genes 10 (10) (2019). Available from: https://doi.org/10.3390/genes10100828.

[75] G. Zhang, et al., N6-methyladenine DNA modification in Drosophila, Cell 161 (4) (2015) 893–906. Available from: https://doi.org/10.1016/j.cell.2015.04.018.

[76] S. Urieli-Shoval, Y. Gruenbaum, J. Sedat, A. Razin, The absence of detectable methylated bases in *Drosophila melanogaster* DNA, FEBS Lett. 146 (1)

(1982) 148–152 [Online]. Available from: <http://www.ncbi.nlm.nih.gov/htbin-post/Entrez/query?db = m&form = 6&dopt = r&uid = 0006814955>.
[77] Q. Xie, et al., N(6)-methyladenine DNA modification in glioblastoma, Cell 175 (5) (2018) 1228–1243. Available from: https://doi.org/10.1016/j.cell.2018.10.006. e20.
[78] M. Kulis, A.C. Queiros, R. Beekman, J.I. Martin-Subero, Intragenic DNA methylation in transcriptional regulation, normal differentiation and cancer, Biochim. Biophys. Acta 1829 (11) (2013) 1161–1174. Available from: https://doi.org/10.1016/j.bbagrm.2013.08.001.
[79] B. Yao, et al., Active N(6)-methyladenine demethylation by DMAD regulates gene expression by coordinating with polycomb protein in neurons, Mol. Cell 71 (5) (2018) 848–857. Available from: https://doi.org/10.1016/j.molcel.2018.07.005. e6.
[80] R.A. Waterland, R.L. Jirtle, Transposable elements: targets for early nutritional effects on epigenetic gene regulation, Mol. Cell Biol. 23 (15) (2003) 5293–5300 [Online]. Available from: <http://www.ncbi.nlm.nih.gov/entrez/query.fcgi?cmd = Retrieve&db = PubMed&dopt = Citation&list_uids = 12861015>.
[81] R.A. Waterland, R.L. Jirtle, Early nutrition, epigenetic changes at transposons and imprinted genes, and enhanced susceptibility to adult chronic diseases, Nutrition 20 (1) (2004) 63–68. Available from: https://doi.org/10.1016/j.nut.2003.09.011.
[82] G.L. Rainsford, S.H. Schuman, The family in crisis. A case study of overwhelming illness and stress, JAMA: J. Am. Med. Assoc. 246 (1) (1981) 60–63. Available from: https://doi.org/10.1001/jama.246.1.60.
[83] D. Liu, et al., Maternal care, hippocampal glucocorticoid receptors, and hypothalamic-pituitary-adrenal responses to stress, Science 277 (5332) (1997) 1659–1662 [Online]. Available from: <http://www.ncbi.nlm.nih.gov/entrez/query.fcgi?cmd = Retrieve&db = PubMed&dopt = Citation&list_uids = 9287218>.
[84] D. Francis, J. Diorio, D. Liu, M.J. Meaney, Nongenomic transmission across generations of maternal behavior and stress responses in the rat, Science 286 (5442) (1999) 1155–1158 [Online]. Available from: <http://www.ncbi.nlm.nih.gov/entrez/query.fcgi?cmd = Retrieve&db = PubMed&dopt = Citation&list_uids = 10550053>.
[85] I.C. Weaver, et al., Epigenetic programming by maternal behavior, Nat. Neurosci. 7 (8) (2004) 847–854 [Online]. Available from: <http://www.ncbi.nlm.nih.gov/entrez/query.fcgi?cmd = Retrieve&db = PubMed&dopt = Citation&list_uids = 15220929>.
[86] I.C. Weaver, et al., The transcription factor nerve growth factor-inducible protein a mediates epigenetic programming: altering epigenetic marks by immediate-early genes, J. Neurosci. 27 (7) (2007) 1756–1768 [Online]. Available from: <http://www.ncbi.nlm.nih.gov/entrez/query.fcgi?cmd = Retrieve&db = PubMed&dopt = Citation&list_uids = 17301183>.
[87] I.C. Weaver, et al., The methylated-DNA binding protein MBD2 enhances NGFI-A (egr-1)-mediated transcriptional activation of the glucocorticoid receptor, Philos. Trans. R. Soc. Lond. B Biol. Sci. 369 (1652) (2014). Available from: https://doi.org/10.1098/rstb.2013.0513.
[88] F.A. Champagne, I.C. Weaver, J. Diorio, S. Dymov, M. Szyf, M.J. Meaney, Maternal care associated with methylation of the estrogen receptor-alpha1b promoter and estrogen receptor-alpha expression in the medial preoptic area of female offspring, Endocrinology 147 (6) (2006) 2909–2915 [Online]. Available from: <http://www.ncbi.nlm.nih.gov/entrez/query.fcgi?cmd = Retrieve&db = PubMed&dopt = Citation&list_uids = 16513834>.
[89] I.C. Weaver, et al., Reversal of maternal programming of stress responses in adult offspring through methyl supplementation: altering epigenetic marking later in life, J. Neurosci. Res. Support Non-U.S. Gov't 25 (47) (2005) 11045–11054. Available from: https://doi.org/10.1523/JNEUROSCI.3652-05.2005.

[90] M. Szyf, Prospects for medications to reverse causative epigenetic processes in neuropsychiatry disorders, Neuropsychopharmacology 42 (1) (2017) 367–368. Available from: https://doi.org/10.1038/npp.2016.219.
[91] M. Szyf, Prospects for the development of epigenetic drugs for CNS conditions, Nat. Rev. 14 (7) (2015) 461–474. Available from: https://doi.org/10.1038/nrd4580.
[92] T.L. Roth, F.D. Lubin, A.J. Funk, J.D. Sweatt, Lasting epigenetic influence of early-life adversity on the BDNF gene, Biol. Psychiatry 65 (9) (2009) 760–769. doi: S0006-3223(08)01530-8 [pii] 10.1016/j.biopsych.2008.11.028.
[93] C. Murgatroyd, et al., Dynamic DNA methylation programs persistent adverse effects of early-life stress, Nat. Neurosci. 12 (12) (2009) 1559–1566. doi: nn.2436 [pii] 10.1038/nn.2436.
[94] I.C. Weaver, M.J. Meaney, M. Szyf, Maternal care effects on the hippocampal transcriptome and anxiety-mediated behaviors in the offspring that are reversible in adulthood, Proc. Natl. Acad. Sci. U. S. A. 103 (9) (2006) 3480–3485 [Online]. Available from: <http://www.ncbi.nlm.nih.gov/entrez/query.fcgi?cmd = Retrieve&db = PubMed&dopt = Citation&list_uids = 16484373>.
[95] M. Frank, R. Kemler, Protocadherins, Curr. Opin. Cell Biol. 14 (5) (2002) 557–562. doi: S0955067402003654 [pii].
[96] P.O. McGowan, et al., Broad epigenetic signature of maternal care in the brain of adult rats, PLoS One 6 (2) (2011) e14739. Available from: https://doi.org/10.1371/journal.pone.0014739.
[97] P.O. McGowan, et al., Epigenetic regulation of the glucocorticoid receptor in human brain associates with childhood abuse, Nat. Neurosci. 12 (3) (2009) 342–348. doi: nn.2270 [pii] https://doi.org/10.1038/nn.2270.
[98] G.C. Ruppenthal, G.L. Arling, H.F. Harlow, G.P. Sackett, S.J. Suomi, A 10-year perspective of motherless-mother monkey behavior, J. Abnorm. Psychol. 85 (4) (1976) 341–349 [Online]. Available from: <http://www.ncbi.nlm.nih.gov/entrez/query.fcgi?cmd = Retrieve&db = PubMed&dopt = Citation&list_uids = 821983>.
[99] S.J. Suomi, M.L. Collins, H.F. Harlow, G.C. Ruppenthal, Effects of maternal and peer separations on young monkeys, J. Child. Psychol. Psychiatry 17 (2) (1976) 101–112 [Online]. Available from: <http://www.ncbi.nlm.nih.gov/entrez/query.fcgi?cmd = Retrieve&db = PubMed&dopt = Citation&list_uids = 819450>.
[100] N. Provencal, et al., The signature of maternal rearing in the methylome in rhesus macaque prefrontal cortex and T cells, J. Neurosci. 32 (44) (2012) 15626–15642. Available from: https://doi.org/10.1523/JNEUROSCI.1470-12.2012.
[101] R. Massart, et al., The signature of maternal social rank in placenta deoxyribonucleic acid methylation profiles in rhesus monkeys, Child. Dev. 88 (3) (2017) 900–918. Available from: https://doi.org/10.1111/cdev.12640.
[102] M. Schmidt, et al., Fetal glucocorticoid receptor (Nr3c1) deficiency alters the landscape of DNA methylation of murine placenta in a sex-dependent manner and is associated to anxiety-like behavior in adulthood, Transl. Psychiatry 9 (1) (2019) 23. Available from: https://doi.org/10.1038/s41398-018-0348-7.
[103] R. Massart, et al., Early life adversity alters normal sex-dependent developmental dynamics of DNA methylation, Dev. Psychopathol. (2016) 1–14. Available from: https://doi.org/10.1017/S0954579416000833.
[104] S.J. Marzi, et al., Analysis of DNA methylation in young people: limited evidence for an association between victimization stress and epigenetic variation in blood, Am. J. Psychiatry 175 (6) (2018) 517–529. Available from: https://doi.org/10.1176/appi.ajp.2017.17060693.
[105] N. Perroud, et al., Increased methylation of glucocorticoid receptor gene (NR3C1) in adults with a history of childhood maltreatment: a link with the severity and

type of trauma, Transl. Psychiatry 1 (2011) e59. Available from: https://doi.org/10.1038/tp.2011.60.
[106] D. Cicchetti, E.D. Handley, Methylation of the glucocorticoid receptor gene, nuclear receptor subfamily 3, group C, member 1 (NR3C1), in maltreated and non-maltreated children: associations with behavioral undercontrol, emotional lability/negativity, and externalizing and internalizing symptoms, Dev. Psychopathol. 29 (5) (2017) 1795–1806. Available from: https://doi.org/10.1017/s0954579417001407.
[107] L. Tozzi, et al., Epigenetic changes of FKBP5 as a link connecting genetic and environmental risk factors with structural and functional brain changes in major depression, Neuropsychopharmacology 43 (5) (2018) 1138–1145. Available from: https://doi.org/10.1038/npp.2017.290.
[108] S.H. Parade, et al., Change in FK506 binding protein 5 (FKBP5) methylation over time among preschoolers with adversity, Dev. Psychopathol. 29 (5) (2017) 1627–1634. Available from: https://doi.org/10.1017/s0954579417001286.
[109] M.B. Harms, et al., Early life stress, FK506 binding protein 5 gene (FKBP5) methylation, and inhibition-related prefrontal function: a prospective longitudinal study, Dev. Psychopathol. 29 (5) (2017) 1895–1903. Available from: https://doi.org/10.1017/s095457941700147x.
[110] R. Yehuda, et al., Holocaust exposure induced intergenerational effects on FKBP5 methylation, Biol. Psychiatry 80 (5) (2016) 372–380. Available from: https://doi.org/10.1016/j.biopsych.2015.08.005.
[111] T. Klengel, et al., Allele-specific FKBP5 DNA demethylation mediates gene-childhood trauma interactions, Nat. Neurosci. 16 (1) (2013) 33–41. Available from: https://doi.org/10.1038/nn.3275.
[112] E. Unternaehrer, et al., Childhood maternal care is associated with DNA methylation of the genes for brain-derived neurotrophic factor (BDNF) and oxytocin receptor (OXTR) in peripheral blood cells in adult men and women, Stress (2015) 1–11. Available from: https://doi.org/10.3109/10253890.2015.1038992.
[113] E.J. Kraaijenvanger, Y. He, H. Spencer, A.K. Smith, P.A. Bos, M.P.M. Boks, Epigenetic variability in the human oxytocin receptor (OXTR) gene: a possible pathway from early life experiences to psychopathologies, Neurosci. Biobehav. Rev. 96 (2018) 127–142. Available from: https://doi.org/10.1016/j.neubiorev.2018.11.016.
[114] J.P. Gouin, et al., Associations among oxytocin receptor gene (OXTR) DNA methylation in adulthood, exposure to early life adversity, and childhood trajectories of anxiousness, Sci. Rep. 7 (1) (2017) 7446. Available from: https://doi.org/10.1038/s41598-017-07950-x.
[115] L. Provenzi, R. Giorda, S. Beri, R. Montirosso, SLC6A4 methylation as an epigenetic marker of life adversity exposures in humans: a systematic review of literature, Neurosci. Biobehav. Rev. (2016). Available from: https://doi.org/10.1016/j.neubiorev.2016.08.021.
[116] S. King, et al., The ice storm: an opportunity to study the effects of prenatal stress on the baby and the mother, Sante mentale au Quebec 25 (1) (2000) 163–185 [Online]. Available from: <http://www.ncbi.nlm.nih.gov/pubmed/18253576>. La tempete de verglas: une occasion d'etudier les effets du stress prenatal chez l'enfant et la mere.
[117] D.P. Laplante, A. Brunet, N. Schmitz, A. Ciampi, S. King, Project Ice Storm: prenatal maternal stress affects cognitive and linguistic functioning in 5 1/2-year-old children, J. Am. Acad. Child. Adolesc. Psychiatry 47 (9) (2008) 1063–1072. Available from: https://doi.org/10.1097/CHI.0b013e31817eec80.
[118] D.J. Walder, D.P. Laplante, A. Sousa-Pires, F. Veru, A. Brunet, S. King, Prenatal maternal stress predicts autism traits in 6(1/2) year-old children: Project Ice Storm,

Psychiatry Res. 219 (2) (2014) 353–360. Available from: https://doi.org/10.1016/j.psychres.2014.04.034.
[119] L. Cao-Lei, et al., DNA methylation mediates the impact of exposure to prenatal maternal stress on BMI and central adiposity in children at age 13(1/2) years: Project Ice Storm, Epigenetics (2015). Available from: https://doi.org/10.1080/15592294.2015.1063771.
[120] L. Cao-Lei, K.N. Dancause, G. Elgbeili, D.P. Laplante, M. Szyf, S. King, DNA methylation mediates the effect of maternal cognitive appraisal of a disaster in pregnancy on the child's C-peptide secretion in adolescence: Project Ice Storm, PLoS One 13 (2) (2018) e0192199. Available from: https://doi.org/10.1371/journal.pone.0192199.
[121] F. Veru, K. Dancause, D.P. Laplante, S. King, G. Luheshi, Prenatal maternal stress predicts reductions in CD4+ lymphocytes, increases in innate-derived cytokines, and a Th2 shift in adolescents: Project Ice Storm, Physiol. Behav. 144 (2015) 137–145. Available from: https://doi.org/10.1016/j.physbeh.2015.03.016.
[122] A.M. Turcotte-Tremblay, R. Lim, D.P. Laplante, L. Kobzik, A. Brunet, S. King, Prenatal maternal stress predicts childhood asthma in girls: Project Ice Storm, BioMed. Res. Int. 2014 (2014) 201717. Available from: https://doi.org/10.1155/2014/201717.
[123] L. Cao-Lei, et al., DNA methylation signatures triggered by prenatal maternal stress exposure to a natural disaster: Project Ice Storm, PLoS One 9 (9) (2014) e107653. Available from: https://doi.org/10.1371/journal.pone.0107653.
[124] L. Cao-Lei, K.N. Dancause, G. Elgbeili, D.P. Laplante, M. Szyf, S. King, Pregnant women's cognitive appraisal of a natural disaster affects their children's BMI and central adiposity via DNA methylation: Project Ice Storm, Early Hum. Dev. 103 (2016) 189–192. Available from: https://doi.org/10.1016/j.earlhumdev.2016.09.013.
[125] L. Cao-Lei, F. Veru, G. Elgbeili, M. Szyf, D.P. Laplante, S. King, DNA methylation mediates the effect of exposure to prenatal maternal stress on cytokine production in children at age 13(1/2) years: Project Ice Storm, Clin. Epigenetics 8 (2016) 54. Available from: https://doi.org/10.1186/s13148-016-0219-0.
[126] M.J. Garabedian, C.A. Harris, F. Jeanneteau, Glucocorticoid receptor action in metabolic and neuronal function, F1000Research 6 (2017) 1208. Available from: https://doi.org/10.12688/f1000research.11375.1.
[127] C.M. Pariante, C.B. Nemeroff, A.H. Miller, Glucocorticoid receptors in depression, Isr. J. Med. Sci. 31 (12) (1995) 705–712 [Online]. Available from: <http://www.ncbi.nlm.nih.gov/entrez/query.fcgi?cmd = Retrieve&db = PubMed&dopt = Citation&list_uids = 8543464>.
[128] A. Cintra, et al., Glial and neuronal glucocorticoid receptor immunoreactive cell populations in developing, adult, and aging brain, Ann. N. Y. Acad. Sci. 746 (1994) 42–61. Available from: https://doi.org/10.1111/j.1749-6632.1994.tb39210.x. discussion 61-3, Nov 30.
[129] B. Liu, T.N. Zhang, J.K. Knight, J.E. Goodwin, The glucocorticoid receptor in cardiovascular health and disease, Cells 8 (10) (2019). Available from: https://doi.org/10.3390/cells8101227.
[130] R.A. Lee, C.A. Harris, J.C. Wang, Glucocorticoid receptor and adipocyte biology, Nucl. Receptor Res. 5 (2018). Available from: https://doi.org/10.32527/2018/101373.
[131] G. Caratti, L. Matthews, T. Poolman, S. Kershaw, M. Baxter, D. Ray, Glucocorticoid receptor function in health and disease, Clin. Endocrinol. (Oxf.) 83 (4) (2015) 441–448. Available from: https://doi.org/10.1111/cen.12728.
[132] R. Patel, J. Williams-Dautovich, C.L. Cummins, Minireview: new molecular mediators of glucocorticoid receptor activity in metabolic tissues, Mol. Endocrinol. 28 (7) (2014) 999–1011. Available from: https://doi.org/10.1210/me.2014-1062.

[133] L. Escoter-Torres, G. Caratti, A. Mechtidou, J. Tuckermann, N.H. Uhlenhaut, S. Vettorazzi, Fighting the fire: mechanisms of inflammatory gene regulation by the glucocorticoid receptor, Front. Immunol. 10 (2019) 1859. Available from: https://doi.org/10.3389/fimmu.2019.01859.
[134] A.A. Bartlett, H.E. Lapp, R.G. Hunter, Epigenetic mechanisms of the glucocorticoid receptor, Trends Endocrinol. Metab. 30 (11) (2019) 807–818. Available from: https://doi.org/10.1016/j.tem.2019.07.003.
[135] A. Crudo, et al., Prenatal synthetic glucocorticoid treatment changes DNA methylation States in male organ systems: multigenerational effects, Endocrinology 153 (7) (2012) 3269–3283. Available from: https://doi.org/10.1210/en.2011-2160.
[136] N. Provencal, et al., Glucocorticoid exposure during hippocampal neurogenesis primes future stress response by inducing changes in DNA methylation, Proc. Natl. Acad. Sci. U. S. A. (2019). Available from: https://doi.org/10.1073/pnas.1820842116. Aug 9.
[137] N. Provencal, et al., Association of childhood chronic physical aggression with a DNA methylation signature in adult human T cells, PLoS One 9 (4) (2014) e89839. Available from: https://doi.org/10.1371/journal.pone.0089839.
[138] A.S. Sundman, F. Pértille, L. Lehmann Coutinho, E. Jazin, C. Guerrero-Bosagna, P. Jensen, DNA methylation in canine brains is related to domestication and dog-breed formation, PLoS One 15 (10) (2020) e0240787. Available from: https://doi.org/10.1371/journal.pone.0240787.

CHAPTER 3

Conflict-driven evolution*

Eugene V. Koonin[1], Yuri I. Wolf[1] and Mikhail I. Katsnelson[2]
[1]National Center for Biotechnology Information, National Library of Medicine, National Institutes of Health, Bethesda, MD, United States
[2]Institute for Molecules and Materials, Radboud University, Nijmegen, The Netherlands

Introduction

Biological evolution has produced systems endowed with organizational complexity that has no precedent outside the domain of biology [1–7]. Indeed, it has been proposed that complexity could be used as a signature of "artifacts," that is, a defining criterion for identification of life (in potential extraterrestrial habitats) [8]. There is no universal definition of complexity [9–11], even if "when we see it, we know it." The complementary criteria of complexity that seem to be biologically relevant include the number of nucleotide sites that are subject to selection or number of genes in a genome, the number of functional components in an organism or a functional systems, or hierarchical organization of biological systems [5–8,12]. Arguably, the most general definition, pathway complexity, is derived from algorithmic complexity in mathematics and represents the number of steps that are required to generate an object. Pathway complexity exceeding a certain threshold has been postulated to be attainable only through biological evolution [8].

Many attempts to elucidate the origins of biological complexity have been made, from the vantage points of either biology or physics. According to the classic biological narratives, starting with Darwin's famous scenario for the evolution of the eye [13], complex features are gradually evolving adaptations. More recently, however, the adaptationist perspective on complexity has been questioned, and various scenarios for nonadaptive emergence of complex traits have been proposed [14–19]. These models take a population-genetic perspective on the origin of genomic complexity genomic embellishments, for example, integrated mobile genetic elements (MGE), introns or duplicated genes, accrue in organisms with small effective

* This article is dedicated to Professor Eviatar Nevo, on the occasion of his 90th anniversary

New Horizons in Evolution
DOI: https://doi.org/10.1016/B978-0-323-90752-1.00004-3

population size because, in such populations, the power of purifying selection is insufficient to purge these sequences from the genome. Such genomic embellishments, then, can gain function, in particular, under the subfunctionalization model, where duplicated genes differentially lose ancestral subfunctions. This nonadaptive model has been conceptualized as "constructive neutral evolution" which holds that complexity emerges and is fixed in evolution due to the mutual dependence of the components of an evolving cellular system resulting from subfunctionalization [20–23]. The neutralist models notwithstanding, other population-genetic analyses suggest that complexity, after all, can emerge as adaptation. This possibility is brought about by analysis of models of genome evolution in prokaryotes which shows that, on average, acquired genes are beneficial to the organism, likely, due to the increase in functional plasticity in microbes with diverse gene repertoires [24,25].

Neither adaptive nor neutral models of the evolution of complexity capture one of its key features, the pronounced nonuniformity of the emergence of complex traits. Indeed, new levels of complexity appear in evolution abruptly as notably exemplified by such momentous events as the origin of eukaryotes, plants, or animals which all occurred over very short time intervals, on the geological scale. To explain these "sudden" surges of complexity in evolution, the concepts of punctuated equilibrium [26,27] and major and minor evolutionary transitions [28,29] have been developed. These, however, remain largely qualitative explanatory frameworks, without an underlying general theory.

A complementary approach from physics links the complexity surges with self-organized criticality (SOC), an intrinsic property of dynamical systems with many degrees of freedom and nonlinear behavior that undergo series of "avalanches" separated in time by stasis periods. The distinctive feature of the SOC dynamics is the power law distribution of avalanche sizes such that most of the avalanches are small and large ones are rare—and hence the rarity of major changes in system complexity [30–36]. The SOC dynamics clearly resembles the punctuated equilibrium concept of evolution, with extended periods of stasis separated by rapid evolutionary transitions. Indeed, Bak and colleagues, the inventors of SOC, developed models that were inspired by biological systems and aimed at mimicking their evolution [32,33,35,36]. In the classic Bak–Sneppen model [32,37], links between organisms that are assumed to reflect ecological connections and are represented by physical proximity in the model space drive the coevolution of the entire virtual community.

Evolution of such systems results in self-organization into a critical quasiequilibrium that is punctuated by "avalanches" of extinction (extinction of the least fit organisms results in disruption of the local environment, cause extinction of the neighbor organisms), and the avalanche size is power-law distributed. Although these phenomena have been studied for the past 25 years, there is no theory allowing one to derive the specific conditions leading to SOC. Importantly, however, a connection has been discovered between competing interactions and frustration in spin glasses and SOC [38,39]. Indeed, SOC has been shown to be an emergent property of spin glasses with a large (diverging) number of neighbors [39].

Spin glasses are the best studied class of nonergodic systems that can be informally defined as physical systems with a memory of past states such that the current state depends the history of the system, and the future states are not predictable exactly [40,41]. Nonergodicity results from the competition between short-range and long-range interactions which results in frustration causes the formation of complex patterns [42]. Spin glasses and some other glass-like phases share with biological systems two fundamental properties: (1) memory of past states resulting in nonergodicity and (2) complexity, and thus, potentially, could serve as simple physical models for understanding biology. Laughlin, Pines, and colleagues have outlined the parallels between the modern physical theory of glass-like states and biological theory [43,44]. Primarily, they emphasized the apparent glass-like properties of biological macromolecules, and the role of competing interactions and frustration in protein folding has been subsequently explored in detail [45,46]. In our previous work, we instead focused on the role of frustration as a primary driver of evolutionary dynamics [47].

Here, we continue this line of discourse, by exploring how various types of conflicts that pervade biology bring about frustration states, emergence of complexity, and evolutionary transitions. We also focus on a particular type of ubiquitous biological conflicts, those between genetic parasites and their hosts, and discuss the evolutionary entanglement between MGE and host defense systems that promotes remarkable diversification and complexification on both sides of the conflict.

Competing interactions and frustration drive biological evolution

Competing interactions, frustrations, and SOC necessarily appear in glass-like phases with more than six dimensions [48]. In evolving biological systems, the fitness landscape, that is, the configuration space, typically is of an extremely high dimensionality. Indeed, the numerous genes in a genome or, if an individual gene is considered, the numerous nucleotide sites represent distinct dimensions [49]. Therefore the appearance of competing interactions and frustration in biological evolution is inevitable, resulting in nonergodicity and SOC, and as the ultimate consequence, evolutionary transitions that follow the punctuated equilibrium regime (Fig. 3.1).

Competing interactions permeate all levels of biological organization and all types of biological processes (Table 3.1). The lowest organizational level where specific biological complexity is apparent is represented by nucleic acid and protein molecules that fold into unique three-dimensional structures essential for their functions [50]. The beginning of life can be identified with the emergence of catalytically active RNA molecules (ribozymes) endowed with RNA polymerase activity within the hypothetical, primordial RNA World [51]. As one would expect, experimental selection for ribozyme RNA polymerases has shown that this activity is achievable (thus far, to a limited extent) only by RNA

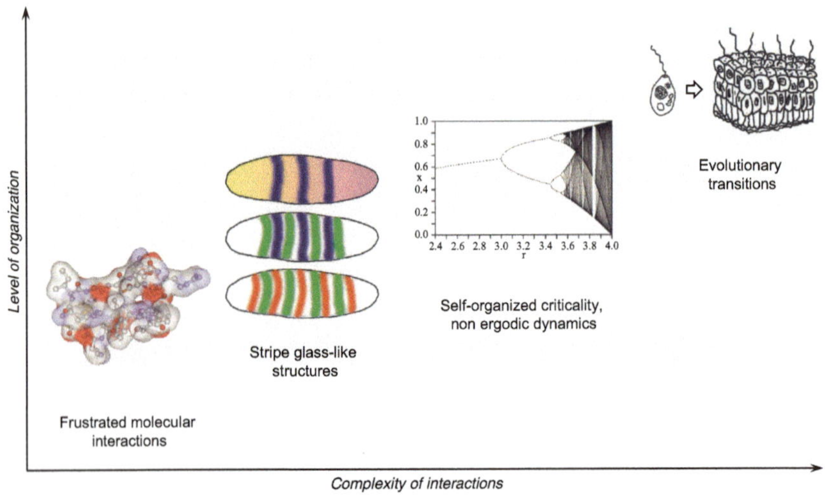

Figure 3.1 Competing interactions, frustrated states, and evolution of complexity. The figure schematically depicts the path from competing molecular interactions resulting in frustration to the evolution of increasing levels of biological complexity through SOC phenomena. *Reproduced, with permission, from Y.I. Wolf, M.I. Katsnelson, E.V. Koonin. Physical foundations of biological complexity, Proc Natl Acad Sci U S A. 115: (2018) E8678-E8687 [47].*

Table 3.1 Competing interactions and frustrated states in biological evolution.

System	Competing interactions causing frustration	Emergent properties
RNA	Short-range interactions (local hydrogen bonding within stems, stacking) versus long-range interactions (long-distance hydrogen bonding, salt bridges)	Complex 3-dimensional structures of RNA molecules including ribozymes
Proteins	Short-range interactions (Van der Waals) versus long-range interactions (hydrogen bonds, salt bridges) between amino acid side chains	Regular patterns and stable conformations in protein structures
Macromolecular complexes	Interactions within a subunit versus interactions between subunits	Elaborate organization of nucleoproteins (ribosomes, chromatin) and protein complexes (e.g. proteasome or electron transfer chains)
(Proto)cells[a]	**Membranes (compartmentalization, confinement of metabolies) versus channels and pores (transport of chemicals)**	**Compartments and cellular trafficking machinery dependent on electrochemical gradients**
Autonomous replicators (hosts) and semiautonomous replicators (parasites)	**"Cooperator" replicators versus "cheater" replicators**	**Self- versus non-self-discrimination and defense; increasingly complex, large genomes of increasing size; protocells**
Autonomous reproducers (hosts) versus semiautonomous replicators (parasites)	**Host cells and viruses**	**Evolutionary arms race; defense and counter-defense systems; evolutionary entanglement between parasites and defense systems**

(Continued)

Table 3.1 (Continued)

System	Competing interactions causing frustration	Emergent properties
Autonomous reproducers (hosts) versus semiautonomous replicators (parasites)	Host cells versus transposons	Intragenomic DNA replication control; diverse evolutionary innovations through hijacking of transposons
Eukaryotic cells	**Host (archaeal) cells versus endosymbiont (alpha-proteobacteria, proto-mitochondria)**	**Complex organization of eukaryotic cells**
Communities of unicellular organisms	**Individual cells versus cellular collectives**	**Exchange of "common goods" and signals; quorum sensing; cell proliferation control; programmed cell death, multicellularity**
Multicellular organism	Proliferating versus quiescent cells	Tumorigenesis, aging, death
Populations	**Individual organisms versus groups**	**Interorganismal cooperation; eusocilaity**
Ecosystems	Species occupying different niches	Interspecies conflicts, host–parasite and predator–prey relationships, mutualism, commensalism
Societies[b]	–	–

[a]Those competing interactions and frustrated states that appear to underlie MTE are shown in bold.
[b]We do not specify the conflicts driving the origin and evolution of human societies.

molecules with elaborate, complex secondary and tertiary structures [52–54]. The competition between short-range and long-range interactions is readily observable in RNA and protein molecules, and the frustration caused by this competition appears to define their folding that is crucial for all molecular functions [45,46]. This frustration would have been central to the evolution of the ribozyme polymerases, arguably one of the defining events in the origin of life.

Actually, most of the functions within cells are performed by various macromolecular complexes [55,56]. In such complexes, there is an obvious competition between interactions within macromolecules that are required for the folding of individual subunits and the interactions between subunits that are essential for the complex formation.

Conformation changes in both ribosomal proteins and RNA during ribosome morphogenesis [57,58], in transcription factors upon binding their recognition sites in DNA [59,60], and in virus proteins during virion morphogenesis [61] are only some of the most notable examples. Moreover, allosteric regulation of enzymes, typically, multimeric ones, is based on transitions between macromolecular conformations that differ minimally in terms of free energy but possess distinct biological properties [50]. The existence of multiple conformations with similar free energies in macromolecular complexes is caused by competing interactions, in a close analogy to the polymorphism of simple inorganic solids that is well studied in condensed matter physics.

Evolutionary entanglement between hosts and parasites as a key factor of evolution

From a different perspective that is unique to life, selection pressures that act in opposite directions and produce tradeoffs that ubiquitous in biology [62,63]. The interplay between such competing factors produces complex fitness landscapes with many local peaks and basins of attraction, on which numerous evolutionary strategies play out that are distinct in their specific manifestations but closely similar in terms of fitness. Conflicting selective processes are the key driver of host–parasite coevolution which is one of the major aspects in the entire history of life. Indeed, emergence of genetic parasites appears to be inevitable in all replicator systems because replicator systems that are protected from parasites are evolutionarily unstable [64,65]. Genetic parasites with different reproduction modes and varying degrees of autonomy (viruses, plasmids, transposons, and more) are associated with all cellular life forms [66–69]. Frustration in host–parasite systems is caused by the trade-offs between parasite reproduction, host reproduction, and interactions that stabilize the host–parasite system as a whole. The conflicts between the selective factors operating on each of these levels appear to be major drivers of the evolution of biological complexity [70]. Computer simulations of evolution under a broad range of conditions show that, in well-mixed replicator systems, parasites overwhelm the hosts, eventually resulting in collapse of the entire system, that is, extinction of both the hosts and the parasites [65,71–74]. By contrast, compartmentalization (that, in reality, could involve partitioning of replicator ensembles between vesicles or simply separation in a viscous medium) stabilizes the host–parasite system and leads to diversification and emergence of complexity [71–73]. Compartmentalization in the primordial replicator systems that is essential, in part, to control parasites culminated in the origin of cells. In physical terms, the outcome of host–parasite coevolution under these models studies is pattern formation, a characteristic result of frustration in glass-like states.

Compartmentalization is the simplest and, arguably, the most fundamental effect of host–parasite conflicts but, in all cellular life forms, these conflicts also drive the evolution of versatile host defense systems and counter-defense systems in parasites, another prominent and ubiquitous manifestation of biological complexity [75–78]. In the course of evolution, conflicts between hosts and parasites are resolved into multiple, distinct, stable evolutionary regimes. These regimes span the entire range of host–parasite relationships, from highly aggressive parasites, such as lytic viruses, that kill the host and move to the next one, to cooperative elements, such as many plasmids, that often provide beneficial functionalities to the host [68,69]. This persistent diversity of host–parasite interactions is a major part of biological complexity at the level of ecosystems and the entire biosphere.

It seems to be no exaggeration to state that frustration caused by intergenomic conflicts is the key driver of the evolution of biological complexity in general [70,79]. The genomes of all life forms contain multiple integrated MGE which in many multicellular eukaryotes (animals and plants) account for the majority of the genome sequence [80–82]. The competing interactions causing frustration are particularly obvious in the case of MGE with dual roles, such as toxin–antitoxin (TA), abortive infection (Abi), and restriction-modification (RM) modules in prokaryotes [83–86]. The TA, Abi, and RM systems protect the hosts from more aggressive parasites, in particular, viruses, but their lifestyle positions themselves as a type of "passive" MGE that spread, primarily, by hijacking plasmids. There are several levels of frustration in prokaryotes that harbor TA, Abi, and RM modules (which includes the great majority of bacteria and archaea). The TA and RM systems compete, on the one hand, with viruses which they attack, protecting the host, but on the other hand, with the host itself, in which they induce dormancy or death when it "attempts" to get rid of these elements. Another party to this play are plasmids on which RM, Abi, and TA modules are habitually transferred. The interplay between different forms of competition in these complex networks results in stabilization of the entire multicomponent host–parasite system such that entities with all types of reproduction strategies persist indefinitely. Put another way, host–parasite coevolution underlies ensures the persistence of biological complexity at level of ecosystems.

A major, intrinsic feature of the host–parasite coevolutionary networks is the "guns for hire" phenomenon, whereby active components of MGE, such as nucleases involved in transposition, are hijacked by the host for defense functions, and conversely, host defense systems or their components are recruited by MGE for antidefense or other roles [87–89]. The CRISPR-Cas systems that provide adaptive immunity to archaea and bacteria present a particular striking showcase for the "guns for hire" concept. Phylogenomic analysis shows that most of the constituent parts of the CRISPR-Cas systems evolved from diverse MGE [90]. Reciprocally, defective CRISPR-Cas systems incapable of target DNA cleavage were recruited by Tn7-like transposons in which they enable RNA-guided sequence-specific transposition [91–93]. Multiple other cases of MGE gene recruitment for defense functions and the reciprocal hijacking of defense systems by MGE have been identified including animal immune systems and viruses [89,94]. Thus conflict-driven evolution involves, as an essential component, not only competition but also various forms of cooperation between the conflicting agencies.

Competing interactions drive major innovations and transitions in evolution

Frustration resulting from competing interactions appears to be a major driver of evolutionary innovations and transitions. Maynard Smith and Szathmary developed the concept of major transitions in evolution (MTE) that are defined as a distinct class of evolutionary innovations that engender transitions in individuality, or emergence of new levels of selection [28,29]. Crucially, within the framework of the MTE concept, the transitions are construed not simply as major evolutionary innovations but meet specific criteria that make them analogous to phase transitions in physics. Accordingly, every MTE is a major innovation but not every major innovation is a MTE. To reiterate, the hallmark of MTE is an evolutionary transition in individuality which is accompanied by a change in the level of selection. An obvious example of a MTE is the origin of multicellular organisms from unicellular life forms but MTE, although not numerous, permeate the entire history of life (Table 3.1 and Fig. 3.1). The second signature of MTE, according to Szathmary, is the emergence new mechanisms of information encoding and transmission; thus, for example, multicellularity is accompanied by the rise of epigenetic information [29]. It is easy to see that each MTE involves competing interactions and/or levels of selections (Table 3.1).

Most obviously, evolution of multicellular life forms that occurred on multiple occasions in the history of life, in each case, involves the fundamental conflict between selection factors that act on individual cells and those that affect cellular ensembles or tissues. Obviously, to maintain the integrity of a multicellular organism, individual cell proliferation must be tightly controlled.

The origin of the first cells is an extremely difficult, poorly understood, even if, arguably, most important innovation in the history of life. That said, it is hard to imagine an evolutionary scenario under which the emergence of full-fledged cells was not antedated by a stage of evolution when all genetic information was contained in small genetic elements that would resemble modern virus or transposon genomes [95–97]. The origin of cells must have involved accretion of such elements resulting in large genomes on the scale of modern prokaryotes. Competition between the selective factors that affect individual genetic elements and factors those that act on ensembles of such elements giving rise to cellular genomes is inherent in this scenario for the origin of cells. In this case as in the case of multicellularity, replication of individual elements has suppressed for the ensemble, that is, the cell, to retain its integrity. This type of scenario forms the basis of a recent mathematical model of primordial cell evolution [98].

The origin of eukaryotes, the next MTE after the origin of cells, apparently was triggered by endosymbiotic alpha-proteobacteria that gave rise to the mitochondria. Accordingly, eukaryogenesis involved the inescapable conflict between the "interests" of the endosymbiont and those of the evolving eukaryotic cell that had to be resolved through coordination between the reproductive cycles of the host and the symbiont [99–101]. An analogous conflict is incumbent in the evolution of photosynthesizing algae (and eventually plants) whereby the cyanobacterial endosymbiont gave rise to the chloroplast. During these MTE, the frustration caused by the host–symbiont competition was resolved into stable symbiotic associations, but the conflicts persist and can flare up, for example, in the form of mitochondrial diseases [102] and lysis of mitochondria that in some organisms results in frequent invasion of nonfunctional mitochondrial DNA into the host genome [103].

The subsequent MTE that resulted in the emergence of eusocial animals and societies involved obvious competition between individuals and collectives, or between collectives at different levels of organization. Generally, competing interactions between entities at different organization levels appear to be an important if not the defining driver of evolutionary transitions in individuality that underlie MTE.

The competition between the different levels of individuality in MTE is tightly linked to host–parasite conflicts. Mathematical modeling of multicellularity evolution suggests that defense against viruses might be a major driver of this MTE. Specifically, parasite pressure results in the evolution of programmed cell death, an "altruistic" form of defense that functions only in conjunction with cell aggregation and could promote the emergence of multicellularity [104]. Genetic parasites likely played major roles also in the preceding MTE, in particular, the origin of eukaryotes where endosymbiosis is thought to have triggered massive invasion of self-splicing introns from the endosymbiont into the host genome where they gave rise to the spliceosomal introns [100,101,105]. This onslaught of parasitic elements on the genomes of the emerging eukaryotes could have been the driving force behind the emergence of the signature features of eukaryotic cells, such as the nucleus, the spliceosome and the ubiquitin signaling network, that are essential for the dramatic leap in cellular organization complexity during eukaryogenesis [106]. The sequence of events in eukaryogenesis is difficult to infer with confidence, but protection against the adverse effect of the invading introns, such as production of aberrant proteins from unspliced transcripts, would be a major selective factor underpinning the evolution of these hallmarks of the eukaryotic cellular organization [100,101,106]. As sketchy as our understanding of the origins of the first cells is, it appears certain that conflicts between selfish and "cooperative" genetic elements played a key role at this stage of evolution [95,97]. Thus all MTE appear to be predicated not only competing interactions but, more specifically, on host–parasite conflicts.

Each MTE encompasses a series of major evolutionary innovations but not every major innovation is a MTE. However, the difference between the MTE and other evolutionary innovations appears to be quantitative rather than qualitative. The major innovations that do not involve the emergence of a new level of selection are nevertheless associated with local transitions in individuality, that is, evolution of novel, complex functions through evolutionary fixation of new interactions between gene and their protein products. Important examples are interactions between photosystem components in the case of photosynthesis [107–109] or between methanogenic pathway enzymes in the case of archaeal methanogenesis [110,111]. Arguably, all major evolutionary innovations are based on the emergence of new units of selection (Darwinian individual), even if only MTE are linked to new levels of selection (classes of Darwinian individuals). Therefore competing interactions and frustration appear to be inherent in all such major innovations.

It is important to note that MTE and major evolutionary innovations, in general, are subject to a complementary interpretation as cooperation between Darwinian individuals that results in the emergence of a new level of selection. Competing interactions lead directly to cooperation as clearly demonstrated by the evolution of multicellularity in response to parasite pressure and tradeoffs between hoarding and sharing of "public goods" [70,104]. In short, competition begets cooperation.

Cancer, aging, and death

The conflict between individual cells proliferation and the maintenance of large, stable cell ensembles (tissues and organs) in multicellular life forms is resolved by evolution of multiple layers of controls of cell reproduction, a striking manifestation of biological complexity. Yet, an alternative and common resolution to this conflict caused by impairment of control mechanisms is rampant proliferation of "cheater" cells resulting in tumor formation and, in particular, cancer in animals [112]. The appearance of cheaters in cell collectives is inevitable for the same reasons that emergence of parasites is intrinsic to replicator systems.

Aging and eventual individual death are intrinsic stages of the ontogenesis of multicellular organisms that are, apparently, caused by the same conflict [113]. This is the case because of the inherent conflict between the "vigor" (fitness) of individual cells that amounts to high division rate and cellular cooperation that requires limiting the division rate or even bringing cell division to a halt. The frustration caused by this conflict hampers the elimination of senescent cells that accumulate deleterious somatic mutations in competition with high-fitness cells that form tumors when allowed to divide uncontrollably.

Frustration as the major cause of complexity in nature and specifics of biology

Unifying explanations of universal phenomena are inherently shaky. Nevertheless, it appears likely that complexity emerges only in nonergodic systems, and nonergodicity, in turn, is caused by competing interactions. The competition between short-range and long-range interactions, under additional constraints, results in SOC that allows for the evolution of complexity. This general perspective seems to pertain to all types of complexity in nature, from striped glasses to galaxies (Fig. 3.1). However, contrary to some claims [36], evolution of complexity is not guaranteed by SOC alone. Indeed, SOC does not result in hierarchical complexity because fractal patterns produced by SOC are not genuinely complex inasmuch as, by definition, they appear the same across all spatial and/or temporal scales (and hence can be produced by a simple algorithm). As an alternative to SOC, evolution of biological complexity has been studied in abstract models based on Highly Optimized Tolerance which emerges via robustness tradeoffs, such as competition between specialists and generalists in unstable environments [114,115]. Frustrations caused by competing evolutionary factors, such as adaptation to a broad or a narrow range of conditions, seem to underpin this model as well.

The patterns of complexity at different levels of biological organization, such as macromolecules, cells, multicellular organisms, populations, and ecosystems, substantially differ. Evolutionarily, these hierarchical levels of organization are linked through MTE that, as we argue, are driven by competing interactions and the resulting frustration. Indeed, this appears to be the universal driver of the evolution of hierarchy in nature. The actual mathematical theory of frustration-driven evolution remains to be developed and might require mathematical ideas and techniques beyond those currently available [47].

Conclusions

What are the defining features of life? The famous adage of Dobzhansky, "nothing in biology makes sense except in the light of evolution" [116], although true in itself, certainly, fails as a demarcation criterion because everything in the universe evolves. Indeed, outside the quantum realm, the world is teeming with nonergodic (and hence evolving) systems, and the universe itself can be adequately described only in the light of its evolution over during the ~13.8 billion years elapsed since the Big Bang [117]. However, the hierarchical complexity and elaboration that mark life are unmatched by anything outside biology. The critical distinction between biological systems and inanimate nonergodic, complex ones appears to be the genetic memory mechanism together with the genotype-to-phenotype mapping. This mechanism is based on the replication of digital information carriers (nucleic acids) that preserve and the memory of the patterns emerging from competing interactions at different levels and pass this memory across generations with sufficient fidelity to provide for selection via the phenotype-to-genotype feedback. Attempts to define life at a fundamental level, perhaps, represent a pseudophilosophical folly [118–120]. Nevertheless, complexity emerging from competing interactions and frustration, combined with memory perpetuated via replication of digital information carriers, underlies all life and appears to be unique to biology. Accordingly, any system that possesses these properties would qualify as living.

Declarations

Ethical approval and consent to participate

Not applicable.

Consent for publication

Not applicable.

Availability of supporting data

Not applicable.

Competing interests

The authors declare that they have no competing interests.

Author contributions

EVK, YIW, and MIK wrote the manuscript.

Funding

EVK and YIW are funded through the Intramural Research Program of the National Institutes of Health of the USA. MIK acknowledges financial support from the NWO via the Spinoza Prize.

References

[1] C. Adami, What is complexity? Bioessays 24 (2002) 1085–1094.
[2] C. Adami, The use of information theory in evolutionary biology, Ann. N. Y. Acad. Sci. 1256 (2012) 49–65.
[3] C. Adami, C. Ofria, T.C. Collier, Evolution of biological complexity, Proc. Natl. Acad. Sci. U. S. A. 97 (2000) 4463–4468.
[4] M. Gell-Mann, The Quark and the Jaguar: Adventures in the Simple and the Complex, St. Martin's Griffin, New York, 1995.
[5] D.W. McShea, R.N. Brandon, Biology's First Law: The Tendency for Diversity and Complexity to Increase in Evolutionary Systems, University Chicago Press, Chicago, IL, 2010.
[6] E.V. Koonin, The meaning of biological information, Philos. Trans. A Math. Phys. Eng. Sci. 374 (2016) 20150065.
[7] E.V. Koonin, A non-adaptationist perspective on evolution of genomic complexity or the continued dethroning of man, Cell Cycle 3 (2004) 280–285.
[8] S.M. Marshall, A.R.G. Murray, L. Cronin, A probabilistic framework for identifying biosignatures using Pathway Complexity, Philos. Trans. A Math. Phys. Eng. Sci. 375 (2017) 20160342.
[9] M. Gell-Mann, S. Lloyd, Information measures, effective complexity, and total information, Complexity 2 (1996) 44–52.
[10] R. Badii, A. Politi, Complexity. Hierarchical Structures and Scaling in Physics, Cambridge University Press, Cambridge, 1997.
[11] S. Lloyd, Measures of complexity: a nonexhaustive list, IEEE Control. Syst. Mag. 21 (2001) 7–8.
[12] N.A. Heim, J.L. Payne, S. Finnegan, M.L. Knope, M. Kowalewski, S.K. Lyons, et al., Hierarchical complexity and the size limits of life, Proc. Biol. Sci. 284 (2017) 20171039.
[13] C. Darwin, On the Origin of Species, A.F. Murray, London, 1859.
[14] S.J. Gould, R.C. Lewontin, The spandrels of San Marco and the Panglossian paradigm: a critique of the adaptationist programme, Proc. R. Soc. Lond. B Biol. Sci. 205 (1979) 581–598.
[15] S.J. Gould, The exaptive excellence of spandrels as a term and prototype, Proc. Natl. Acad. Sci. U. S. A. 94 (1997) 10750–10755.
[16] M. Lynch, The origins of eukaryotic gene structure, Mol. Biol. Evol. 23 (2006) 450–468.
[17] M. Lynch, The frailty of adaptive hypotheses for the origins of organismal complexity, Proc. Natl. Acad. Sci. U. S. A. 104 (1) (2007) 8597–8604.
[18] M. Lynch, The Origins of Genome Archiecture, Sinauer Associates, Sunderland, MA, 2007.

[19] M. Lynch, J.S. Conery, The origins of genome complexity, Science 302 (2003) 1401–1404.
[20] A. Stoltzfus, On the possibility of constructive neutral evolution, J. Mol. Evol. 49 (1999) 169–181.
[21] A. Stoltzfus, Constructive neutral evolution: exploring evolutionary theory's curious disconnect, Biol. Direct 7 (2012) 35.
[22] M.W. Gray, J. Lukes, J.M. Archibald, P.J. Keeling, W.F. Doolittle, Cell biology. Irremediable complexity? Science 330 (2010) 920–921.
[23] J. Lukes, J.M. Archibald, P.J. Keeling, W.F. Doolittle, M.W. Gray, How a neutral evolutionary ratchet can build cellular complexity, IUBMB Life 63 (2011) 528–537.
[24] I. Sela, Y.I. Wolf, E.V. Koonin, Theory of prokaryotic genome evolution, Proc. Natl. Acad. Sci. U. S. A. 113 (2016) 11399–11407.
[25] J. Iranzo, J.A. Cuesta, S. Manrubia, M.I. Katsnelson, E.V. Koonin, Disentangling the effects of selection and loss bias on gene dynamics, Proc. Natl. Acad. Sci. U. S. A. 114 (2017) E5616–E5624.
[26] S.J. Gould, Tempo and mode in the macroevolutionary reconstruction of Darwinism, Proc. Natl. Acad. Sci. U. S. A. 91 (1994) 6764–6771.
[27] N. Eldredge, S.J. Gould, On punctuated equilibria, Science 276 (1997) 338–341.
[28] J. Maynard Smith, E. Szathmary, The Major Transitions in Evolution, Oxford University Press, Oxford, 1997.
[29] E. Szathmary, Toward major evolutionary transitions theory 2.0, Proc. Natl. Acad. Sci. U. S. A. 112 (2015) 10104–10111.
[30] P. Bak, C. Tang, K. Wiesenfeld, Self-organized criticality: an explanation of the 1/f noise, Phys. Rev. Lett. 59 (1987) 381–384.
[31] P. Bak, C. Tang, K. Wiesenfeld, Self-organized criticality, Phys. Rev. A Gen. Phys. 38 (1988) 364–374.
[32] P. Bak, K. Sneppen, Punctuated equilibrium and criticality in a simple model of evolution, Phys. Rev. Lett. 71 (1993) 4083–4086.
[33] S. Maslov, M. Paczuski, P. Bak, Avalanches and 1/f noise in evolution and growth models, Phys. Rev. Lett. 73 (1994) 2162–2165.
[34] S. Maslov, Y.C. Zhang, Exactly solved model of self-organized criticality, Phys. Rev. Lett. 75 (1995) 1550–1553.
[35] P. Bak, M. Paczuski, Complexity, contingency, and criticality, Proc. Natl. Acad. Sci. U. S. A. 92 (1995) 6689–6696.
[36] P. Bak, How Nature Works: The Science of Self-Organized Criticality, Springer, New York, 1996.
[37] K. Sneppen, P. Bak, H. Flyvbjerg, M.H. Jensen, Evolution as a self-organized critical phenomenon, Proc. Natl. Acad. Sci. U. S. A. 92 (1995) 5209–5213.
[38] F. Pázmándi, G. Zaránd, G.T. Zimányi, Self-organized criticality in the hysteresis of the Sherrington-Kirkpatrick model, Phys. Rev. Lett. 83 (1999) 1034–1037.
[39] J.C. Andresen, Z. Zhu, R.S. Andrist, H.G. Katzgraber, V. Dobrosavljevic, G.T. Zimanyi, Self-organized criticality in glassy spin systems requires a diverging number of neighbors, Phys. Rev. Lett. 111 (2013) 097203.
[40] S.F. Edwards, P.W. Anderson, Theory of spin glasses, J. Phys. F: Met. Phys. 5 (1975) 965–974.
[41] K.H. Fischer, J.A. Hertz, Spin Glasses, Cambridge Univ Press, Cambridge, 1993.
[42] J.F. Sadoc, R. Mozzeri, Geometric Frustration, Cambridge University Press, Cambridge, 1999.
[43] R.B. Laughlin, D. Pines, The theory of everything, Proc. Natl. Acad. Sci. U. S. A. 97 (2000) 28–31.
[44] R.B. Laughlin, D. Pines, J. Schmalian, B.P. Stojkovic, P. Wolynes, The middle way, Proc. Natl. Acad. Sci. U. S. A. 97 (2000) 32–37.

[45] D.U. Ferreiro, E.A. Komives, P.G. Wolynes, Frustration in biomolecules, Q. Rev. Biophys. 47 (2014) 285–363.
[46] D.U. Ferreiro, E.A. Komives, P.G. Wolynes, Frustration, function and folding, Curr. Opin. Struct. Biol. 48 (2018) 68–73.
[47] M.I. Katsnelson, Y.I. Wolf, E.V. Koonin, Towards physical principles of biological evolution, Phys. Scr. 93 (2018) 043001.
[48] W. Wang, M.A. Moore, H.G. Katzgraber, Fractal dimension of interfaces in Edwards-Anderson spin glasses for up to six space dimensions, arXiv:171204971 (2017).
[49] S. Gavrilets, Fitness Landscapes and the Origin of Species, Princeton University Press, Princeton, NJ, 2004.
[50] A.V. Finkelstein, O.B. Ptitsyn, Protein Physics: A Course of Lectures, Academic Press, New York, 2002.
[51] RNA worlds: from life's origins to diversity, in: J.F. Atkins, R.F. Gesteland, T.R. Cech (Eds.), Gene Regulation, Cold Spring Harbor Laboratory Press, Cold Spring Harbor, NY, 2010.
[52] K.E. McGinness, G.F. Joyce, In search of an RNA replicase ribozyme, Chem. Biol. 10 (2003) 5–14.
[53] D.P. Horning, G.F. Joyce, Amplification of RNA by an RNA polymerase ribozyme, Proc. Natl. Acad. Sci. U. S. A. 113 (2016) 9786–9791.
[54] B. Samanta, G.F. Joyce, A reverse transcriptase ribozyme, Elife 6 (2017) e31153.
[55] A.C. Gavin, M. Bosche, R. Krause, P. Grandi, M. Marzioch, A. Bauer, et al., Functional organization of the yeast proteome by systematic analysis of protein complexes, Nature 415 (2002) 141–147.
[56] R. Krause, C. von Mering, P. Bork, A comprehensive set of protein complexes in yeast: mining large scale protein-protein interaction screens, Bioinformatics 19 (2003) 1901–1908.
[57] Z. Shajani, M.T. Sykes, J.R. Williamson, Assembly of bacterial ribosomes, Annu. Rev. Biochem. 80 (2011) 501–526.
[58] H. Kim, S.C. Abeysirigunawarden, K. Chen, M. Mayerle, K. Ragunathan, Z. Luthey-Schulten, et al., Protein-guided RNA dynamics during early ribosome assembly, Nature 506 (2014) 334–338.
[59] C. Wolberger, Multiprotein-DNA complexes in transcriptional regulation, Annu. Rev. Biophys. Biomol. Struct. 28 (1999) 29–56.
[60] K. Ogata, K. Sato, T.H. Tahirov, Eukaryotic transcriptional regulatory complexes: cooperativity from near and afar, Curr. Opin. Struct. Biol. 13 (2003) 40–48.
[61] J.D. Perlmutter, M.F. Hagan, Mechanisms of virus assembly, Annu. Rev. Phys. Chem. 66 (2015) 217–239.
[62] F.J. Poelwijk, M.G. de Vos, S.J. Tans, Tradeoffs and optimality in the evolution of gene regulation, Cell 146 (2011) 462–470.
[63] M.G. de Vos, A. Dawid, V. Sunderlikova, S.J. Tans, Breaking evolutionary constraint with a tradeoff ratchet, Proc. Natl. Acad. Sci. U. S. A. 112 (2015) 14906–14911.
[64] E.V. Koonin, Y.I. Wolf, M.I. Katsnelson, Inevitability of the emergence and persistence of genetic parasites caused by thermodynamic instability of parasite-free states, Biol. Direct 12 (2017) 31.
[65] E. Szathmary, The evolution of replicators, Philos. Trans. R. Soc. Lond. B Biol. Sci. 355 (2000) 1669–1676.
[66] E.C. Holmes, The Evolution and Emergence of RNA Viruses, Oxford University Press, Oxford, 2009.
[67] E.V. Koonin, The Logic of Chance: The Nature and Origin of Biological Evolution, FT press, Upper Saddle River, NJ, 2011.

[68] E.V. Koonin, P. Starokadomskyy, Are viruses alive? The replicator paradigm sheds decisive light on an old but misguided question, Stud. Hist. Philos. Biol. Biomed. Sci. 59 (2016) 125–134.
[69] M. Jalasvuori, E.V. Koonin, Classification of prokaryotic genetic replicators: between selfishness and altruism, Ann. N. Y. Acad. Sci. 1341 (2015) 96–105.
[70] E.V. Koonin, Viruses and mobile elements as drivers of evolutionary transitions, Philos. Trans. R. Soc. Lond. B Biol. Sci. 371 (2016) 20150442.
[71] N. Takeuchi, P. Hogeweg, Evolution of complexity in RNA-like replicator systems, Biol. Direct 3 (2008) 11.
[72] N. Takeuchi, P. Hogeweg, Evolutionary dynamics of RNA-like replicator systems: a bioinformatic approach to the origin of life, Phys. Life Rev. 9 (2012) 219–263.
[73] N. Takeuchi, P. Hogeweg, E.V. Koonin, On the origin of DNA genomes: evolution of the division of labor between template and catalyst in model replicator systems, PLoS Comput. Biol. 7 (2011)e1002024.
[74] E. Szathmary, L. Demeter, Group selection of early replicators and the origin of life, J. Theor. Biol. 128 (1987) 463–486.
[75] S.J. Labrie, J.E. Samson, S. Moineau, Bacteriophage resistance mechanisms, Nat. Rev. Microbiol. 8 (2010) 317–327.
[76] A. Stern, R. Sorek, The phage-host arms race: shaping the evolution of microbes, Bioessays 33 (2011) 43–51.
[77] K.S. Makarova, Y.I. Wolf, E.V. Koonin, Comparative genomics of defense systems in archaea and bacteria, Nucleic Acids Res. 41 (2013) 4360–4377.
[78] E.V. Koonin, K.S. Makarova, Y.I. Wolf, Evolutionary Genomics of Defense Systems in Archaea and Bacteria, Annu. Rev. Microbiol. 71 (2017) 233–261.
[79] L. Aravind, V. Anantharaman, D. Zhang, R.F. de Souza, L.M. Iyer, Gene flow and biological conflict systems in the origin and evolution of eukaryotes, Front. Cell Infect. Microbiol. 2 (2012) 89.
[80] H.H. Kazazian Jr., Mobile elements: drivers of genome evolution, Science 303 (2004) 1626–1632.
[81] K.R. Oliver, W.K. Greene, Transposable elements: powerful facilitators of evolution, Bioessays 31 (2009) 703–714.
[82] B. Chenais, A. Caruso, S. Hiard, N. Casse, The impact of transposable elements on eukaryotic genomes: from genome size increase to genetic adaptation to stressful environments, Gene 509 (2012) 7–15.
[83] K. Gerdes, S.K. Christensen, A. Lobner-Olesen, Prokaryotic toxin-antitoxin stress response loci, Nat. Rev. Microbiol. 3 (2005) 371–382.
[84] K.S. Makarova, Y.I. Wolf, E.V. Koonin, Comprehensive comparative-genomic analysis of type 2 toxin-antitoxin systems and related mobile stress response systems in prokaryotes, Biol. Direct 4 (2009) 19.
[85] L. Van Melderen, Toxin-antitoxin systems: why so many, what for? Curr. Opin. Microbiol. 13 (2010) 781–785.
[86] L. Van Melderen, M. Saavedra De Bast, Bacterial toxin-antitoxin systems: more than selfish entities? PLoS Genet. 5 (2009) e1000437.
[87] E.V. Koonin, K.S. Makarova, Y.I. Wolf, Evolutionary entanglement of mobile genetic elements and host defence systems: guns for hire, Nat Rev Genet 21 (2020) 119–131.
[88] E.V. Koonin, K.S. Makarova, Mobile genetic elements and evolution of CRISPR-Cas systems: all the way there and back, Genome Biol. Evol. 9 (2017) 2812–2825.
[89] E.V. Koonin, K.S. Makarova, Y.I. Wolf, M. Krupovic, Evolutionary entanglement of mobile genetic elements and host defence systems: guns for hire, Nat. Rev. Genet. 21 (2020) 119–131.

[90] G. Faure, S.A. Shmakov, W.X. Yan, D.R. Cheng, D.A. Scott, J.E. Peters, et al., CRISPR-Cas in mobile genetic elements: counter-defence and beyond, Nat. Rev. Microbiol. 17 (2019) 513–525.
[91] J.E. Peters, K.S. Makarova, S. Shmakov, E.V. Koonin, Recruitment of CRISPR-Cas systems by Tn7-like transposons, Proc. Natl. Acad. Sci. U. S. A. 114 (2017) E7358–E7366.
[92] J. Strecker, A. Ladha, Z. Gardner, J.L. Schmid-Burgk, K.S. Makarova, E.V. Koonin, et al., RNA-guided DNA insertion with CRISPR-associated transposases, Science 365 (2019) 48–53.
[93] T.S. Halpin-Healy, S.E. Klompe, S.H. Sternberg, I.S. Fernandez, Structural basis of DNA targeting by a transposon-encoded CRISPR-Cas system, Nature 577 (2020) 271–274.
[94] E.V. Koonin, M. Krupovic, Evolution of adaptive immunity from transposable elements combined with innate immune systems, Nat. Rev. Genet. 16 (2015) 184–192.
[95] E.V. Koonin, W. Martin, On the origin of genomes and cells within inorganic compartments, Trends Genet. 21 (2005) 647–654.
[96] E.V. Koonin, T.G. Senkevich, V.V. Dolja, The ancient Virus World and evolution of cells, Biol. Direct 1 (2006) 29.
[97] E.V. Koonin, On the origin of cells and viruses: primordial virus world scenario, Ann. N. Y. Acad. Sci. 1178 (2009) 47–64.
[98] N. Takeuchi, P. Hogeweg, K. Kaneko, The origin of a primordial genome through spontaneous symmetry breaking, Nat. Commun. 8 (2017) 250.
[99] T.M. Embley, W. Martin, Eukaryotic evolution, changes and challenges, Nature 440 (2006) 623–630.
[100] W. Martin, E.V. Koonin, Introns and the origin of nucleus-cytosol compartmentalization, Nature 440 (2006) 41–45.
[101] P. Lopez-Garcia, D. Moreira, Selective forces for the origin of the eukaryotic nucleus, Bioessays 28 (2006) 525–533.
[102] A. Suomalainen, B.J. Battersby, Mitochondrial diseases: the contribution of organelle stress responses to pathology, Nat. Rev. Mol. Cell Biol. 19 (2018) 77–92.
[103] D. Bensasson, D. Zhang, D.L. Hartl, G.M. Hewitt, Mitochondrial pseudogenes: evolution's misplaced witnesses, Trends Ecol. Evol. 16 (2001) 314–321.
[104] J. Iranzo, A.E. Lobkovsky, Y.I. Wolf, E.V. Koonin, Virus-host arms race at the joint origin of multicellularity and programmed cell death, Cell Cycle 13 (2014) 3083–3088.
[105] E.V. Koonin, Intron-dominated genomes of early ancestors of eukaryotes, J. Hered. 100 (2009) 618–623.
[106] E.V. Koonin, The origin of introns and their role in eukaryogenesis: a compromise solution to the introns-early versus introns-late debate? Biol. Direct 1 (2006) 22.
[107] J. Raymond, O. Zhaxybayeva, J.P. Gogarten, S.Y. Gerdes, R.E. Blankenship, Whole-genome analysis of photosynthetic prokaryotes, Science 298 (2002) 1616–1620.
[108] A.Y. Mulkidjanian, E.V. Koonin, K.S. Makarova, S.L. Mekhedov, A. Sorokin, Y.I. Wolf, et al., The cyanobacterial genome core and the origin of photosynthesis, Proc. Natl. Acad. Sci. U. S. A. 103 (2006) 13126–13131.
[109] B. Nowicka, J. Kruk, Powered by light: phototrophy and photosynthesis in prokaryotes and its evolution, Microbiol. Res. 186-187 (2016) 99–118.
[110] W.F. Martin, F.L. Sousa, Early microbial evolution: the age of anaerobes, Cold Spring Harb. Perspect. Biol. 8 (2015) a018127.
[111] D.Y. Sorokin, K.S. Makarova, B. Abbas, M. Ferrer, P.N. Golyshin, E.A. Galinski, et al., Discovery of extremely halophilic, methyl-reducing euryarchaea provides

insights into the evolutionary origin of methanogenesis, Nat. Microbiol. 2 (2017) 17081.
[112] D. Hanahan, R.A. Weinberg, Hallmarks of cancer: the next generation, Cell 144 (2011) 646–674.
[113] P. Nelson, J. Masel, Intercellular competition and the inevitability of multicellular aging, Proc. Natl. Acad. Sci. U. S. A. 114 (2017) 12982–12987.
[114] J.M. Carlson, J. Doyle, Highly optimized tolerance: a mechanism for power laws in designed systems, Phys. Rev. E Stat. Phys Plasmas Fluids Relat. Interdiscip. Top. 60 (1999) 1412–1427.
[115] T. Zhou, J.M. Carlson, J. Doyle, Evolutionary dynamics and highly optimized tolerance, J. Theor. Biol. 236 (2005) 438–447.
[116] T. Dobzhansky, Nothing in biology makes sense except in the light of evolution, Am. Biol. Teach. 35 (1973) 125–129.
[117] S. Weinberg, Dreams of a Final Theory: The Scientist's Search for the Ultimate Laws of Nature, Vintage, New York, 1994.
[118] E.N. Trifonov, Vocabulary of definitions of life suggests a definition, J. Biomol. Struct. Dyn. 29 (2011) 259–266.
[119] E.V. Koonin, Defining life: an exercise in semantics or a route to biological insights? J. Biomol. Struct. Dyn. 29 (2012) 603–605.
[120] J.W. Szostak, Attempts to define life do not help to understand the origin of life, J. Biomol. Struct. Dyn. 29 (2012) 599–600.

CHAPTER 4

Evolutionary perspectives on cancer and aging

Walter F. Bodmer[1] and Daniel J.M. Crouch[2]
[1]Cancer and Immunogenetics Laboratory, Weatherall Institute of Molecular Medicine, John Radcliffe Hospital, University of Oxford, Oxford, United Kingdom
[2]JDRF/Wellcome Diabetes and Inflammation Laboratory, Wellcome Centre for Human Genetics, University of Oxford, Old Road Campus, Oxford, United Kingdom

Introduction

It is widely accepted that cancer is a somatic evolutionary process in which the steps are genetic, or relatively stable epigenetic, changes, each step successively providing some advantage for the development and outgrowth of the cancer (e.g., see Refs. [1,2]). Evolutionary pressures have pushed the somatic "error rate", especially the mutation rate, down to a level where for most organisms, cancer is no longer of any selective significance. These evolutionary pressures have, largely as a by-product, reduced cancer incidence at early ages in humans to a level where it is not directly a significant factor for natural selection, at least up to the present time. If humans suffered the same error rates as mice, then we would all die of cancer, certainly before the age of 5! [3]. Thus cancer in humans is mostly a disease of old age and has become significant in modern human populations largely because of the advances in health prevention and disease treatment over the last 100–150 years. Cancer is a thus a concomitant of senescence, and not a primary contributor to senescence.

For most cancers, a major step in this evolutionary process is the cellular escape from control over the process by which regularly dividing adult tissue stem cells differentiate into cells with a programmed, limited life span. The most aggressive cancers are often those that have lost most of their capacity to differentiate.

The major features of aging in complex eukaryotes are most notably extensive tissue malfunction as a result of the dysregulation of differentiation (for recent reviews see Refs. [4–7]) and, in this respect there appears to be some parallel between the processes of aging and the development of cancers. There is much discussion on the genetic and epigenetic

New Horizons in Evolution
DOI: https://doi.org/10.1016/B978-0-323-90752-1.00008-0

changes that could explain the aging-associated changes, and the factors, including dietary influences, which could influence the rates at which these changes are produced and how the changes could be ameliorated.

Most adult tissues are formed by adult tissue stem cells that persistently divide and produce differentiated daughter cells that form the mature tissue. These differentiated cells no longer divide, and eventually die by apoptosis, namely programmed cell death. When complex multitissue organisms pass the reproductive age, they suffer from increasing malfunction of many organs and there has so far been no overall satisfactory theory to explain this aging process.

Any replicating system in which heritable variants with differing replicative potentials can arise is subject to a Darwinian evolutionary process, and these systems must include the continually replicating adult (i.e., not embryonic, pluripotent) tissue stem cells that control most of the tissue integrity of most relatively long lived, multicellular, complex organisms, including humans. Only the limited number of tissues, such as neurons in the brain, where there is no apparent adult tissue turnover, will not be subject to such a Darwinian selective process. Once an organism has passed the reproductive age there is inevitably greatly reduced counter selection against mutations, or stable epigenetic changes that affect gene expression, in a manner that allows tissue stem cells to grow to their own advantage at the expense of the effect on the organism through reduced maintenance of the organism's integrity. This "selfish" selective advantage of the adult tissue stem cell is, we propose, mediated by a reduced ability to differentiate and by reduced apoptosis, leading to an increased growth rate of the stem cell but with reduced function.

Adult tissue stem cells and the extent to which there is somatic selection in their evolution has been discussed in Refs. [8–10]. There is, however, hardly any mention of the possibility that the degenerative changes in most aging tissues could be the result of somatic clonal selection for genetic or epigenetic variants that, to an increasing extent, escape some of the constraints of differentiation and apoptosis and so lead to tissue malfunction.

We suggest, therefore, that cancer can be viewed as an extreme form of aging, with respect to the tissue from which the cancer originates, and that similar violations of differentiation programs by adult stem cells are responsible for many normal aging phenomena.

As Williams [11] said in his much-quoted, perceptive article on human aging, "We are not machines that simply wear out, because our parts continually reproduce and refresh themselves throughout our lifetime". However, as Lopez-Otin et al. [7] said in their comprehensive review of "The Hallmarks of Ageing", "Stem cell exhaustion unfolds as the integrative consequence of multiple types of ageing-associated damages and likely constitutes one of the ultimate culprits of tissue and organismal ageing" (e.g., see also Ref. [12]). What is missing from this statement is an explanation of *why* all this damage leading to exhaustion should somehow accumulate disproportionately in the stem cells.

We propose that it is Darwinian somatic selection leading to the gradual dysregulation of the stem cell differentiation process, and eventually to persistent tissue malfunction, that is a major determinant of human aging and its associated wide range of age-related diseases. This proposal is based on the following assumptions: (1) natural selection at the level of the whole organism effectively stops, or is at least greatly weakened, after reproduction is finished, (2) long-term maintenance of most adult vertebrate tissues is undertaken by adult tissue stem cells, but (3) after some period of time, adult tissue stem cells become exhausted and are then replaced by "fresh" stem cells.

Background supporting data

It was first shown more than 20 years ago that normal human skin contains large numbers of clones of keratinocytes carrying TP53 mutations [13,14]. Because these mutations are often due to thymine dimers, it was assumed that they were caused by UV in sunlight on exposed areas of the skin. The fact that the mutations are mostly missense mutations with functional effects strongly implies that the TP53 mutations give the clones a growth advantage, just like that found in many cancers and precancerous growths, including colorectal adenomas. When there is a tumor adjacent to such a clone, it carries a different TP53 mutation, suggesting that in the cancer, the TP53 mutation is not a tumor initiating mutation but provides growth persistence (or immortalization) to the tumor (see e.g., Ref. [15]), after other mutations have been selected. Recently, a more extensive study of normal skin biopsies, using DNA sequencing, found evidence for frequent clones carrying a wide range of the sort of driver mutations found in cancers, notably those in NOTCH1, a gene whose function is closely tied to the maintenance of adult stem cells in epithelial tissues [16]. There is similar extensive evidence for somatic clonal selection in hematological cells from older individuals (see e.g., Refs. [10,17–20]).

Weissmann's group showed that older mice have 5 times as many hematopoietic stem cells (HSCs) in their bone marrow as young or middle-aged mice [21], but that these "old" HSCs were much less efficient than "young" HSCs at engrafting the bone marrow of irradiated mice. This clearly suggested somatic selection for clones that had lost much of their differentiating potential and as a result were faster and/or more persistent in their continued growth. More recently, it has become clear that elderly humans without cancers may have only a few dominant HSC clones in their normal bone marrow [22,23] and that these HSCs often carry mutations in DNMT3A and TET2, genes that are key regulators of DNA methylation, which is the main epigenetic mechanism for gene expression control in differentiated cells. Mutations are also found in genes controlling hematopoiesis and in TP53 [6]. This is clear evidence for somatic selection of dominant clones that have outgrown their contemporaries. A comprehensive analysis of gene expression in highly purified HSCs from young and old mice showed increased promoter methylation associated with differentiation-promoting genes in the latter and decreased methylation with genes associated with HSC maintenance [24]. These results are best explained by somatic selection for methylation changes in the older mice, changes that give rise to HSC clones with higher rates of proliferation at the expense of effective differentiation.

A strikingly similar observation has been made in the fruit fly, *Drosophila melanogaster*, by Biteau et al. [25]. They showed a decline in the ability of intestinal stem cells (ISCs) to differentiate properly in old (40 days) compared with young (3 days) flies, driven by the JNK gene. This decline in differentiating ability is accompanied by an increased rate of ISC proliferation.

An interesting, analogous phenomenon has been described by Andrew Wilkie and his colleagues [26,27], which they call "selfish selection of spermatogonial mutations". In normal, especially older, males they find clones of spermatogonial cells, effectively the adult tissue stem cells for sperm production, that carry, for example, mutations in the Kras oncogene that are known to cause severe inherited diseases. These mutations appear to be advantageous for the outgrowth of the spermatogonia that carry them but severely disadvantageous to the individuals who inherit these mutations through fertilization by mutated sperm. They suggest that this form of selfish selection can explain much of the apparent increase in male germ line mutation rates with age.

Origins of somatic genetic variation

Mutation in dividing somatic cells is an unavoidable phenomenon. However efficient the DNA repair processes, every cell when it divides will transmit at least a few new mutations to its daughter cells. There is therefore, as might be expected and as we have already discussed, abundant evidence for the existence of somatic genetic variation in a wide variety of human tissues (see Risques and Kennedy [28] for a recent extensive review). Stable DNA methylation changes, which underlie the whole cellular differentiation process, and which can also be the basis for selected gene expression changes in cancers, are an alternative source of genetic variation that is analogous to mutation. Such methylation changes are likely to be very important functional changes at the somatic level as they occur at a higher rate and can directly influence the level of expression of a relevant gene. Recent evidence suggests rates of gene methylation in HSCs increase with age [24].

While extreme differences in the rates of mutation or methylation due to severe germ line mutations in, for example, certain DNA repair genes will be expected to increase the rate of somatic mutations and methylation changes that we suggest underlie the aging process, the normal background error rates will, we propose, be enough to lead to normal rates of aging. This is entirely analogous to the fact that normal rates of mutation can lead to tumor development, so that genetic instability is not required for tumor initiation [29,30].

Mutations can either be neutral, meaning they have no functional effect that influences their survival or rate of proliferation, or be selectively disadvantageous leading to a decreased rate of proliferation or survival, or be selectively advantageous resulting in a higher rate of proliferation and/or survival. Cells with neutral mutations can, just by chance, increase in frequency relative to nonmutant cells. While the probability of this is low for any particular mutation, since there is so much of the DNA that can give rise to mutations, clonal variation can arise from neutral genetic variation, but this variation, by its nature, is generally not likely to lead to any relevant functional change. Variants that are disadvantageous to stem cell proliferation will be selected against and so are not expected to play any role in the aging process. Even quite small selective disadvantages give rise to exceedingly low probabilities of survival. It is therefore only the variants that favor increased stem cell proliferation that are likely to give rise to functionally affected clones. Even quite small selective advantages give hugely increased chances of survival relative to neutral clones (for the classical population genetics analysis behind these arguments see, e.g., Cavalli-Sforza and Bodmer [31,32]).

Mechanisms of somatic selection in adult tissue stem cells

The key to our model for aging is not just to suggest an effect of selection on the aging adult tissue stem cells (see, e.g., Beerman et al. [10]), but to propose the underlying basis for the mechanism of action of the selection and the reason for its detrimental effect on adult tissue functions

Thus we suggest that the common feature in all the examples of age-related functional changes in adult tissue stem cells that we have discussed is somatic selection for mutations (or stable epigenetic changes) that give rise to an increased rate of proliferation and persistence of adult tissue stem cells at the expense of normal functional differentiation and, eventually, apoptosis. It is this process that leads to long-term tissue malfunction. Once an organism has passed the reproductive age, by which is meant the age range during which they have offspring, there is little, if any, counter selection at the organismal level to this inevitable selfish Darwinian cellular level process that leads to the gradual erosion of the differentiating capacity of adult tissue stem cells through selection for increased stem cell proliferation. Weaker selective effects at the organismal level, mediated through the benefits that individuals are able to bestow on their younger relatives by helping to care for them [33,34], may help to explain the relative rarity of tumorigenesis as compared to the normal aging process during postreproductive life.

The cancer model of Tomlinson and Bodmer [35] uses the known cellular composition of the colon as its biological framework. This is based on epithelial stem cells at the base of the colonic crypts that line the surface of the colon. These stem cells control a crypt's complete turnover about every 4–6 days and give rise to transit amplifying cells, which differentiate into terminally differentiated cells that die by apoptosis and are removed into the lumen of the intestine. The mathematical model assumes that the overall proliferation rate of stem cells can be represented simply by $\alpha = \alpha_3 - \alpha_2 - \alpha_1$, where α_3 is the rate of cell division, α_2 is the rate at which daughter cells become differentiated and α_1 the rate of apoptosis. In a normal steady state $\alpha = 0$, and so, following this model, if nonmutant stem cells have a negligible apoptosis rate $\alpha_1 = 0$, then $\alpha_3 = \alpha_2$. This can either be achieved by asymmetric cell division, where when a stem cell divides, one daughter always remains a stem cell while the other always becomes a transit amplifying cell destined to differentiate, or if, as now seems more likely, stem cells on average produce one transit amplifying cell and one stem cell but any given division may produce two daughter stem cells or two transit amplifying daughter cells. In that case the rate at which stem cells produce two daughter stem cells must equal the rate at which they produce two transit amplifying cells. The steady state can then be perturbed by a mutation that decreases the rate at which two transit amplifying cells are produced, so that more daughter stem cells are produced on average than daughter transit amplifying cells. In that case $\alpha_3 > \alpha_2$ and so $\alpha > 0$ and the stem cells have an increased proliferation rate relative to the steady state maintained by nonmutant stem cells. That is the basis for assuming that mutations which decrease the rate at which a stem cell produces daughter transit amplifying and then differentiated cells may be the basis for selection for increased rates of proliferation of the stem cells at the expense of normal rates of differentiation. This then disturbs the normal steady state ratio between stem and differentiated cells and so is a basis for tissue malfunction.

Why should these effects be concentrated in adult stem cells, as opposed to transit amplifying cells? Transit amplifying cells complete a limited number of divisions before differentiating and then undergoing apoptosis, and they will therefore require a major restructuring of their cellular growth pathways to reactivate a proliferative phenotype.

Differentiated cell populations could, in principle, gain a selective advantage through the acquisition of mutations causing delayed apoptosis, but these mutations could only be passed on to a limited number of daughter cells and would thus have only short-term effects.

These arguments against a role for mutations that affect only the function of transit amplifying or of differentiated cells lead naturally to the suggestion that the predominant cause of aging lies in the adult stem cells *themselves* evolving a resistance to apoptosis or, more likely initially, to differentiation. This resistance to differentiation, leading to increased proliferation, will result in greater numbers of mutant stem cells, as explained above, and will confer a selective advantage on them. Johnston et al. [36,37], developing the model of Tomlinson and Bodmer [35] by incorporating homeostatic feedback, showed that successive genetic or stable epigenetically based changes that either decrease apoptosis or decrease differentiation rates in stem cells lead to a series of increasingly large, stable benign tumors, with increasingly large numbers of stem cells and differentiated cells, the latter still having their life spans limited by apoptosis. The incorporation of feedback enables homeostatic regulation of the stem cells and buffering against stochastic physiological variation. This enables the stem cell number to vary around a fixed mean over time, leading to corresponding minor variation in the balance between stem cell division and differentiation. Cancers, resulting from exponential, continuous, growth only arise after the cumulative effect of a number of such steps exceeds the limitations of benign growth. In this sense, the aging adult tissue stem cells give rise to the analog of an adenoma rather than a carcinoma. Decreases in the rates of differentiation, and concomitant increases in the rates of stem cell proliferation, that either preserve or do not greatly change the total numbers of cells present in a region of tissue and so maintain basic functionality, will confer smaller selective disadvantages to the organism as a whole than those which cause an overall large growth in numbers and greater disruption of tissue differentiation, as happens in cancers.

As apoptosis occurs infrequently in normally functioning stem cells and is presumed to occur mainly in response to signs of stem cell exhaustion, failure to differentiate properly is very likely to be the candidate for the selective effect of initial mutations. Resistance to apoptosis will confer no significant selective advantage to cells that are not yet displaying signs of exhaustion. For the *eventual persistence* of the resulting abnormally functioning tissue, there must, however, be further selection specifically against apoptosis of the stem cells carrying the mutations that lead to defective differentiation. This further selection will ensure against full replacement of the overgrown population of abnormally differentiated mutant or epigenetically altered cells by fully functional, normally differentiated tissue derived from a fresh stem cell, originating from an underlying pool of potential fully functional stem cells. Such replacement will occur in response to signs of exhaustion, including a failure to differentiate at the normal healthy rate. This can explain the presence of, for example, TP53 mutations promoting immortality in aging adult stem cells, analogous to the presence of TP53 mutations in late larger colorectal adenomas [38]. Failure of stem cell differentiation, associated with persistent increased proliferation, thus acts as a strongly positively selected phenotype at the level of the stem cell, and a weakly negatively selected phenotype at the level of the organism. It is just such changes that we propose are the candidates for many of the causes of normal aging.

In contrast to the aging situation, during life up to the end of the reproductive period there is strong selection at the higher level of the individual whole organism, for the integrity of adult tissues over many cycles of stem cell proliferation. As a result, there are mechanisms that differentially protect the adult tissue stem cell from accumulating various forms of damage, including DNA damage, and that ensure the replacement of a stem cell once it has exceeded a certain level of "exhaustion" (e.g., see Ref. [12]). One feature of this protective process seems to be the existence of two levels of adult tissue stem cells. One level is functional at any given time and is responsible for the clonality of individual colorectal crypts [39], while the second has a slower proliferation rate (some say that cells at this level are quiescent [40]). The second level "quiescent" stem cells are poised to replace the first level stem cells as soon as the need arises, namely when a currently fully functional stem cell reaches exhaustion. There is direct evidence for these two levels of stem cells in the human gut [41–44]. It is the counteraction of the exhaustion of individual stem cells, presumably by countering apoptosis, that gives rise to the persistence of dysregulated differentiated tissue, as discussed above. With this background, we will now discuss in more detail how the model of Johnston et al. [36], with some modification can, we suggest, explain how selection works on adult tissue stem cells.

Quantitative model

An extension to the model of Johnston et al. [36], which can explain how long-term persistence of individual cells can occur and why this is only likely to happen to cells that have already acquired a failure to differentiate at the full healthy rate, is shown schematically in Fig. 4.1. This shows a graphical depiction of the differentiation process and its relationship to the various parameters. We retain the same general notation as in the original Johnston et al. [36] model, which relates specifically to cell types within the colonic crypt, as already discussed. We represent the number of quiescent or "second layer" stem cells as N_0, the number of functional stem cells as N_1, the number of semidifferentiated still dividing cells as N_2 and the number of fully differentiated cells as N_3. For each stem cell type we use the parameters α_1, α_2, and α_3 (for quiescent stem cells) and β_1, β_2, and β_3 (for functional stem cells) to represent the apoptosis, differentiation and renewal rates, respectively. For transit amplifying or semidifferentiated cells we use the equivalent parameters δ_1, δ_2, and δ_3, and for fully differentiated cells we define a single parameter γ, representing their apoptosis rate.

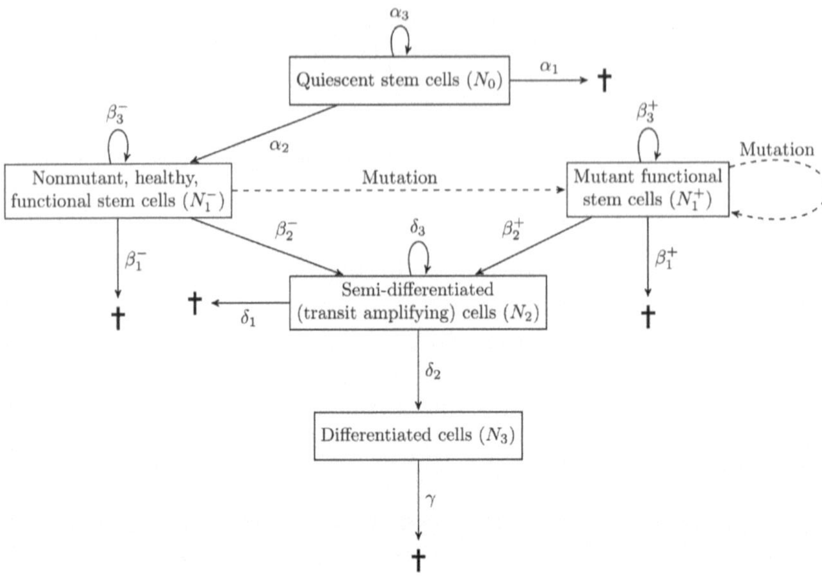

Figure 4.1 Schematic representation of a model for stem cell proliferation and differentiation. Crucifix signs indicate cell death.

Thus as depicted in Fig. 4.1, the individual cells of the quiescent stem cell population can, when they divide, die (α_1), become a healthy (non-mutant) functional stem cell (α_2) or self- renew (α_3). Similarly, the healthy functional stem cell can die (β_1), become a transit amplifying semidifferentiated cell, which can self-renew and then become fully differentiated (β_2), or renew (β_3). The same applies to the semidifferentiated cells (δ_1 for dying, δ_2 for differentiating and δ_3 for self-renewal), while differentiated cells can only die by apoptosis (γ).

We assume that when the population of stem or semidifferentiated cells increases, the rate at which new cells are produced also increases, but instead of assuming a linear dependence of per-capita rate on population size we assume that there is a maximum per-capita rate represented by feedback terms in the following equations. This feedback is an essential feature of our model, as it is required to obtain stable states of the relative proportions of the various cell types, where the change in these proportions depends on the magnitude of the effects of mutations on the properties of the functional stem cells. A simple model with no feedback control necessarily results in exponential growth of adult tissue stem cells and so cannot explain what happens with aging other than giving rise to a cancer. The rates of change of the various cell type counts are as specified by Johnston et al. [36] using the following differential equations:

$$\frac{dN_0}{dt} = \left(\alpha_3 - \alpha_2 - \alpha_1 - \frac{k_0 N_0}{1 + m_0 N_0}\right) N_0, \tag{4.1}$$

$$\frac{dN_1}{dt} = \left(\beta_3 - \beta_2 - \beta_1 - \frac{k_1 N_1}{1 + m_1 N_1}\right) N_1 + \alpha_2 N_0 + \frac{k_0 N_0^2}{1 + m_0 N_0}, \tag{4.2}$$

$$\frac{dN_2}{dt} = \left(\delta_3 - \delta_2 - \delta_1 - \frac{k_2 N_2}{1 + m_2 N_2}\right) N_2 + \beta_2 N_1 + \frac{k_1 N_1^2}{1 + m_1 N_1} \tag{4.3}$$

$$\frac{dN_3}{dt} = -\gamma N_3 + \delta_2 N_2 + \frac{k_2 N_2^2}{1 + m_2 N_2}. \tag{4.4}$$

In these equations, the nonlinear homeostatic feedback to control cell numbers is captured by the terms $k_0 N_0^2/(1 + m_0 N_0)$, $k_1 N_1^2/(1 + m_1 N_1)$ and $k_2 N_2^2/(1 + m_2 N_2)$. These effectively impose carrying capacities on the three types of cell, which can be overwhelmed when the cells replicate too rapidly. If $m_0 = m_1 = m_2 = 0$ feedback is linear. Cell numbers beyond the limit imposed by feedback are forced to differentiate into the downstream cell type. Nonlinearity is essential to capture the ability of cells to overwhelm homeostatic control, as this is necessary to produce exponential cancerous growths, as shown by Johnston et al. [36].

The model can be expanded to allow for continual replacement of mutant (+) dysregulated stem cells in the functional layer from a supply of "fresh" nonmutant (−) stem cells produced by the quiescent level of stem cells. It can then be shown that the equilibrium for the total nonquiescent stem cell population size, when the mutant is able to spread, is

$$\tilde{N}_1 = \tilde{N}_1^+ + \tilde{N}_1^- = \frac{\beta^+}{(k_1 - m_1 \beta^+)}, \tag{4.5}$$

where N^+ is the mutant stem cell population size, N^- is the nonmutant stem cell population size, and β^+ is the net proliferation rate of the mutant stem cell population. This result, surprisingly, does not depend on any properties specific to the nonmutant functional stem cells. However, if the mutant survives, this does not cause the elimination of the nonmutant type. When a new mutant is introduced into a population consisting

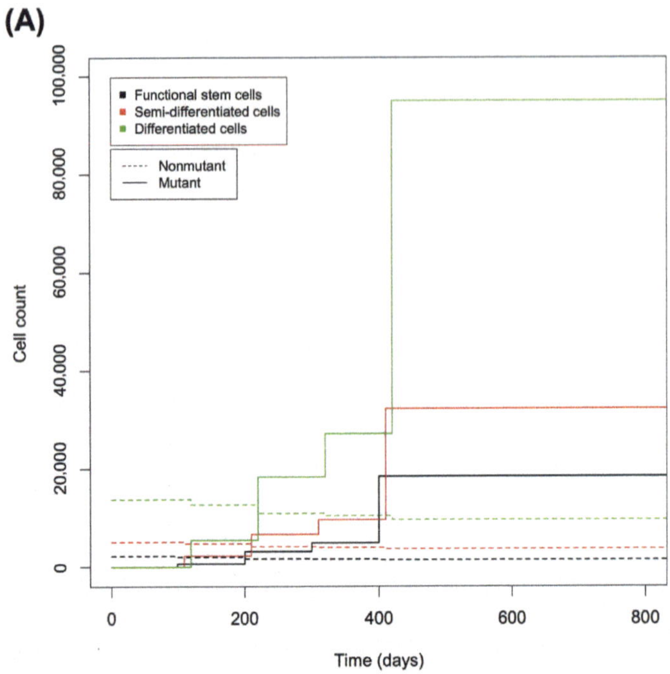
(A)
Functional stem cells
Semi-differentiated cells
Differentiated cells
Nonmutant
Mutant
Cell count
0
20,000
40,000
60,000
80,000
100,000
0
200
400
600
800
Time (days)

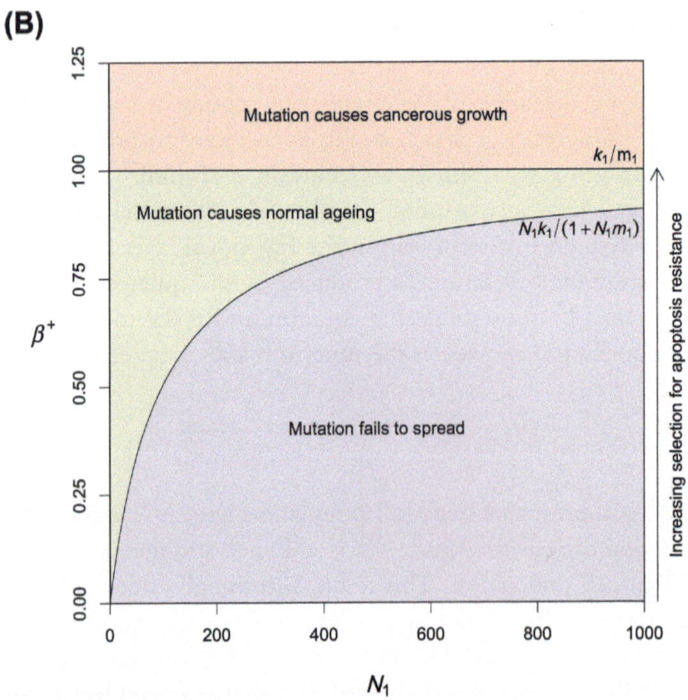
(B)
Mutation causes cancerous growth
k_1/m_1
Mutation causes normal ageing
$N_1k_1/(1+N_1m_1)$
Mutation fails to spread
Increasing selection for apoptosis resistance
β^+
0.00
0.25
0.50
0.75
1.00
1.25
0
200
400
600
800
1000
N_1

of an existing mutant type and a healthy nonmutant type, and if it can increase in frequency, the new mutant will entirely replace the original mutant, consistent with the observation of individual mutant HSC clones dominating sections of aged bone marrow. The new mutation could either be a different mutation in the nonmutant stem cell population $\left(N_1^-\right)$ or an additional mutation in the already mutated stem cell population $\left(N_1^+\right)$.

The effects of mutations on tissue function will be intrinsic to the mechanisms by which mutations affect differentiation and will also depend on the increased number of resulting differentiated and semidifferentiated cells, N_3 and N_2, which themselves depend on N_1^+, N_1^- and N_0, the numbers of mutant and nonmutant functional stem cells and quiescent stem cells, respectively. A numerical example of an outcome of the model based on the above equations is shown in Fig. 4.2A. This shows equilibrium values for N_1^+ and N_1^- after a series of mutations changing the differentiation $\left(\beta_2^+\right)$ and apoptosis $\left(\beta_1^+\right)$ rate parameters in the functional stem cell population. Consistent with our proposed model, the first three mutations confer a reduced differentiation phenotype, while the fourth and final mutation confers almost complete resistance to apoptosis, in effect immortalization. The numbers of semidifferentiated and differentiated cells descending from mutant and nonmutant stem cells are also shown. Equilibrium states are calculated from modified equations based on Johnston et al. [37]. The final equilibrium state shows that, now, most of the cells are mutant, with a large number of presumably poorly functioning differentiated cells.

Figure 4.2 (A) An illustrative example of the outcome of the model based on Eqs. (4.1)–(4.4) for a sequence of functional stem cell mutations ($\{\beta_1^+, \beta_2^+\}$ = {0.01, 0.1}, {0.01, 0.025}, {0.01, 0.01}, {0.01, 0.005}, and {0, 0.005}) occurring successively every 100 days. The initial values are those for the nonmutant rates, $\beta_1^- = 0.01$ and $\beta_2^- = 0.1$, which do not vary. Day zero is a point after the end of the organism's reproductive lifespan, when selection for healthy tissue functioning has been greatly diminished. Each vertical line marks a new mutation. Its effect on the numbers of each cell type is given by the horizontal lines, which show stable equilibrium cell counts for each new mutation. Remaining parameters were set to $k_0 = 0.1$, $m_0 = 0.1$, $\alpha_1 = 0.01$, $\alpha_2 = 0.05$, $\alpha_3 = 1$, $k_1 = 0.01$, $m_1 = 0.01$, $\beta_3^+ = 1$, $\beta_3^- = 1$, $k_2 = 0.001$, $m_2 = 0.001$, $\delta_1 = 0.1$, $\delta_2 = 0.5$, $\delta_3 = 1$ and $\gamma = 0.5$ and did not vary, on the assumption that the mutations only affected the parameters of the functional stem cell population, N_1. (B) Parameter space leading to normal aging effects (stable growth) and cancerous effects (exponential growth) versus functional stem cell population size. Parameters k_1 and m_1 were both set to 0.01.

Regions of stability in the β parameter space for the system with saturating feedback in the functional stem cell population are shown in Fig. 4.2B, plotted against the functional stem cell population size, N_1. If $\beta^+ < N_1 k_1/(1 + N_1 m_1)$, then the mutant stem cells cannot sustain their number and the mutant contribution to the tissue region becomes extinct resulting in a nonmutant healthy stable tissue region (blue region). If $\beta^+ > k_1/m_1$ the growth saturation limit is exceeded, and the cell populations grow with no bound as in a cancer [36,37] (red region). The green region corresponds to normal aging, where mutant functional stem cells can sustain their growth but do not overwhelm homeostatic nonlinear feedback so as to grow exponentially and turn into a cancer. (For further details of the mathematical treatment of this model see Bodmer and Crouch [45]).

Discussion

We have shown how adult tissue stem cell growth based on selection, after the end of the reproductive period, for increased stem cell proliferation at the cost of reduced differentiation and eventually elimination of apoptosis, can lead to immortalization of aberrantly differentiating adult tissue stem cells and stable malfunctioning mature tissue, but not cancer. We suggest that this can explain the aging process and reduced efficacy of the adult tissue stem cells, for those tissues that are continually turning over. This is compatible with observations of increased numbers of somatic clones with mutations that are often found in cancers, as a function of increasing age. In the particular case of HSCs this also fits with the evidence of changes in their patterns of differentiation associated with increasing age, and also associated with oligoclonality of the HSCs.

Our quantitative model is a key part of our case for the role of somatic selection against differentiation and apoptosis in adult tissue stems cells as a major feature of the aging process. Essential features of the model are that it includes the minimal number of plausible cell compartments that is known to model adult tissues, using crypts in the normal colon as our biological example, and that stable states cannot be established without some form of feedback. The aim of such a model is not so much to provide a basis for estimating relevant parameters, such as division and differentiation rates, but to provide a quantitatively and biologically plausible mathematical description of how such selection could work in a way that is consistent with the biological observations we have reviewed. This is most directly demonstrated by Fig. 4.2B, which shows the limited range of values of β^+, the net mutant functional stem cell reproduction rate, in relation to N_1, the number of functional stem cells, that can lead to functional differentiated tissue without giving rise to a cancer. Here it is clear that, under our model, there is an upper bound on the amount of normal aging that can possibly occur within a single tissue region, measured in terms of excess functional stem cells. In the case of aging, it is most probably just a single mutation in an aging clone, while cancer only occurs with the accumulation of several mutations in a clone.

The quantitative implementation of the model confirms that failure of functional adult stems cells to differentiate at the normal rate increases their rate of renewal, and so leads generally to stable overgrowth of cells. The model, however, only leads to a cancer when differentiation rates reduce to below a certain threshold. Within growths associated with mutations, or stable epigenetic expression changes, that increase the net functional stem cell rate of proliferation, β^+, the proportions of the various cell types are different to those observed within healthy tissue, exhibiting an excess of semidifferentiated and differentiated cells. These predictions are entirely consistent with observations of aged tissue.

Clearly for those tissues where there is no apparent turnover connected with tissue stem cells, for example, neurons in the brain, aging may be mostly due to wear and tear with age connected with constant use, just like the parts of a car. However, even such tissues are likely to be materially affected by declining function of the stem cells for supporting tissues and cell types, for example, the glial cells in the brain.

Spatial expansion of mutant clones at the expensive of other cells has been shown to occur in various epithelial structured tissues [16,46–48]. Although growths of poorly differentiating stem cells descending from a single mutant clone are unlikely to account for the aging of an entire epithelial tissue such as the skin or gut, the rate of successful selective increases of such mutations becomes much higher after the end of the reproductive life of the organism. Under such circumstances, the relevant mutations will occur multiple times in different tissue regions, thereby accounting for significant damage. A mutation causing a growth which took over a very large part of a tissue would most likely be one overwhelming any kind of homeostatic feedback and therefore, in effect, be cancerous, as our model shows (Fig. 4.2A).

On the assumption that, as we have argued, the somatic selection of variant clones with dysregulated differentiation is inevitable and may be a key to human aging, the basic scientific problem is the lack of understanding of the normal processes of tissue homeostasis before the onset of organismal aging, and of the changes in these processes with organismal aging. A systematic study of stem cells and their immediate progeny in a variety of tissues and at different organismal ages, analogous to the study of HSCs by Xie et al. [23], should help to reveal the changes that are associated with aging. Analyses of mRNA in single cells should reveal the balance of different cell types at different stages of differentiation, while direct and methylation DNA sequencing should reveal the genes in which there are somatic mutations or stable methylation changes in expression in older organisms compared to younger organisms. It will then be interesting to see to what extent the genes in which there are significant changes correspond to the genes in which there are driver mutations or stable methylation changes in cancers from the relevant tissue, as is the case for the somatic mutations found in HSCs from normal healthy elderly individuals [23]. In agreement with this possibility, Slack et al. [49] have suggested that drugs that attack some of the cancer promoting pathways, for example, those associated with mutations in Kras and PTEN, might counter aging effects. This idea is consistent with the notion that abnormally functioning adult stem cell clones may accumulate genetic or epigenetic changes by a process of somatic selection for driver mutations or stable methylation changes that is similar to that found in cancers.

Conflict of interest

The authors declare no conflicts of interest.

Author contributions

WFB conceived the project. WFB and DJMC wrote the manuscript.

Acknowledgments

We are pleased to be able to make this contribution to a publication in honor of Eviatar Nevo's 90th birthday, recognizing the extraordinary range and quality of his contributions to the study of evolution. This paper is an edited version of Bodmer and Crouch [45].

This work was supported by JDRF Grant 5-SRA-2015-130-A-N and Wellcome Trust Grants 107212/Z/15/Z and 203131/Z/16/Z.

References

[1] W. Bodmer, The somatic evolution of cancer—The Harveian Oration of 1996, J. R. Coll. Physicians Lond. 31 (1997) 82–89.
[2] W.F. Bodmer, Cancer and evolution, Isr. J. Ecol. Evol. 52 (2006) 233–245. 2006/01/01.
[3] E.F. Bashford, J.A. Murray, The incidence of cancer in mice of known age, in: Proceedings of the Royal Society B, vol. 81, 1909, pp. 310–313.
[4] L.B. Boyette, R.S. Tuan, Adult stem cells and diseases of aging, J. Clin. Med. 3 (2014) 88–134.
[5] L. Fontana, L. Partridge, Promoting health and longevity through diet: from model organisms to humans, Cell 161 (2015) 106–118.
[6] M.A. Goodell, T.A. Rando, Stem cells and healthy aging, Science 350 (2015) 1199–1204.
[7] C. Lopez-Otin, M.A. Blasco, L. Partridge, M. Serrano, G. Kroemer, "The hallmarks of aging, Cell 153 (2013) 1194–1217.
[8] I.L. Weissman, Stem cells are units of natural selection for tissue formation, for germline development, and in cancer development, Proc. Natl. Acad. Sci. U. S. A. 112 (2015) 8922–8928.
[9] R.A.J. Signer, S.J. Morrison, Mechanisms that regulate stem cell aging and life span, Cell Stem Cell 12 (2013) 152–165.
[10] I. Beerman, D. Bhattacharya, S. Zandi, M. Sigvardsson, I.L. Weissman, D. Bryder, et al., "Functionally distinct hematopoietic stem cells modulate hematopoietic lineage potential during aging by a mechanism of clonal expansion, Proc. Natl. Acad. Sci. U. S. A. 107 (2010) 5465–5470.
[11] G.C. Williams, "Pleiotropy, natural-selection, and the evolution of senescence, Evolution 11 (1957) 398–411.
[12] D.J. Rossi, C.H.M. Jamieson, I.L. Weissman, "Stems cells and the pathways to aging and cancer, Cell 132 (2008) 681–696.
[13] A.S. Jonason, S. Kunala, G.J. Price, R.J. Restifo, H.M. Spinelli, J.A. Persing, et al., "Frequent clones of p53-mutated keratinocytes in normal human skin, Proc. Natl. Acad. Sci. U. S. A. 93 (1996) 14025–14029.
[14] Z.P. Ren, A. Hedrum, F. Ponten, M. Nister, A. Ahmadian, J. Lundeberg, et al., "Human epidermal cancer and accompanying precursors have identical p53 mutations different from p53 mutations in adjacent areas of clonally expanded non-neoplastic keratinocytes, Oncogene 12 (1996) 765–773.
[15] J.R. Jenkins, K. Rudge, G.A. Currie, "Cellular immortalization by a Cdna clone encoding the transformation-associated phosphoprotein P53, Nature 312 (1984) 651–654.
[16] I. Martincorena, P.J. Campbell, Somatic mutation in cancer and normal cells (vol. 351, aaf5401, 2016) Science 353 (2016) 132.
[17] R.H. Cho, H.B. Sieburg, C.E. Muller-Sieburg, "A new mechanism for the aging of hematopoietic stem cells: aging changes the clonal composition of the stem cell compartment but not individual stem cells, Blood 111 (2008) 5553–5561.
[18] C. Benz, M.R. Copley, D.G. Kent, S. Wohrer, A. Cortes, N. Aghaeepour, et al., "Hematopoietic stem cell subtypes expand differentially during development and display distinct lymphopoietic programs, Cell Stem Cell 10 (2012) 273–283.
[19] G. Genovese, A.K. Kahler, R.E. Handsaker, J. Lindberg, S.A. Rose, S.F. Bakhoum, et al., "Clonal hematopoiesis and blood-cancer risk inferred from blood DNA sequence, N. Engl. J. Med. 371 (2014) 2477–2487.
[20] S. Jaiswal, P. Fontanillas, J. Flannick, A. Manning, P.V. Grauman, B.G. Mar, et al., "Age-related clonal hematopoiesis associated with adverse outcomes, N. Engl. J. Med. 371 (2014) 2488–2498.

[21] S.J. Morrison, A.M. Wandycz, K. Akashi, A. Globerson, I.L. Weissman, "The aging of hematopoietic stem cells, Nat. Med. 2 (1996) 1011–1016.

[22] M. Buscarlet, S. Provost, Y.F. Zada, A. Barhdadi, V. Bourgoin, G. Lepine, et al., "DNMT3A and TET2 dominate clonal hematopoiesis and demonstrate benign phenotypes and different genetic predispositions, Blood 130 (2017) 753–762.

[23] M.C. Xie, C. Lu, J.Y. Wang, M.D. McLellan, K.J. Johnson, M.C. Wendl, et al., "Age-related mutations associated with clonal hematopoietic expansion and malignancies, Nat. Med. 20 (2014) 1472–1478.

[24] D.Q. Sun, M. Luo, M. Jeong, B. Rodriguez, Z. Xia, R. Hannah, et al., "Epigenomic profiling of young and aged HSCs reveals concerted changes during aging that reinforce self-renewal, Cell Stem Cell 14 (2014) 673–688.

[25] B. Biteau, C.E. Hochmuth, H. Jasper, "JNK activity in somatic stem cells causes loss of tissue homeostasis in the aging Drosophila gut, Cell Stem Cell 3 (2008) 442–455.

[26] G.J. Maher, A. Goriely, A.O.M. Wilkie, Cellular evidence for selfish spermatogonial selection in aged human testes, Andrology 2 (2014) 304–314.

[27] G.J. Maher, S.J. McGowan, E. Giannoulatou, C. Verrill, A. Goriely, A.O.M. Wilkie, Visualizing the origins of selfish de novo mutations in individual seminiferous tubules of human testes, Proc. Natl. Acad. Sci. U. S. A. 113 (2016) 2454–2459.

[28] R.A. Risques, S.R. Kennedy, Aging and the rise of somatic cancer-associated mutations in normal tissues, PLoS Genet. 14 (2018) e1007108.

[29] W. Bodmer, Genetic instability is not a requirement for tumor development, Cancer Res. 68 (2008) 3558–3560.

[30] I. Tomlinson, W. Bodmer, Selection, the mutation rate and cancer: ensuring that the tail does not wag the dog, Nat. Med. 5 (1999) 11–12.

[31] L.L. Cavalli-Sforza, W.F. Bodmer, The Genetics of Human Populations, Dover, Mineola, NY, 1999.

[32] L.L. Cavalli-Sforza, W.F. Bodmer, The Genetics of Human Populations, W.H. Freeman, San Francisco, 1971.

[33] W.D. Hamilton, Genetical evolution of social behaviour II, J. Theor. Biol. 7 (1964) 17–52.

[34] W.D. Hamilton, Genetical evolution of social behaviour I, J. Theor. Biol. 7 (1964) 1–16.

[35] I.P.M. Tomlinson, W.F. Bodmer, Failure of programmed cell-death and differentiation as causes of tumors—some simple mathematical-models, Proc. Natl. Acad. Sci. U. S. A. 92 (1995) 11130–11134.

[36] M.D. Johnston, C.M. Edwards, W.F. Bodmer, P.K. Maini, S.J. Chapman, "Mathematical modeling of cell population dynamics in the colonic crypt and in colorectal cancer, Proc. Natl. Acad. Sci. U. S. A. 104 (2007) 4008–4013.

[37] M.D. Johnston, C.M. Edwards, W.F. Bodmer, P.K. Maini, S.J. Chapman, "Examples of mathematical modeling: tales from the crypt, Cell Cycle 6 (2007) 2106–2112.

[38] M. Pignatelli, G.W.H. Stamp, G. Kafiri, D. Lane, W.F. Bodmer, "Over-expression of P53 nuclear oncoprotein in colorectal adenomas, Int. J. Cancer 50 (1992) 683–688.

[39] M.R. Novelli, J.A. Williamson, I.P.M. Tomlinson, G. Elia, S.V. Hodgson, I.C. Talbot, et al., "Polyclonal origin of colonic adenomas in an XO/XY patient with FAP, Science 272 (1996) 1187–1190.

[40] T.H. Cheung, T.A. Rando, "Molecular regulation of stem cell quiescence, Nat. Rev. Mol. Cell Biol. 14 (2013) 329–340.

[41] S. Roth, R. Fodde, "Quiescent stem cells in intestinal homeostasis and cancer, Cell Commun. Adhes. 18 (2011) 33–44.

[42] H. Tian, B. Biehs, S. Warming, K.G. Leong, L. Rangell, O.D. Klein, et al., A reserve stem cell population in small intestine renders Lgr5-positive cells dispensable (vol. 478, pg. 255, 2011) Nature 482 (2012).

[43] K.S. Yan, L.A. Chia, X.N. Li, A. Ootani, J. Su, J.Y. Lee, et al., The intestinal stem cell markers Bmi1 and Lgr5 identify two functionally distinct populations, Proc. Natl. Acad. Sci. U. S. A. 109 (2012) 466−471.
[44] S.J. Buczacki, H.I. Zecchini, A.M. Nicholson, R. Russell, L. Vermeulen, R. Kemp, et al., Intestinal label-retaining cells are secretory precursors expressing Lgr5, Nature 495 (2013) 65−69.
[45] W.F. Bodmer, D.J.M. Crouch, Somatic selection of poorly differentiating variant stem cell clones could be a key to human ageing, J. Theor. Biol. 489 (2020) 110153.
[46] N.A. Wright, Epithelial stem cell repertoire in the gut: clues to the origin of cell lineages, proliferative units and cancer, Int. J. Exp. Pathol. 81 (2000) 117−143.
[47] A.M. Nicholson, C. Olpe, A. Hoyle, A.S. Thorsen, T. Rus, M. Colombe, et al., Fixation and spread of somatic mutations in adult human colonic epithelium, Cell Stem Cell 22 (2018) 909−918. e8.
[48] I. Martincorena, "Somatic mutation and clonal expansions in human tissues, Genome Med. 11 (2019) 35.
[49] C. Slack, N. Alic, L. Partridge, "Could cancer drugs provide ammunition against aging? Cell Cycle 15 (2016) 153−155.

CHAPTER 5

Evolutionary medicine—*Apolipoprotein L1* in human health and disease

Etty Kruzel-Davila[1] and Karl Skorecki[1,2]
[1]Department of Nephrology, Rappaport Faculty of Medicine and Research Institute, Technion-Israel Institute of Technology, Rambam Health Care Campus, Haifa, Israel
[2]Azrieli Faculty of Medicine, Bar-Ilan University, Safed, Israel

Introduction

For many years, it has been known that the burden of progressive chronic kidney disease (CKD) is not distributed proportionally across global population groups [1]. Populations of subSaharan African ancestry experience an approximately fourfold higher prevalence of end-stage kidney disease (ESKD) [2]. This distribution has been documented best in US-based surveys comparing the burden of kidney disease in the African American community with that of other populations [2,3]. The estimated lifetime risk for the development of ESKD approaches 8% among individuals of recent African ancestry, whereas the risk among individuals of European ancestry is 2%–3% [3].

Using population genetics tools, two groups first reported the results of a modified version of a genome-wide association study (GWAS) to identify a genomic region containing risk variants strongly linked to ESKD risk loci [4,5]. This modified approach is termed mapping by admixture linkage disequilibrium (MALD) [6,7]. Linkage disequilibrium (LD) refers to the nonrandom association of alleles at different loci. Loci are said to be in LD when the frequency of association of their different alleles deviates from that expected by random association. MALD is suitable for population-based discovery of disease risk alleles in admixed populations with a strong genetic locus variant underlying a large difference in disease susceptibility between the parent populations comprising the admixture. The basic principle of MALD relies on genotyping the small proportion of single-nucleotide polymorphic (SNP) markers that differ in frequency across populations of different ancestries [7]. When

New Horizons in Evolution
DOI: https://doi.org/10.1016/B978-0-323-90752-1.00002-X

abrupt recent admixture occurs between populations that had been separated previously by geographic or other boundaries, the resulting admixed population inherits large blocks of chromosomal regions of one ancestry or the other. These regions can be identified by SNP markers that show substantially different allele frequencies between ancestral populations. This admixture-generated LD enables linkage-based disease gene discovery in GWAS, with the ability to calculate logarithm of odds scores for large regions that usually contain many dozens of genomic loci. Using MALD enabled the investigators to identify enriched African ancestry SNP markers located in chromosome 22 among African American with kidney disease compared with healthy controls and their genome-wide average [4,5]. Subsequently, the combination of a strong signal for natural selection in the region of chromosome 22 in the African ancestry population-based on integrated haplotype score data [8–10], and the genotyping data from Yoruba (African) individuals in the 1000 Genomes Project [11], yielded the discovery of two variants residing within the *apolipoprotein L1 (APOL1)* gene on chromosome 22, as causative for the increased risk for ESKD in people of African ancestry [10,12]. These two variants at the *APOL1* gene, account for more than 70% of the increased risk for nondiabetic CKD in individuals of African ancestry. The ancestral nonrisk allele has been designated G0 and the derived two-risk alleles (renal risk variants- RRV), have been designated as G1 (encoding S342G and I384M substitutions) and G2 (encoding N388 and Y389 deletions), respectively [10,12]. Translating this information into the actual burden of disease, it is estimated that more than 70 million people world-wide (including approximately 6 million African Americans) have the high-risk *APOL1* genotype comprising two parental risk alleles [13–15].

The two G1 SNPs that together constitute G1 are in almost perfect, but not 100%, LD. About 1% of haplotypes with the G342 allele do not have the M384 allele [16]. The G342 allele, in the absence of M384, is thought to have the same effect on disease risk as the two variants together [17,18]. G0 is thought of as the group of haplotypes that does not contain G1 or G2 [19]. Interestingly, G1 and G2 are in perfect negative LD, that is, they never occur together on the same parental chromosome. The perfect negative LD indicates that these mutations arose and spread relatively recently. Subsequent studies date the selective sweep that resulted in their high frequency to less than 10,000 years ago [10,20]. The alleles frequency of these variants is very high, reaching to approximately

51% of African Americans having at least one risk allele and 13% having two-risk alleles [12–15,17,20–22]. These variants are among the most powerful common risk variants yet identified. For most common heritable diseases with nonMendelian inheritance, the genetic contribution represents the summation of the infinitesimal contribution of many common variants, each with a small effect size, scattered across the genome [23]. In general, there is an inverse relationship between the deleterious mutant allele frequency and the effect size. Common variants generally have weak effects with a low odds ratio (OR), usually well below 1.2 [24]. By contrast, Mendelian variants are rare but have a large effect size. Genes variants under strong positive selection, such as those that confer pathogen resistance in the single heterozygous state but susceptibility to a common disease in the homozygous or compound heterozygous state, are an exception to this rule, as has been best described for the relationship between the sickle cell anemia gene and malaria [25,26]. Genetic variants that have increased to a high frequency in this manner can show an unusual combination of high frequency and high penetrance. The ORs for the association of APOL1 with various forms of progressive nondiabetic CKD suit this paradigm and are among the highest ever described for a common disease association, ranging from 7 to 89 for various etiologies of nondiabetic ESKD [10,12,17,27–35]. This rare exception of high frequency alleles with a high OR for disease association raises questions about the selection pressure driving this process [20,36].

While this represents among the highest ORs ever reported for common nonmonogenic disease risk causal variants, only a subset of individuals who carry two APOL1 risk alleles actually develops kidney disease. For example, among people with the two-risk allele genotype, the lifetime risk for kidney disease for focal segmental glumerulosclerosis (FSGS) is 4%, whereas the risk increases to over 50% in untreated individuals with HIV infection [13,17]. However, examination of 337 autopsy kidneys collected at the University of Mississippi Medical Center from people without overt clinical kidney disease demonstrated that APOL1 risk alleles are associated with exaggerated age-related nephron loss, along with enlargement of the remaining glomeruli [37]. Such silent kidney disease might itself underlies known extra-renal morbidities related to cardiovascular disease. Moreover, these findings support the contention that clinically silent pathology has the potential to evolve to progressive CKD when exposed to a second hit. Environmental factors and genetic background may alter kidney disease risk [38]. In this regard, environmental modifiers have been proven to be the most significant mediators, while a recent GWAS failed to detect polymorphisms associated with ESKD beyond APOL1 [39]. The biologic rationale for environmental modifiers, that serve as a second hit which transform genotypic risk to clinical disease is supported by the following data: robust induction of APOL1 by inflammatory mediators such as interferons, patients with APOL1 high-risk genotype treated by interferon developed collapsing FSGS [40], and roles of interacting risk (HIV) and protective (JC polyoma) viruses in APOL1 nephropathy [15,17,33,38]. The most powerful environmental influence to date, is untreated HIV infection, with an OR of 29−89 to develop HIVAN in individuals with 2 APOL1 risk alleles and OR of 2−5 in individuals with one G1 allele [17,33]. Other genetic or environmental factors are yet to be discovered to fill the unexplained gap of preserved kidney function in individuals at genotypic risk and may translate to preventive and therapeutic implications.

A glimpse into the *Trypanosoma*—APOL1 arms race, explaining the high frequencies of *APOL1* renal risk variants

APOL1 is one member of a series of adjacent paralogous innate immunity genes (*APOL1-6*), arising in human ancestors as a result of gene duplication in primates, after the divergence of the primate lineage from other mammals. New World monkeys do not harbor APOL1, thus dating the APOL1 duplication 30 million years ago after the split between Old and New World monkeys. *APOL1* has been lost or pseudogenized in multiple primate species in the Old World lineage, for example the chimpanzees, suggesting that not only is *APOL1* dispensable in certain cases but it may even be detrimental [18,41−43]. APOL1 is the only member of the APOL family that encodes a protein containing an N-terminal signal peptide. Such signal peptide sequences are needed for protein secretion into the bloodstream [41,42,44], thereby enabling its function as a trypanolytic factor, which render humans to have innate resistance to some members of the *Trypanosoma* family of parasites [45,46]. In addition to the signal peptide, APOL1 comprises several domains that are essential for its function as a trypanolytic factor: N-terminal domain which forms the colicin-like domain, that has the potential to perforate membranes, a BH3-only subdomain located within the N-terminal domain that may affect apoptosis and autophagy, membrane-addressing domain that contain a putative pH-dependent apparatus enabling APOL1 activation in acidic pH and the C-terminus domain that includes a coiled coil domain with a leucine zipper motif that is important for the interactions with other proteins and maybe lipids and contains the G1 & G2 mutations [13–15,18,21,22,45].

Human African trypanosomiasis (HAT) is a disease caused by *trypanosome* parasites that infect humans and other animals residing in subSaharan Africa. In humans, the parasite leads to sleeping sickness, and it can be fatal if untreated. There are two overall clinical presentations of the disease: an acute form occurring mostly in East Africa, caused by *T.b. rhodesiense* (3% of HAT cases) and a more chronic form occurring mainly in West and Central Africa caused by *T.b. gambiense* (97% of HAT cases) [47]. APOL1 is the native weapon which protects against many different species of *trypanosome* parasites in humans [45]. Circulating APOL1 can lyse *T. brucei* and protects human beings against African sleeping sickness. Despite their close evolutionary proximity to human beings, chimpanzees are sensitive to *trypanosoma* infection because they have lost the *APOL1* gene, as delineated above [43], whereas human beings are resistant to *T. brucei* because of the presence of this gene [45]. APOL1 circulates in two different serum complexes that have been identified as trypanosome lytic factors (TLFs) [48]. After TLF1 and/or TLF2 uptake by the *trypanosoma*, APOL1 is released from the carrier complexes and inserts into endosomal membranes at an acidic pH [45,46,49–51]. Trypanolysis is linked to transmembrane ionic flux, osmotic swelling of the lysosome [45,46,52], and mitochondrial membrane permeabilization [51]. The carboxy-terminal domain of APOL1 that contains the G1 and G2 mutations is a long amphipathic α-helix that is required for trypanolytic activity [46,53]. Pursuant to this arms race, two *trypanosome* subspecies have evolved independent mechanisms to become resistant to APOL1: *Trypanosoma brucei rhodesiense* (*T. br. Rhodesiense*), which predominates in East Africa, and *Trypanosoma brucei gambiense* (*T. br. Gambiense*), which is more common in West Africa [48,54]. *T. br. Rhodesiense*, has evolved a virulence factor called serum-resistant activity (SRA). SRA binds the C-terminus of APOL1 in an acidic endolysosome organelle, thereby conferring resistance to APOL1-mediated lysis [49,50,52,55]. The *T. br. gambiense* subspecies evolved more diverse and complex mechanisms of defense against APOL1: reduced APOL1 uptake due to mutations in trypanosomal receptor for TLF, increased cysteine protease activity and stiffening of endosomal/lysosomal membranes [54]. As a subsequent step in this evolutionary arms race, the APOL1 protein harboring the G1 and G2 variants, which are located at the interaction site of APOL1 and SRA, restores trypanolytic activity by reducing SRA-binding affinity. The dominant selective advantage of these variants has been attributed to this effect [10,18]. Notably, G1 has reduced trypanolytic activity against *T. br. rhodesiense* and retains its ability to bind SRA compared to G2 [10,18], suggesting that the G2 variant has evolved in response to the expression of SRA in *T. br. rhodesiense*, whereas the G1 variant might have evolved in response to other selection pressures extending beyond *T. br. rhodesiense*. Until recently, an unresolved mystery regarding the nonoverlapping endemic areas for *T. br. rhodesiense*, located in modern East Africa and areas of high G1 frequencies in West Africa, perturbed researchers in the APOL1 field [10,18,56]. Recent population-based data has provided compelling potential explanation for the perplexing distribution of APOL1 risk alleles and the different trypanolytic activity of the two APOL1 risk variants against *T. br. rhodesiense*. Cooper et al. tested for the presence of APOL1 risk alleles in individuals living in Uganda with and without *T. br. rhodesiense* infection and in Guineans with and without *T. br. gambiense* infection [57]. They found evidence that the G2 allele (but not the G1 allele) was protective against *T. br. rhodesiense*, as expected per biologic studies. They did not observe an effect of either allele on infection (seropositivity) with *T. br. gambiense*, but they found that G1 protected against symptomatic disease whereas unexpectedly, G2 increased susceptibility to a severe clinical course. This association was replicated in a subsequent study [58]. These results enhance and sharpen the previously simplistic presentation of an overall protective effect of both G1 and G2 against *T.b. rhodesiense* as the predominant explanation for the rise to high allele frequency of these derived variants [10,18]. The protective properties of G1 against active sleeping sickness in individuals infected with *T. br. gambiense* seems to explain the widespread distribution of this variant in West Africa. On the other hand, the increased risk individuals with G2 mutations face when infected with this parasite, may explain why G2 is generally less common than G1, even though G2 has a greater trypanolytic activity against *T. br. rhodesiense* parasites. Several reports have described the enigmatic entity of trypanotolerance: asymptomatic, aparasitemic seropositive individuals infected by *T. b. gambiense*. Several hypotheses have been provided, including alternation of the innate immune system. Specifically, cytokine response and host polymorphisms in key mediators of inflammatory response have been suggested [59,60]. Cooper et al. invoke the potential role of G1 as a contributor to this trypanotolerance. However, the mechanism mediating this protective effect was not defined. Given the absence of trypanolytic activity of G1 sera on *T. b. gambiense* [10], other mechanisms that interact with components of the innate immune system might mediate this G1 protective effect, and await clarification. The significant differences between G1 and G2 uncovered in this study imply that these two gene variants may also have variable mechanisms leading to CKD [61]. Currently, the data support a model where G2 is an older mutation that confer protection against *T. br. rhodesiense*, while G1 is a more recent mutation that arose in West Africa and spread very quickly due to its protection from active *T. br. gambiense* infection. As Theodosius Dobzhansky once said, "nothing in biology makes sense except in the light of evolution." It still remains unclear why G2 is nevertheless fairly common in West Africa, despite increasing the risk of suffering symptomatic trypamosomiasis caused by *T. br. gambiense*. One explanation for this could be that G2 may protect against other as yet unidentified pathogens. No doubt, the APOL1 evolution story will continue to evolve and unravel additional roles of APOL1 RRV as innate immune proteins.

The mode of inheritance paradox

Cohorts of nondiabetic CKD patients show a highly significant OR under an APOL1 two-risk allele recessive mode of inheritance. Several studies have described a 7- to 10-fold increased risk of hypertension-attributed kidney disease, a 10- to 17-fold increase in FSGS, and a 29- to 89-fold increase in HIVAN [10,12,17,33]. Some reports have indicated a mild risk effect of a single G1 risk allele [17,33] and a younger age at initiation of dialysis [62,63]. APOL1 duplication can partially alter susceptibility to kidney disease in G1 heterozygous individuals [64]. Nevertheless, a recessive mode of disease risk inheritance has been demonstrated repeatedly and prominently. This mode of inheritance does not seem to reconcile with the dispensability of APOL1 for kidney development and health in human and most nonhuman primates [18,42,65,66] and with experimental overexpression models demonstrating gain-of-function injury [18,40,67–85].

Several proposals have been put forth to explain this paradox: (1) A dose-dependent gain-of injury with a threshold inflection point is required. However, APOL1 protein expression in kidney biopsies was decreased in individuals with CKD regardless of APOL1 genotype [86,87]. Similarly, APOL1 plasma level was not increased in carriers of 2

APOL1 risk alleles with CKD [88,89]. The absence of correlation with circulating and endogenous (kidney) APOL1 level on one hand and on the other hand the robust induction of APOL1 by interferon [40] suggest that the dose of APOL1 can only partially explain the discrepancy of gain of injury and recessive mode of inheritance; (2) A multimerization model has been suggested as a solution to this enigma [90]. Recently, Shah et al. reported that WT APOL1 is mostly monomeric, whereas APOL1 risk variants can form large oligomers, leading to opening of the mitochondrial permeability transition pore and cell death [80]. This difference in propensity of different variants to oligomerize could help to explain APOL1 risk variants gain-of-function biology despite a recessive mode of inheritance; (3) Loss of inhibition model: Disrupted putative autoinhibitory domain in the SRA-binding region of the RRV may cause loss-of-function of the inhibitory domain, leading to overall gain-of-function of the APOL1 protein. Alternatively, an as yet unidentified interacting human protein/lipid that attenuates toxicity by binding the C-terminal domain, similar to the APOL1 and SRA interaction system in *T. br. rhodesiense*, may play a role. Given the preserved differential toxicity of the RRV compared to GO in various model organism that do not normally express APOL1, namely, yeast, flies and mice, the hypothesis about an intrinsic autoinhibitory loss-of-function of the C-terminal is appealing and may explain why different mutations in the C-terminus lead to the same gain-of-function, which is actually a loss of inhibition effect; and (4) G0 has a protective role, therefore loss of protection may explain the recessive mode of inheritance and kidney protection in the heterozygous state [15,21]. This hypothesis was supported by several models, including podocytes and parietal epithelial cell lines, Zebrafish, and mouse model of HIV-associated nephropathy [91–94]. Notwithstanding the foregoing, these various hypotheses are not mutually exclusive.

In conclusion, while the mechanisms mediating kidney injury have not been clarified yet, and many unanswered questions remain, more pieces of the puzzle join to shed light on the complex evolutionary story of APOL1. The discovery of APOL1 provides an exciting opportunity for personalized medicine in CKD. Various platform for high-throughput screening for drug discovery and antisense oligonucleotide approaches are being developed to alter APOL1 toxicity and reduce APOL1 expression. These avenues would hopefully yield new potential therapies that would mitigate the risk of kidney disease and its complications in individuals at genotypic risk [81] (Fig. 5.1).

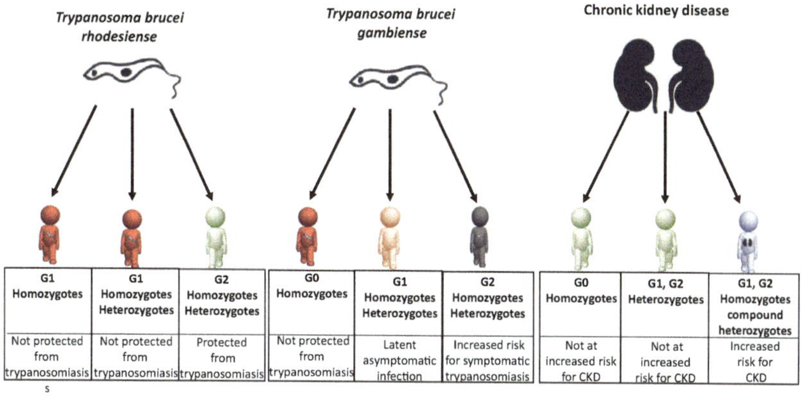

Figure 5.1 Individuals carrying different variants of the *APOL1* gene are protected against sleeping sickness and kidney disease to different extents. Individuals with two copies of wild-type APOL1 (G0 homozygotes) are not protected against sleeping sickness caused by T. b. rhodesiense (red figure, left column) or T. b. gambiense (red figure, middle column), but they do not have an increased risk of chronic kidney disease (CKD) (yellow figure, right column). Individuals with one copy of wild-type APOL1 and one copy of the G1 variant (G1 heterozygotes), and individuals with two copies of the G1 variant (G1 homozygotes) are not protected against sleeping sickness caused by T. b. rhodesiense (second red figure, left column) and are more likely to have latent asymptomatic infection by T. b. gambiense (pink figure, middle column). Individuals with one copy of wild-type APOL1 and one copy of the G2 variant (G2 heterozygotes), and individuals with two copies of the G2 variant (G2 homozygotes) are protected against sleeping sickness caused by T. b. rhodesiense (green figure, left column) but are at increased risk of developing symptomatic infection by T. b. gambiense (gray figure, middle column). Like G0 homozygotes, G1 heterozygotes and G2 heterozygotes do not have an increased risk of CKD (second yellow figure, right column). However, G1 homozygotes, G2 homozygotes and compound heterozygotes (individuals with both G1 and G2) all have an increased risk of CKD (blue figure, right column). *Reprinted with permission from E. Kruzel-Davila, K. Skorecki, The double-edged sword of evolution, Elife 6 (2017).*

References

[1] K.T. Mills, et al., A systematic analysis of worldwide population-based data on the global burden of chronic kidney disease in 2010, Kidney Int. 88 (2015) 950–957.

[2] A.J. Collins, et al., United States Renal Data System 2011 Annual Data Report: atlas of chronic kidney disease & end-stage renal disease in the United States, Am. J. Kidney Dis. 59 (2012) A7, e1.

[3] M.E. Grams, et al., Lifetime incidence of CKD stages 3–5 in the United States, Am. J. Kidney Dis. 62 (2013) 245.

[4] J.B. Kopp, et al., MYH9 is a major-effect risk gene for focal segmental glomerulosclerosis, Nat. Genet. 40 (2008) 1175–1184.

[5] W.H. Kao, et al., MYH9 is associated with nondiabetic end-stage renal disease in African Americans, Nat. Genet. 40 (2008) 1185–1192.

[6] R. Chakraborty, K.M. Weiss, Admixture as a tool for finding linked genes and detecting that difference from allelic association between loci, Proc. Natl. Acad. Sci. U S A 85 (1988) 9119–9123.
[7] M.W. Smith, S.J. O'Brien, Mapping by admixture linkage disequilibrium: advances, limitations and guidelines, Nat. Rev. Genet. 6 (2005) 623–632.
[8] S.R. Grossman, et al., A composite of multiple signals distinguishes causal variants in regions of positive selection, Science 327 (2010) 883–886.
[9] B.F. Voight, et al., A map of recent positive selection in the human genome, PLoS Biol. 4 (2006) e72.
[10] G. Genovese, et al., Association of trypanolytic ApoL1 variants with kidney disease in African Americans, Science 329 (2010) 841–845.
[11] L. Clarke, et al., The 1000 Genomes Project: data management and community access, Nat. Methods 9 (2012) 459–462.
[12] S. Tzur, et al., Missense mutations in the APOL1 gene are highly associated with end stage kidney disease risk previously attributed to the MYH9 gene, Hum. Genet. 128 (2010) 345–350.
[13] P.D. Dummer, et al., APOL1 kidney disease risk variants: an evolving landscape, Semin. Nephrol. 35 (2015) 222–236.
[14] D.J. Friedman, M.R. Pollak, Apolipoprotein L1 and kidney disease in African Americans, Trends Endocrinol. Metab. (2016).
[15] E. Kruzel-Davila, et al., APOL1 nephropathy: a population genetics and evolutionary medicine detective story, Semin. Nephrol. 37 (2017) 490–507.
[16] S. Limou, et al., Sequencing rare and common APOL1 coding variants to determine kidney disease risk, Kidney Int. (2015).
[17] J.B. Kopp, et al., APOL1 genetic variants in focal segmental glomerulosclerosis and HIV-associated nephropathy, J. Am. Soc. Nephrol. 22 (2011) 2129–2137.
[18] R. Thomson, et al., Evolution of the primate trypanolytic factor APOL1, Proc. Natl. Acad. Sci. U S A 111 (2014) E2130–E2139.
[19] H. Lannon, et al., Apolipoprotein L1 (APOL1) risk variant toxicity depends on the haplotype background, Kidney Int. 96 (2019) 1303–1307.
[20] D.J. Friedman, M.R. Pollak, Genetics of kidney failure and the evolving story of APOL1, J. Clin. Invest. 121 (2011) 3367–3374.
[21] E. Kruzel-Davila, et al., APOL1 nephropathy: from gene to mechanisms of kidney injury, Nephrol. Dial. Transpl. (2015).
[22] D.J. Friedman, M.R. Pollak, APOL1 and kidney disease: from genetics to biology, Annu. Rev. Physiol. 82 (2020) 323–342.
[23] G. Gibson, Rare and common variants: twenty arguments, Nat. Rev. Genet. 13 (2012) 135–145.
[24] T.A. Manolio, Bringing genome-wide association findings into clinical use, Nat. Rev. Genet. 14 (2013) 549–558.
[25] A.C. Allison, The distribution of the sickle-cell trait in East Africa and elsewhere, and its apparent relationship to the incidence of subtertian malaria, Trans. R. Soc. Trop. Med. Hyg. 48 (1954) 312–318.
[26] M. Cyrklaff, et al., Hemoglobins S and C interfere with actin remodeling in *Plasmodium falciparum*-infected erythrocytes, Science 334 (2011) 1283–1286.
[27] D.J. Friedman, et al., Population-based risk assessment of APOL1 on renal disease, J. Am. Soc. Nephrol. 22 (2011) 2098–2105.
[28] N. Papeta, et al., APOL1 variants increase risk for FSGS and HIVAN but not IgA nephropathy, J. Am. Soc. Nephrol. 22 (2011) 1991–1996.
[29] M.S. Lipkowitz, et al., Apolipoprotein L1 gene variants associate with hypertension-attributed nephropathy and the rate of kidney function decline in African Americans, Kidney Int. 83 (2013) 114–120.

[30] M.C. Foster, et al., APOL1 variants associate with increased risk of CKD among African Americans, J. Am. Soc. Nephrol. 24 (2013) 1484–1491.
[31] A. Parsa, et al., APOL1 risk variants, race, and progression of chronic kidney disease, N. Engl. J. Med. 369 (2013) 2183–2196.
[32] D.M. Fine, et al., APOL1 risk variants predict histopathology and progression to ESRD in HIV-related kidney disease, J. Am. Soc. Nephrol. 23 (2012) 343–350.
[33] A.N. Kasembeli, et al., APOL1 risk variants are strongly associated with HIV-associated nephropathy in black South Africans, J. Am. Soc. Nephrol. (2015).
[34] C.A. Peralta, et al., APOL1 genotype and race differences in incident albuminuria and renal function decline, J. Am. Soc. Nephrol. (2015).
[35] M.E. Grams, et al., Race, APOL1 risk, and eGFR decline in the general population, J. Am. Soc. Nephrol. (2016).
[36] W.G. Wasser, et al., Population genetics of chronic kidney disease: the evolving story of APOL1, J. Nephrol. 25 (2012) 603–618.
[37] W.E. Hoy, et al., APOL1 risk alleles are associated with exaggerated age-related changes in glomerular number and volume in African-American adults: an autopsy study, J. Am. Soc. Nephrol. 26 (2015) 3179–3189.
[38] B.I. Freedman, K. Skorecki, Gene–gene and gene–environment interactions in apolipoprotein L1 gene-associated nephropathy, Clin. J. Am. Soc. Nephrol. 9 (2014) 2006–2013.
[39] C.D. Langefeld, et al., Genome-wide association studies suggest that APOL1-environment interactions more likely trigger kidney disease in African Americans with nondiabetic nephropathy than strong APOL1-second gene interactions, Kidney Int. 94 (2018) 599–607.
[40] B. Nichols, et al., Innate immunity pathways regulate the nephropathy gene apolipoprotein L1, Kidney Int. 87 (2015) 332–342.
[41] H. Monajemi, et al., The apolipoprotein L gene cluster has emerged recently in evolution and is expressed in human vascular tissue, Genomics 79 (2002) 539–546.
[42] E.E. Smith, H.S. Malik, The apolipoprotein L family of programmed cell death and immunity genes rapidly evolved in primates at discrete sites of host–pathogen interactions, Genome Res. 19 (2009) 850–858.
[43] T. Mikkelsen, L.D. Hillier, E. Eichler, M. Zody, et al., Initial sequence of the chimpanzee genome and comparison with the human genome, Nature 437 (2005) 69–87.
[44] P.N. Duchateau, et al., Apolipoprotein L, a new human high density lipoprotein apolipoprotein expressed by the pancreas. Identification, cloning, characterization, and plasma distribution of apolipoprotein L, J. Biol. Chem. 272 (1997) 25576–25582.
[45] L. Vanhamme, et al., Apolipoprotein L-I is the trypanosome lytic factor of human serum, Nature 422 (2003) 83–87.
[46] D. Perez-Morga, et al., Apolipoprotein L-I promotes trypanosome lysis by forming pores in lysosomal membranes, Science 309 (2005) 469–472.
[47] P.G. Kennedy, Clinical features, diagnosis, and treatment of human African trypanosomiasis (sleeping sickness), Lancet Neurol. 12 (2013) 186–194.
[48] E. Pays, et al., The molecular arms race between African trypanosomes and humans, Nat. Rev. Microbiol. 12 (2014) 575–584.
[49] N.A. Stephens, S.L. Hajduk, Endosomal localization of the serum resistance-associated protein in African trypanosomes confers human infectivity, Eukaryot. Cell 10 (2011) 1023–1033.
[50] J.M. Bart, et al., Localization of serum resistance-associated protein in *Trypanosoma brucei rhodesiense* and transgenic *Trypanosoma brucei brucei*, Cell Microbiol. 17 (2015) 1523–1535.

[51] G. Vanwalleghem, et al., Coupling of lysosomal and mitochondrial membrane permeabilization in trypanolysis by APOL1, Nat. Commun. 6 (2015) 8078.
[52] E. Pays, et al., The trypanolytic factor of human serum, Nat. Rev. Microbiol. 4 (2006) 477–486.
[53] M.P. Molina-Portela, et al., Distinct roles of apolipoprotein components within the trypanosome lytic factor complex revealed in a novel transgenic mouse model, J. Exp. Med. 205 (2008) 1721–1728.
[54] P. Uzureau, et al., Mechanism of *Trypanosoma brucei gambiense* resistance to human serum, Nature 501 (2013) 430–434.
[55] H.V. Xong, et al., A VSG expression site-associated gene confers resistance to human serum in *Trypanosoma rhodesiense*, Cell 95 (1998) 839–846.
[56] W.Y. Ko, et al., Identifying Darwinian selection acting on different human APOL1 variants among diverse African populations, Am. J. Hum. Genet. 93 (2013) 54–66.
[57] A. Cooper, et al., APOL1 renal risk variants have contrasting resistance and susceptibility associations with African trypanosomiasis, Elife 6 (2017).
[58] J.W. Kabore, et al., Candidate gene polymorphisms study between human African trypanosomiasis clinical phenotypes in Guinea, PLoS Negl. Trop. Dis. 11 (2017) e0005833.
[59] B. Bucheton, et al., Human host determinants influencing the outcome of *Trypanosoma brucei gambiense* infections, Parasite Immunol. 33 (2011) 438–447.
[60] H. Ilboudo, et al., Unravelling human trypanotolerance: IL8 is associated with infection control whereas IL10 and TNFalpha are associated with subsequent disease development, PLoS Pathog. 10 (2014) e1004469.
[61] E. Kruzel-Davila, K. Skorecki, The double-edged sword of evolution, Elife 6 (2017).
[62] Z. Kanji, et al., Genetic variation in APOL1 associates with younger age at hemodialysis initiation, J. Am. Soc. Nephrol. 22 (2011) 2091–2097.
[63] S. Tzur, et al., APOL1 allelic variants are associated with lower age of dialysis initiation and thereby increased dialysis vintage in African and Hispanic Americans with non-diabetic end-stage kidney disease, Nephrol. Dial. Transpl. 27 (2012) 1498–1505.
[64] R. Ruchi, et al., Copy number variation at the APOL1 locus, PLoS One 10 (2015) e0125410.
[65] P. Poelvoorde, et al., Distribution of apolipoprotein L-I and trypanosome lytic activity among primate sera, Mol. Biochem. Parasitol. 134 (2004) 155–157.
[66] D.B. Johnstone, et al., APOL1 null alleles from a rural village in India do not correlate with glomerulosclerosis, PLoS One 7 (2012) e51546.
[67] X. Lan, et al., APOL1 risk variants enhance podocyte necrosis through compromising lysosomal membrane permeability, Am. J. Physiol. Ren. Physiol. 307 (2014) F326–F336.
[68] X. Lan, et al., Protein domains of APOL1 and its risk variants, Exp. Mol. Pathol. 99 (2015) 139–144.
[69] X. Lan, et al., Vascular smooth muscle cells contribute to APOL1-induced podocyte injury in HIV milieu, Exp. Mol. Pathol. 98 (2015) 491–501.
[70] D. Cheng, et al., Biogenesis and cytotoxicity of APOL1 renal risk variant proteins in hepatocytes and hepatoma cells, J. Lipid Res. 56 (2015) 1583–1593.
[71] O.A. Olabisi, et al., APOL1 kidney disease risk variants cause cytotoxicity by depleting cellular potassium and inducing stress-activated protein kinases, Proc. Natl. Acad. Sci. U S A (2015).
[72] L. Ma, et al., APOL1 renal-risk variants induce mitochondrial dysfunction, J. Am. Soc. Nephrol. 28 (2017) 1093–1105.
[73] Y. Fu, et al., APOL1-G1 in nephrocytes induces hypertrophy and accelerates cell death, J. Am. Soc. Nephrol. 28 (2017) 1106–1116.

[74] E. Kruzel-Davila, et al., APOL1-mediated cell injury involves disruption of conserved trafficking processes, J. Am. Soc. Nephrol. 28 (2017) 1117–1130.
[75] P. Beckerman, et al., Transgenic expression of human APOL1 risk variants in podocytes induces kidney disease in mice, Nat. Med. 23 (2017) 429–438.
[76] S.S. Hayek, et al., A tripartite complex of suPAR, APOL1 risk variants and alphavbeta3 integrin on podocytes mediates chronic kidney disease, Nat. Med. 23 (2017) 945–953.
[77] D. Granado, et al., Intracellular APOL1 risk variants cause cytotoxicity accompanied by energy depletion, J. Am. Soc. Nephrol. (2017).
[78] H. Wen, et al., APOL1 risk variants cause podocytes injury through enhancing endoplasmic reticulum stress, Biosci. Rep. (2018).
[79] J.Y. Zhang, et al., UBD modifies APOL1-induced kidney disease risk, Proc. Natl. Acad. Sci. U S A 115 (2018) 3446–3451.
[80] S.S. Shah, et al., APOL1 kidney risk variants induce cell death via mitochondrial translocation and opening of the mitochondrial permeability transition pore, J. Am. Soc. Nephrol. 30 (2019) 2355–2368.
[81] M. Aghajan, et al., Antisense oligonucleotide treatment ameliorates IFN-gamma-induced proteinuria in APOL1-transgenic mice, JCI Insight 4 (2019).
[82] V. Kumar, et al., Disrupted apolipoprotein L1-miR193a axis dedifferentiates podocytes through autophagy blockade in an APOL1 risk milieu, Am. J. Physiol. Cell Physiol. 317 (2019) C209–C225.
[83] J. Chun, et al., Recruitment of APOL1 kidney disease risk variants to lipid droplets attenuates cell toxicity, Proc. Natl. Acad. Sci. U S A 116 (2019) 3712–3721.
[84] K. Okamoto, et al., APOL1 risk allele RNA contributes to renal toxicity by activating protein kinase R, Commun. Biol. 1 (2018) 188.
[85] J.H. Ryu, et al., APOL1 renal risk variants promote cholesterol accumulation in tissues and cultured macrophages from APOL1 transgenic mice, PLoS One 14 (2019) e0211559.
[86] S.M. Madhavan, et al., APOL1 localization in normal kidney and nondiabetic kidney disease, J. Am. Soc. Nephrol. 22 (2011) 2119–2128.
[87] S.M. Madhavan, et al., APOL1 variants change C-terminal conformational dynamics and binding to SNARE protein VAMP8, JCI Insight 2 (2017).
[88] L.A. Bruggeman, et al., Plasma apolipoprotein L1 levels do not correlate with CKD, J. Am. Soc. Nephrol. 25 (2014) 634–644.
[89] J. Kozlitina, et al., Plasma levels of risk-variant APOL1 do not associate with renal disease in a population-based cohort, J. Am. Soc. Nephrol. (2016).
[90] S. Limou, et al., APOL1 toxin, innate immunity, and kidney injury, Kidney Int. (2015).
[91] B.R. Anderson, et al., In vivo modeling implicates APOL1 in nephropathy: evidence for dominant negative effects and epistasis under anemic stress, PLoS Genet. 11 (2015) e1005349.
[92] V. Kumar, et al., Disruption of APOL1-miR193a axis induces disorganization of podocyte actin cytoskeleton, Sci. Rep. 9 (2019) 3582.
[93] L.A. Bruggeman, et al., APOL1-G0 protects podocytes in a mouse model of HIV-associated nephropathy, PLoS One 14 (2019) e0224408.
[94] J. Mikulak, et al., Impact of APOL1 polymorphism and IL-1beta priming in the entry and persistence of HIV-1 in human podocytes, Retrovirology 13 (2016) 63.

CHAPTER 6

Network analyses of the impact of visual habitat structure on behavior, demography, genetic diversity, and gene flow in a metapopulation of collared lizards (*Crotaphytus collaris collaris*)

Amy K. Conley[1,2], Jennifer L. Neuwald[1,3] and Alan R. Templeton[1]
[1]Department of Biology, Washington University, St. Louis, MO, United States
[2]New York Natural Heritage Program, College of Environmental Science and Forestry, State University of New York, Albany, NY, United States
[3]Department of Biology and Graduate Degree Program in Ecology, Colorado State University, Fort Collins, CO, United States

Introduction

Networks consist of nodes and edges between nodes. In this paper we utilize several types of networks to address a host of interacting behavioral, ecological, and evolutionary factors within a single biological system. This work lies in the domain of landscape genetics, a discipline that attempts to explain the spatial patterns of genetic variation and gene flow through landscape composition [1]. Gene flow, in turn, can be regarded as arising from an interaction of dispersal behavior and mating behavior [2]. The behavior of an organism is influenced by its perception of the environment and environmental constraints, thereby creating the potential for an interaction between behavior and landscape on gene flow patterns and the distribution of genetic variation over space. Landscape genetics typically focuses on landscape features directly, and only invokes behavioral interactions with the landscape to infer possible explanations for the observed genetic impact of a particular landscape feature [3,4]. An understanding of the interaction between behavior and landscape on dispersal

New Horizons in Evolution
DOI: https://doi.org/10.1016/B978-0-323-90752-1.00005-5

and gene flow is particularly important in conservation biology [5]. Gene flow is often critical for preventing local populations from losing their genetic variation and adaptive flexibility and becoming susceptible to inbreeding depression. The importance of adaptive flexibility and dispersal is increasing in this era of global climate change as both may be needed by many species to adjust to rapid shifts in habitat conditions.

An ideal system for studying the interaction of social behavior with landscape features are metapopulations of the eastern collared lizard (*Crotaphytus collaris collaris*) that inhabit glades in the Ozarks, a highland region found mostly in southern Missouri and northern Arkansas. Glades are open areas with exposed bedrock, often with a southern to western exposure [6,7]. Ozark glades are fire-maintained, and the suppression of fires led to degradation of glades such that 75% of the glade populations were estimated to be extinct by 1980 and collared lizards were listed as "very vulnerable to extirpation for the state" [8]. The dense woody vegetation promoted by fire suppression leads to reduced growth rates, delayed age of maturity, and reduced clutch size and frequency in collared lizards, thereby strongly depressing reproductive rates [9]. Due to glade restoration, fire management, and translocation programs, there has been a strong recovery of collared lizards in the Missouri Ozarks, although they are still listed as a "species of conservation concern" (Missouri Natural Heritage) [10].

Ozark glades are imbedded in a matrix of oak–hickory or pine woodlands. Previous studies have highlighted the impact of this matrix on dispersal and gene flow. Collared lizards show little to no dispersal through an unburned woodland characterized by a dense, woody understory, and little ground cover due to accumulated dead leaves. Burning the woodland eliminates much of the woody understory and results in a rich herbaceous ground cover. Lizards are capable of high levels of dispersal and gene flow, colonization of new glades, and the establishment of a true metapopulation with local extinction balanced by recolonization after woodland burning [6,11,12]. The focus on the matrix in the previous work is typical of many landscape genetic studies on organisms that live in fragmented habitat islands, and potential factors *within* a habitat island that could influence dispersal are typically ignored [13].

Modeling dispersal requires an understanding of emigration, interpatch movement, and immigration [14]. Most current landscape ecology tools focus on the interpatch stage [15]. Fewer studies consider the phases of the dispersal process related to departure and to settlement, which are likely to be influenced by variation in local resource abundance and social parameters [16–19].

Collared lizards are highly social animals with adults of both sexes displaying territorial behavior [20]. Females show less intrasexual aggression than males, and female territories often overlap [20]. Male territories overlap little with adjacent males, but often overlap with multiple female territories [20,21]. Yearling males tend not to defend territories, adopt behaviors that minimize aggression from adults, and retreat from territorial adult males showing aggressive behavior [20,22]. Most intra- and intersexual signaling is visual [23,24], although male intrasexual signaling can escalate into fighting, with bite-force being a strong predictor of outcome [25].

Given that social interactions among collared lizards are mediated by vision, we hypothesized that the distribution and sizes of patches of exposed bedrock within a glade that allow visual contact would have an impact on their social behavior. Directly assessing social forces within individual subpopulations is time-consuming at large scales, and data sets that simultaneously address landscape level dispersal patterns and small-scale local dynamics are rare [26]. We approach this practical constraint by first performing detailed behavioral studies on six glade populations that were chosen to (1) differ in size and (2) have either clustered or evenly distributed patches of exposed bedrock. We test the association between bedrock distribution patterns and social structure as described through home range distribution patterns, social network statistics, and intensity of displays of aggression.

These behavioral studies identified a measure of visual openness associated with network features of social structure. Visual openness of a glade could be estimated from the extent and pattern of patches of open bedrock that could be quantified from remotely sensed data and thereby easily measured within hundreds of glades in our study area. We then hypothesized that quantification of intra-glade bedrock distribution could serve as a proxy for variation in social structure networks and be an important determinant of dispersal and gene flow patterns and local population sizes. We test this hypothesis by making use of extensive mark/recapture data [6] that allows accurate estimates of local population size and dispersal, and genetic survey data [11] that allows us to test if models of dispersal that incorporate intra-glade visual openness will more closely describe observed patterns of genetic differentiation than models that do not consider intra-glade visual openness through network analyses and computer simulations.

Methods

Study system

Our study focuses primarily upon a metapopulation of collared lizards on Stegall Mountain, in the Peck Ranch Conservation Area of the Missouri Department of Conservation. The current lizard population on this mountain is the result of a restoration program. Lizards were introduced onto three glades between 1984 and 1989, following glade restoration that began in 1982. The initial glade populations were fragmented from one another until controlled and ongoing woodland burning began in 1994. Immediately after prescribed burning began, lizards began to disperse between occupied glades and to colonize new glades [6,11]. By 2000, the metapopulation had achieved a stable equilibrium with respect to population size, the number of glades occupied, and rates of glade extinction/recolonization. Also by 2000 there was significant and stable genetic differentiation among glade populations [11]. All behavioral and genetic studies were executed in the stable period, and all the data from previous studies analyzed for this study were limited to the stable period. In this manner, we avoid confoundment with the temporally dynamic history of these populations during the earlier phases of this restoration project.

Because behavioral studies are time and personnel intensive, we only studied six glades behaviorally. To maximize our power with this limited sample, we decided to choose glades that fell into three categories of population size/area (large, medium, and small). Within each size category, we chose a pair of glades that differed greatly in the distribution of exposed bedrock in our qualitative judgment, with one glade being even and the other clustered. Another factor influencing choice of glades was accessibility given that the behavioral work would require returning to the same glades multiple times. Most of the glades on Stegall Mountain can only be reached by long hikes through rugged terrain with no trails. We decided to only use as focal glades those that were within a 30 minute hike or less from the nearest road. Four glades on Stegall Mountain satisfied these criteria: SM-1, SM-2, SM-9, and SM-10 (Fig. 6.1). Two small glades near roads were added outside Stegall Mountain for the behavioral studies: Stalcup Hollow, located nearby in the Peck Ranch Conservation Area, and a glade in St. Francois State Park, located on the route from St. Louis to the Peck Ranch. Both of these glades were also restored, stocked with translocated lizards (1987 for Stalcup Hollow, 1995 for St. Francois

Figure 6.1 A topographic map of Stegall Mountain showing the glades. The glades are shown in yellow except for the four glades used in the behavioral studies. Contour line altitudes above sea level are given in feet. Roads are indicated by dashed lines.

State Park), and under burn management, just as the glades on Stegall Mountain, but in addition were easily accessible due to nearby roads.

Behavioral methods

Collared lizards in Missouri emerge between April and May. Adults are active until mid-August to mid-September, when they retreat to hibernacula for the winter. Hatchlings often remain active into October. We sampled all collared lizard populations on the focal glades from June to August of 2010 and May to August of 2011.

Two to five observers sampled glades twice a week (weather permitting) during the observation period. We captured all lizards by lasso on the first observation day of the week for morphological measurements. In addition, we toe-clipped new individuals for genetic samples and permanent identification. We painted all lizards with an identifying mark and released them at their point of first observation. In the original mark/recapture study of the Stegall populations, there was no indication that marking lizards with nontoxic paint impaired their survival or affected mating success. Markings were on the back and the tail, not the underside or near the dewlap, which are most prominent during displays. Depending on the rate of shedding these marks lasted 1–20 days. We refreshed the marks every week during the weekly lassoing, or as needed. We recorded the positions of all lizards prior to capture.

On the second observation day each week, we only lassoed lizards that were unmarked. If new, we took all the morphological measurements and toe-clip samples. If the lizard was recaptured due to shedding, we renewed the paint mark and released the lizard. We made observations on these released lizards only after a waiting period of at least 20 minutes. Otherwise, we made all second day observations of position and behavior without contact with the animals.

We recorded lizard locations on printed maps created by ground sourcing of glade features and landmarks with a handheld GPS Unit (Trimble June-S). Observers surveyed the glade over the course of 2–4 hours, recording the locations (1) before lassoing an unmarked individual, (2) upon first observing a marked individual, and (3) upon re-encountering a marked individual after a time period of greater than 30 minutes. During focal observation periods (after at least a 20 minute waiting period if the lizard had been handled), the observer recorded the lizard's location every 5 minutes or whenever it moved, for a total period of 20 minutes. Most activity occurred on open bedrock, including hunting as they are sit-and-wait predators that perch on a rock until they see a prey item. Sometimes, in pursuit of prey, a lizard would run into tall grass and scrub and not be visible. Such lizards were not given a position until they returned to view. We also recorded behaviors during the 20-minute period. Behaviors included frequency of pushups, dewlap extensions, arched backs, males and females approaching each other and running away, chasing each other, climbing over the other lizard, resting side by side, and copulation.

We digitized field-mapped points in ArcGIS10. We estimated home range sizes from the area of the minimum convex polygon created from the collective observation points of each lizard. We excluded lizards with fewer than three location points from analysis. We calculated home range boundaries using a kernel density estimator in Geospatial Modeling Environment [27]. We adjusted the kernel smoothing parameters for each lizard so that the area encompassed by the 95% volume kernel density isopleth was equal to the area of the minimum bounding polygon (+/ − 1%), a method considered to be the most accurate for home ranges of herpetofauna [28]. We quantified interaction patterns using social networks based on observed home range distributions. Individual lizards of known gender were the nodes, and the edges of these networks in UCINET [29] were proportional to the area of their 95% kernel density polygons overlap. We visualized networks using NetDraw [30].

We calculated the clustering coefficient and eigen centrality [31] of each glade network. The clustering coefficient measures how connected an individual is to others in the network. Eigen centrality quantifies a node's importance based on the adjacency matrix of the graph. Ego-centric measures of eigen centrality and clustering coefficient consider only the connections surrounding a specific individual, and allow us to examine how evenly connection and influence are distributed throughout the population. We calculated ego-centric eigen centrality and ego-centric clustering coefficient in UCINET. We compared variances in ego-centric measures across individuals within a glade using an F test in R. The variances are an indicator of inequality among individuals in their territorial and social dominance.

We quantified territorial behavior and its variance for male lizards through a territorial score (the sum of all observed incidences of pushups, dewlap extensions, and mating). We analyzed territorial behavior on all males, (18 from clustered, 26 from even habitats). Previous studies have demonstrated that yearling males generally do not have territories nor display aggressive behavior [20,22]. We therefore measured a subset of territorial males that had nonzero territorial scores (8 clustered, 12 even). We assessed differences between habitat classes in mean and variance of territorial scores by 5000 random permutations of male territorial scores across clustered and even populations. We compared the observed differences in mean and variance to the resulting permutation distributions to determine significance.

Quantification of bedrock distribution

In addition to an on-the-ground qualitative assessment of bedrock distribution in the six focal glades, we also quantified the bedrock distributions of every glade on Stegall Mountain using remotely sensed data. We created classified raster data from aerial photographs of Stegall Mountain from the National Agriculture Imagery Program digital orthophoto quarter quads (DOQQ, 1-m resolution). We used Image Classification Analyst (ArcGIS10) to identify and isolate patches of exposed bedrock based on ground-sourced training samples and converted the image to raster format for fragmentation analysis. We use the statistic, radius of gyration ("GYRATE_AM"), to quantify the patterns of bedrock distribution of each glade [32]. The radius of gyration in this study represents the average distance a lizard randomly placed on bedrock could move across bedrock before encountering a boundary defined by nonbedrock substrate. The correlation length is the area weighted mean radius of gyration and represents the extensiveness of all bedrock patches within a glade. Larger correlation length values indicate a more connected, less subdivided landscape. We carried out these analyses using FragStats [33].

Because larger glades tend to have longer correlation lengths, we regressed the log transformed correlation length against the log of total bedrock area (Fig. 6.2), and used the residual to measure clustering. Positive residual correlation lengths indicate that bedrock was more evenly spaced than on average, whereas negative residual correlation lengths indicate that bedrock was more clustered.

Predicting population size from glade area and local bedrock clustering

Population sizes were estimated for all glades on Stegall Mountain every year during the period of sampling of the stable metapopulation phase through mark/recapture techniques with high sample coverage [6]. Although these years were characterized by stability of total population size, individual glades showed fluctuations in size. These fluctuations, expected from small population size, were smoothed by calculating the average population size from 2000 to 2003 for each glade on Stegall Mountain as these were the years of the most intensive sampling. Areas for all glades on Stegall Mountain were measured from aerial photographs using ArcGIS, and similarly the residual correlation lengths were measured for all glades on Stegall Mountain. Because the residual correlation lengths have already been adjusted for total bedrock area within a glade, glade bedrock area and residual correlation length represent orthogonal aspects of glade structure with potentially different effects on population size. Least square regressions were executed using average population size

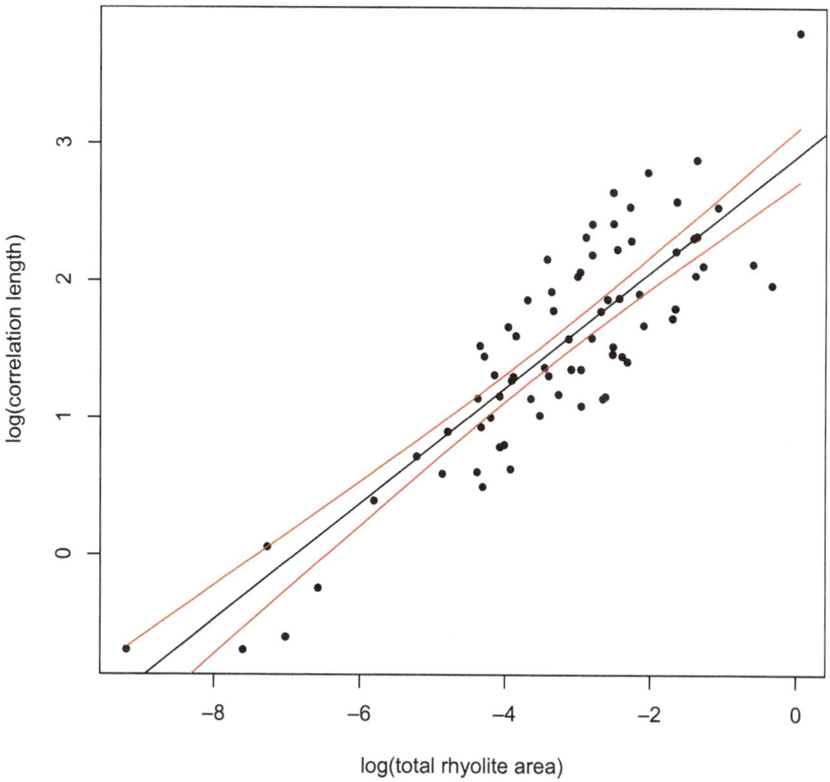

Figure 6.2 Regression of log correlation length of patches of exposed bedrock within glades on Stegall Mountain and St. Francois State park against the log total area of exposed bedrock on those glades. The two curved lines surrounding the linear regression line mark 95% confidence intervals of the predicted relationship.

during the stable metapopulation phase as the dependent variable against area and residual correlation length, considered individually and together with an interaction term.

Measuring and testing dispersal

Populations on Stegall Mountain were surveyed for lizards from 1984 to 2006 coupled with mark/recapture studies [6]. A dispersal event was inferred when an individual was recaptured on a glade other than its glade of previous capture, or if genetic assignment tests assigned a newly caught individual to a population other than the glade upon which it was caught [11]. Linear regression was used to test the relationship between local residual correlation length and number of dispersers originating from a glade during the stable phase. Because males and females have different dispersal characteristics [6], separate regressions were performed for each gender.

The mark/recapture data also identify the glade to which a disperser immigrates. A paired *t*-test was used to test the null hypothesis of random dispersal with respect to glade characteristics by calculating the difference between the glade immigrated to versus the glade emigrated from for each dispersing individual with respect to the following variables: residual correlation length, glade area, glade population size, and glade density (population size divided by glade area).

Circuit analysis was used to examine how local habitat variation influences patterns of dispersal. Circuit theory models dispersal between populations as current flowing through an electrical circuit composed of nodes and resistors [34]. Populations are nodes, with the current produced related to the number of dispersers from the population, and the resistance to current flowing between nodes is determined by the characteristics of the landscape. Circuit analysis better reflects the landscape as experienced by dispersing individuals by incorporating alternative movement pathways and the influence of matrix heterogeneity into predictions of overall connectivity, unlike least-cost-path methods or isolation-by-distance [35–38]. Circuit analysis has proven to be a flexible and robust method that is extensively used in conservation biology [39]. We adjusted the amount of current produced at each node to reflect the number of dispersers leaving a glade, and adjusted the permeability of the matrix between the nodes to reflect the permeability of the habitat to a dispersing lizard.

We tested two models of current and two matrix permeability models for a total of four unique models of dispersal. The two models of current are:

Null model/homogenous

Under this current model, local habitat variation (residual correlation length) does not influence dispersal from a glade. Instead, dispersal is only influenced by the glade area, with larger glades producing more dispersers. Because large values of current can swamp the program, glade area was scaled by dividing the area of each glade by the area of the smallest glade (Glade 43, Fig. 6.1, with an area of 311.82 m^2), creating a range of values from 1 to 166 for standardized current based on area alone. Glade area is highly correlated with glade population size, so area was chosen rather than population size to assess the ability of the model to be run entirely using variables inferred from remote sensing data.

Local habitat model

Under this current model, local residual correlation length influences dispersal from a glade. Glade area was weighted by multiplying area by 1 + the residual correlation length. This weighting implies that greater clustering (negative residual correlation lengths) would produce fewer dispersers, as indicated by our initial dispersal analysis. For computational efficiency, the values were standardized by dividing all adjusted areas by the value of the smallest area (Glade 43), creating a range of values from 1 to 198. Values were rounded to the nearest integer.

The two models of matrix permeability are related to topography. They are:

Null model/flat

Under this model, the terrain between glades plays no role in patterns of dispersal. This was modeled with a simple raster of the study area with a uniform resistance of 1. This results in a pattern of isolation by distance.

Slope resistant model

Under this matrix permeability model, lizards are less likely to disperse along steeper slopes, as indicated by previous analyses [6,11]. To create slope resistant habitat maps, slope was calculated from a digital elevation model, obtained from the Missouri Spatial Data Information Service, at 1:24,000 scale, with a 30-m resolution with a 1000-m buffer. The maximum inclination at each point in the raster was calculated using ArcGIS Spatial analyst; the resulting raster values corresponded to the slope in degrees. This raster was processed by adding 1 to each value to avoid areas of zero resistance. The final slope resistance raster values ranged from 1.02 to 43.52, with higher values corresponding to greater resistance.

We then simulated the four models using Circuitscape Version 3.5.8 for Mac. Each glade population in a model was represented as a valued point location; point values encoded both position and predicted current produced under the null or local habitat model. Points were created using the Polygon to Point tool in ArcGIS that created a single point in the centroid of each glade polygon. Glade polygons were created from tracing glade outlines from an aerial photograph, 2009 NAIP DOQQ (1-m) (*Missouri NAIP* n.d.). To simulate realistic dispersal patterns without prohibitively increasing computational complexity, glades were coded as "short circuit" regions with 0 resistance. This allowed for current, originating from the centroid, to emanate from any part of the polygon. Biologically, a lizard could disperse from any edge of the glade.

To model independent current sources, one must also provide positions of ground, where current leaves the circuit (where dispersers settle). Grounds and current sources cannot overlap in the program. To allow for this, ground points were placed just outside the perimeter of glade polygons at an average distance of 6 m (2–17 m), so as not to overlap with current sources, but not so far as to alter patterns of current flow. Biologically, this meant that dispersers had an equal probability of settling at any encountered glade.

All habitat data was projected in the North American Datum 1983 UTM for Zone 15N. Data were exported to Circuitscape using the Export to Circuitscape Tool with a cell size of $10 \times 10\ m^2$, a scale that provided adequate resolution without overwhelming the memory capacity of the Circuitscape Solver.

To generate predicted patterns of dispersal, cumulative current maps were generated in Circuitscape under all four models. The resulting rasters were imported into ArcGIS 10, and projected into the 1983 North American DATUM UTM 15N projection. The base exponential of current values for each cell in the raster was taken to transform current values from -2 to 5 to a nonnegative range (0–9) that preserved relative values of current, for use later in analysis.

Genetic sampling and analyses

To test the association between bedrock clustering and genetic variation, we included samples surveyed genetically on Stegall Mountain during 2000–03, the years of greatest sampling effort during the stable period. These years, which included the most intense sampling of the study, resulted in a sample size of 132 populations. Six microsatellite loci were used [11] (DRYAD entry: http://doi:10.5061/dryad.ps736). We used pairwise f_{st} calculated using GenoDive (version 2.0b23) [40] as genetic distances (weighted edges) between glade populations (nodes) during 2000–03. The overall f_{st} during this period was significant and stable at 0.04 [11]. The association between pairwise genetic distances and pairwise resistance distances based on different models of dispersal was tested using partial Mantel tests, controlling for patterns driven by isolation by distance

[1,41,42]. A matrix of straight-line distances between the centroids of each polygon served as the control matrix in partial Mantel tests carried out in GenoDive (Meirmans and Van Tienderen, 2004). The resistance distance between each pair of glades under the four dispersal models was calculated using the four cumulative current maps produced under the four dispersal models as conductance surfaces in an isolation by resistance analysis. Resistance distances incorporate both geographic distance between populations as well as the traversability of the interpatch matrix. Pairwise resistance distances were calculated based on an 8-neighbor connection model, and these constitute the edges in a network of glade populations as nodes.

Results

Observed variation in bedrock distribution

Levels of resource clustering on the six focal glades ranged from a residual correlation length of −0.403 on the smallest clustered glade to 0.890 on the largest even glade. The quantification supports our original on-the-ground assessments. For each glade size class, the residual correlation length was larger on the glade classified as "even" than on the glade classified as "clustered" (0.890 for even vs 0.091 for clustered large glades, 0.360 for even vs 0.006 for clustered medium glades, and 0.191 for even vs −0.403 for clustered small glades).

Fig. 6.3 shows the residual correlation distances (the residuals from Fig. 6.2) for the glades on Stegall Mountain, that ranged from −0.794 to 0.890 with a mean of −0.049 and standard deviation of 0.431. These numbers show that a large amount of variation in bedrock clustering exists on this single mountain.

Impact of bedrock clustering on social structure

A total of 171 individual lizards were followed in the behavioral studies, with an average of 12.0 location observations per lizard and an average of 22.6 total observations (locations and behaviors) per lizard. The 2010 populations did not exhibit a significant difference in variance of either clustering coefficient or eigen centrality between habitat classes (Table 6.1). 2010 was the first year of this study, and much initial time and effort was spent on working out protocols, resulting in an abbreviated field season with limited sample sizes. Our 2011 sample was larger, with

Figure 6.3 Stegall Mountain glades classified by clustering of exposed bedrock. The values of the residual correlation length of each glade are given in or next to the glade polygon. Greener colors correspond to glades with more evenly distributed resources, redder glades to more clustered resources.

Table 6.1 Differences in variance of social network measures between habitat classes by year.

Network measure	Year	Even habitat	Clustered habitat	df	*F*	*P*
Clustering coefficient	2010	14,204	6405	14 on 16	0.451	.1415
Eigen centrality	2010	0.058	0.039	14 on 16	0.6753	.466
Clustering coefficient	2011	55,767	131,570	26 on 50	2.359	**.009**
Eigen centrality	2011	0.055	0.113	26 on 60	2.048	**.029**

Notes: The difference in the variance of each statistic between habitat classes is tested by an *F* statistic, with the degrees of freedom (df) given first for the numerator and then the denominator. The probability of the null hypothesis of equal variances is given in the last column. Significant *P*-values ($P < .05$) are indicated in bold.

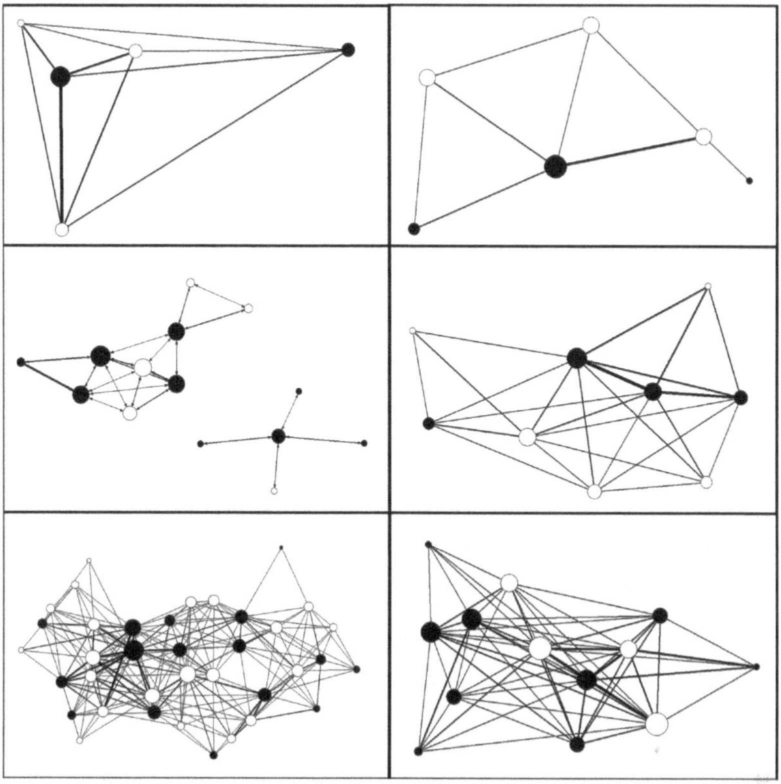

Figure 6.4 Social network graphs estimated from the 2011 samples based on areas of home range overlap for six collared lizard populations in the Missouri Ozarks. Black nodes represent males, white represent females. Node size is weighted by node degree (number of ties) and line thickness is greater for stronger associations (greater areas of home range overlap). Population size increases from top to bottom, with evenly distributed bedrock on the left, clustered bedrock on the right.

the 2011 social networks shown in Fig. 6.4. Individuals on glades with clustered resources exhibited significantly greater variance in clustering coefficient and eigen centrality than individuals on glades with evenly distributed resources (Table 6.1).

To examine how the impact of resource distribution may change with population size, we broke down the data into size classes. The 2010 and 2011 data were pooled to obtain sufficient sample sizes. Habitat class only had a significant effect on the variance of eigen centrality in the largest population class, in which clustered habitat exhibited greater variance in eigen centrality than the evenly distributed habitat (Table 6.2). Variance

Table 6.2 Differences in network monopolization by size and habitat class as measured by the variances of ego-centric eigen centrality and ego-centric clustering coefficient observed in populations from 2010 to 2011.

Glade size	Variable	Even habitat	Clustered habitat	df	*F*	*P*
Small	Clustering coefficient	35,610	315,739	4 on 8	8.86	**.009**
	Eigen centrality	0.249	0.085	4 on 8	0.343	.317
Medium	Clustering coefficient	1511	98,382	12 on 11	65.98	**3.95E-8**
	Eigen centrality	0.064	0.051	12 on 11	0.797	.700
Large	Clustering coefficient	55,277	94,218	25 on 46	1.58	.281
	Eigen centrality	0.043	0.087	23 on 46	2.20	**.041**

Note: Degrees of freedom and *F* tests of variance equality between habitat classes are given, with significant *P*-values ($P < .05$) indicated in bold.

Table 6.3 Effect of resource distribution on male display effort.

Sample and measure of displays per male	Even habitat	*N*	Clustered habitat	*N*	*T*	*P* (one tail)
All males: mean	4.4	18	5.7	26	3109	.311
Territorial males: mean	9.6	12	12.9	8	2273	.227
All males: variance	32.9	18	104.0	26	609	.061
Territorial males: variance	20.8	12	147.3	8	297	**.030**

Notes: Difference in mean and variance of displays per males and per territorial males were calculated using a permutation *T* test based on 5000 iterations on a sample size of *N*. Significant *P*-values ($P < .05$) are indicated in bold.

of clustering coefficient was significantly greater for clustered habitats at both the small and medium population sizes, but not at the largest population size (Table 6.2).

Two groupings of males were examined for differences in level of aggression; all males observed on the glade, and territorial males. The territorial male class excludes males that were never observed to actively display, and only compares the magnitude of display effort among the active males. The only significant effect was that territorial males on clustered habitat exhibited greater variance in aggression scores than territorial males on even habitat (Table 6.3).

Relationship between population size with area and residual correlation length

The regression of population size on residual correlation length by itself was nonsignificant, but glade area by itself was highly significant ($P < .001$) with an R^2 value of 0.672. The model that included both variables had an R^2 value of 0.700, and both area and residual correlation length were significant (Table 6.4). Adding the interaction term raised R^2 to 0.711, but the interaction term was nonsignificant.

Dispersal

One hundred fifty interglade dispersal events were recorded on Stegall Mountain between 2000 and 2006 when the metapopulation was demographically stable [6,11]. Ninety-one dispersers were males and 57 were females (2 hatchling dispersers with sex unknown were removed from analysis). Fig. 6.5 shows a network of the observed dispersal events.

Previous analyses revealed highly significant differences between male and female dispersal patterns [6]. Strong gender effects also emerged when local glade structure was incorporated into the analyses. There was a significant positive relationship between residual correlation length and number of male dispersers produced by a glade (Table 6.5), but no relationship with female dispersal. We compared the goodness of fit of a set of regression models that included population size and residual correlation length as explanatory variables. The model with the lowest Akaike information criterion and highest R^2 value included the explanatory variables population size, residual correlation length, and the interaction between population size and residual correlation length (Table 6.6, Model 1). Table 6.7 shows the estimated regression parameters for this best fit model.

Table 6.4 The regression model of population size as a function of glade area and residual correlation length.

Independent variable	Estimated regression coefficient	*t*	*P*
Constant	1.438	1.594	.118
Area	0.847	10.236	**.000**
Residual correlation length	0.170	2.051	**.046**

Note: A *t* statistic is used to test the null hypothesis that the regression coefficient is 0. Significant *P*-values ($P < .05$) are indicated in bold.

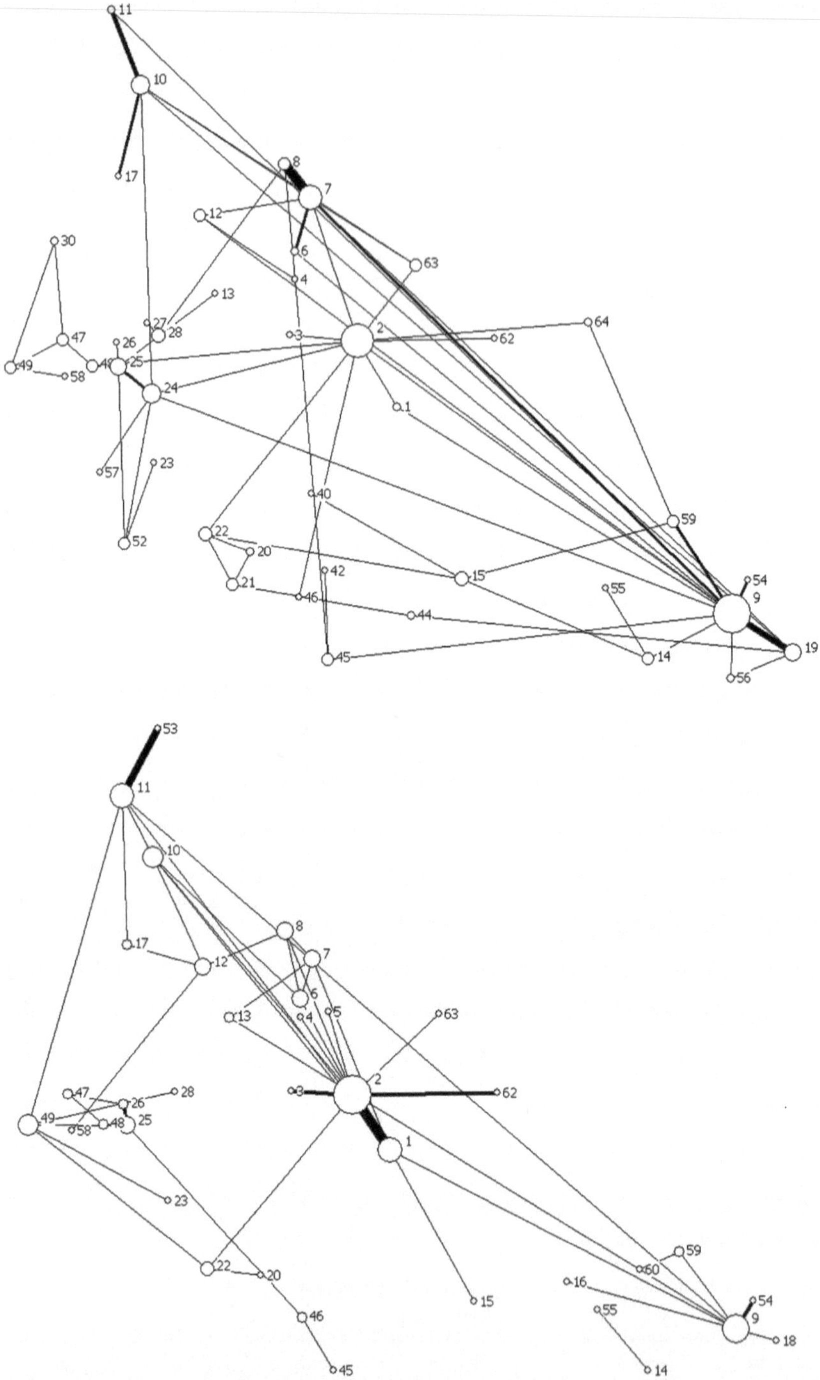

Figure 6.5 Male and female dispersal networks on Stegall Mountain. *Top*: Male dispersal events. *Bottom*: Female dispersal events. Node size is proportional to number of dispersers originating from a glade. Edge width is proportional to number of individuals dispersing between glades.

Table 6.5 Influence of local habitat (log residual correlation length) on number of male and female dispersers on Stegall Mountain between 2000 and 2006 as measured by a linear regression.

Model	R^2	*F*	df	*P* value
Males	0.29	22.16	1 on 52	**1.90E-05**
Females	0.02	1.599	1 on 37	.214

Note: Significant *P*-values ($P < .05$) are indicated in bold.

Table 6.6 Results of different regression models on the number of male dispersers among populations on Stegall Mountain between CE 2000 and 2006.

Explanatory variables	R^2	*F*	df	*P* value	AIC
Model 1: population size × local habitat	0.218	5.92	3 on 50	**.002**	146
Model 2: population size + local habitat	0.127	4.87	2 on 51	**.012**	151
Model 3: population size	0.121	8.32	1 on 52	**.006**	151

Notes: Local habitat refers to the log residual correlation length of exposed bedrock on each glade. Population size refers to the average population size 2000–06. Model 1 includes both of these variables and their interaction, Model 2 includes both of these variables but with no interaction, and Model 3 includes only population size. Significant *P*-values ($P < .05$) are indicated in bold. R^2 is a measure of goodness of fit of the model, with larger values indicating a better fit. AIC is the Akaike information criterion, with lower values indicating a better fit.

Table 6.7 Influence of population size and local habitat quality (as measured by log of correlation length) on number of male collared lizard dispersers on Stegall Mountain between 2000 and 2006 for Model 1 in Table 6.6.

Variable	Estimate	Std error	*P* value
Population size	0.041	0.016	**.012**
Local habitat	−0.369	0.386	.344
Population size × local habitat	0.064	0.025	**.011**

Note: Significant *P*-values ($P < .05$) are indicated in bold.

Table 6.8 shows the results of the paired *t*-test on the difference in several glade characteristics between the glade a dispersing individual went to versus the glade from which it dispersed. The only variable that was significant was the difference in local population size, with lizards of both sexes preferring to go to glades with larger population sizes.

Table 6.8 The results of paired *t*-tests on the difference between several attributes of the glade to which a dispersing individual went versus the glade from which the dispersing individual departed.

		Glade characteristic tested			
Gender	**Test statistics**	**RCL**	**Glade size**	**Population size**	**Population density**
Female	*t*-test	0.586	1.510	2.430	−0.353
	df	43	57	57	57
	p	0.096	0.137	**0.018**	0.725
Male	*t*-test	−0.557	1.490	2.140	0.808
	df	79	97	97	97
	p	0.579	0.140	**0.035**	0.421

Notes: Only dispersing individuals from the stable metapopulation phase on Stegall Mountain (2000–06) were tested, and separate tests were performed on females and males. Tests with significant values ($P \leq .05$) are indicated in bold. *RCL*, residual correlation length; *df*, degrees of freedom.

Relationship between bedrock distribution and within-glade genetic variation

We assessed the effect of bedrock distribution on genetic variation with the genetic data on a sample of 132 glades surveyed on Stegall Mountain from 2000 to 2003 during a phase of stable interglade genetic differentiation [11]. We calculated the effective number of alleles for each population and regressed it with the residual correlation length, finding that population size, residual correlation length, and their interaction all had a significant impact on effective allele number, with the interaction between residual correlation length and population size being negative (Table 6.9).

Association between genetic distance and the predicted resistance distances

We constructed landscape resistance surfaces based on four models (Fig. 6.6). There are substantial differences in the predicted currents between null and local habitat models for both flat and slope-resistant landscapes (Fig. 6.7), indicating power to measure the impact of the local clustering distribution on gene flow. We compared how well each surface of the four surfaces correlated with pairwise f_{st} of the 2000–03 populations using partial Mantel tests [43]. On flat resistance surfaces, local clustering variation had no significant effect on the observed patterns of genetic differentiation (Table 6.10). Among the slope resistant models, local habitat clustering variation had a significant effect on genetic differentiation (Table 6.10).

Table 6.9 Regression of allelic richness on population size, resource clustering and their interaction.

Variable	Estimate	Std error	*t*	*P* value
Population size	0.018	0.003	5.231	**6.68E-07**
Residual correlation length	0.189	0.074	2.568	**.0114**
Population size × residual correlation length	−0.020	0.005	−3.812	**2.13E-04**

Note: Significant *P*-values ($P < .05$) are indicated in bold.

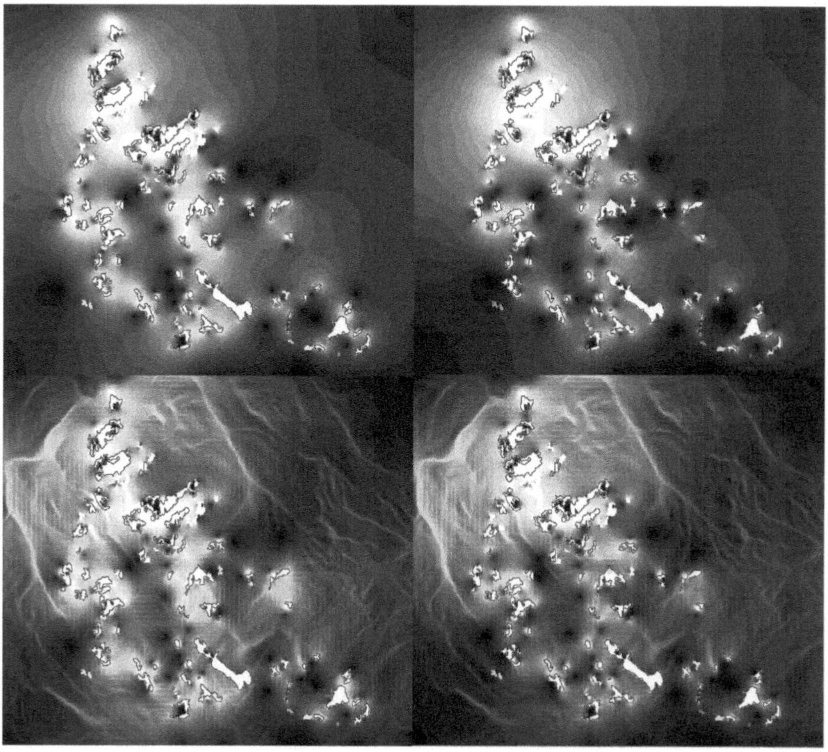

Figure 6.6 Predicted dispersal routes: Cumulative current maps based on circuit analysis incorporating local and landscape parameters. Upper left: Null local model, flat landscape. Upper right: Local habitat model, flat landscape. Lower left: Null local model, slope resistant landscape. Lower right: Local habitat model, slope resistant landscape.

Discussion

The clustering of a key territorial resource, exposed bedrock and its attendant visual openness in this case, has a significant influence on a

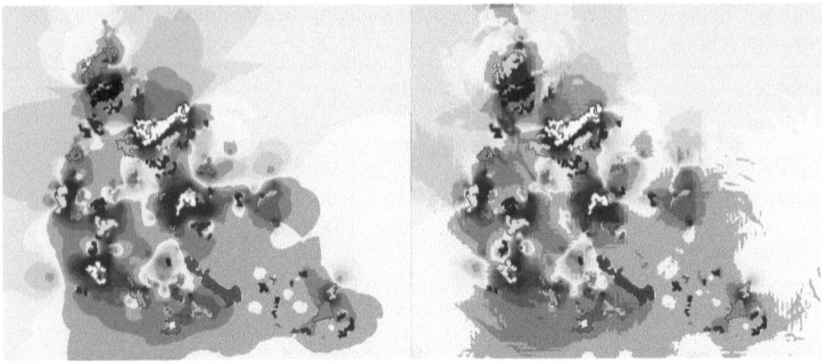

Figure 6.7 The difference in predicted current between null and local habitat models for flat (*left*) and slope-resistant landscapes (*right*). Darker colors correspond to larger differences in predicted current, blue values indicate where local models predicted lower current than null models, orange/red values indicate where local models predicted higher current than null models.

Table 6.10 Association between pairwise genetic distance and resistance distance among 48 populations of collared lizards on Stegall Mountain.

Local model/landscape model	Spearman's *R*	*P* value
Null/flat	0.103	.102
Local habitat/flat	0.103	.091
Null/slope resistant	0.117	.078
Local habitat/slope resistant	**0.138**	**.039**

Notes: Resistance distances were calculated as effective distances across conductance landscapes based on models of landscape resistance (homogenous/flat or correlated with slope) and local dispersal probability (weighted by variation in habitat distribution or homogenous/null). Genetic distances were based on pairwise f_{st} of populations sampled from 2000 to 2003. Significance of association was assessed using partial Mantel tests as measured by a Spearman's R statistic, controlling for isolation by distance, with 10,000 permutations. Significant P-values ($P < .05$) are indicated in bold.

population's social structure, local population size, individual dispersal patterns, and the amount of genetic variation within and among local populations. Populations on clustered habitat displayed significantly greater variance in social network clustering coefficient than populations on evenly distributed habitat in small- and medium-sized glades (Tables 6.1 and 6.2), indicating that a few individuals can be disproportionately socially connected to the rest of the local population. At larger population sizes, variation in the clustering of resources had a significant impact on the potential for a few individuals to monopolize territories (as measured by the variance of eigen centrality) (Table 6.2). This potential is supported by Table 6.3, that shows territorial males have a higher variance in their number of displays in clustered habitats, indicating a high potential for mate monopolization, but in a manner that depends upon population size (Tables 6.1–6.3).

The potential for mate monopolization could influence population demography. Our results indicate that this is indeed the case. Table 6.4 shows that both glade area and the residual correlation length are significant factors in influencing the population size on a glade. Not surprisingly, population size tends to increase with increasing glade size, but it also increases with increasing residual correlation length; that is, increasing evenness. As indicated by the behavioral studies (Tables 6.1–6.3), this means that as the potential for male monopolization increases as resources become more clustered, then population size tends to decrease below the expectations based on area alone. Hence, demographic predictions and conservation strategies can be made more accurately by incorporating both habitat area and internal habitat structure.

A second important demographic parameter is dispersal. Table 6.5 reveals that bedrock resource clustering has a strong association with male dispersal, as does local population size, which in turn strongly interacts with bedrock clustering (Tables 6.6 and 6.7). Table 6.7 shows that larger glades with less fragmented exposed bedrock and less potential for mate monopolization produced more male dispersers than glades with clustered bedrock. It is possible that fewer collared lizard males disperse out of clustered habitats in glades with larger population sizes because the habitat is perceived as high quality. The same clustering which promotes mate competition also facilitates monopolization of multiple females. If yearlings can avoid damage from aggressive males, by remaining on glade edges or adopting socially submissive behavior [20], staying on a habitat proven to attract and support females could be a more effective strategy than risking dispersal.

Given that the choice of dispersal has been made, a dispersing individual also has to decide upon where to settle. Table 6.8 shows that both female and male dispersers make nonrandom choices, preferentially going from glades with smaller population sizes to glades with larger population sizes. Interestingly, glade area and population density did not have a significant effect on this choice, although both are correlated with population size. This strengthens the inference that population size itself is the critical factor. Population size can be considered as an indicator of habitat quality, so this nonrandom choice of the destination glade is consistent with the explanation given above for decreased male dispersal from more clustered glades with larger population sizes. The lack of a significant marginal effect of the residual correlation length on the choice of destination glade (Table 6.8) could arise from the contradictory pressures of males preferring glades with a high potential for mate monopolization (negative residual correlation lengths) versus negative residual correlation lengths being associated with decreased local population size (Table 6.4), which would make the glade less attractive.

Our Circuitscape models indicate that local bedrock clustering and/or the social forces associated with it have a significant impact on landscape-wide dispersal patterns (Fig. 6.6). The differences between the current patterns predicted by our null and local habitat weighted models of dispersal (Fig. 6.7) correspond with the differences between the networks of observed female and male dispersal events (Fig. 6.5). Female dispersal patterns, in areas where it differed distinctly from the male network, more closely resembled the predicted dispersal patterns of the null model of local habitat. This fits with our earlier result that female dispersal was not correlated with variation in residual correlation length. Habitat quality assessment for a female collared lizard may be insensitive to the potential for mate monopolization; if so, it follows that females would exhibit movement patterns predicted purely by matrix heterogeneity and glade size. Conversely, in polygynous systems, male dispersal decisions are more likely to be influenced by the social environment [44]. The movement patterns of male collared lizards support this; our network of observed male dispersal events more closely reflected the predicted patterns of dispersal generated by the local habitat models than the null models.

Given that local bedrock clustering has significant effects on both local population size and dispersal, we would also expect it to influence the pattern and amount of genetic variation both within and among local populations. Table 6.9 shows that there are significant effects of both population size, bedrock clustering, and their interaction on genetic variation within local glade populations as measured by allelic richness. In particular, allelic richness increases with increasing population size and increasing bedrock clustering.

The significant association between observed genetic differentiation and the resistance distances predicted from the slope resistant model incorporating local habitat variation (Table 6.10) indicates that both local and landscape level forces play an important role in patterns of gene flow. While at these scales it is often assumed that gene flow is determined solely by properties of the interpatch matrix, our results indicate that the effect of local habitat features on dispersal interacts with, but is not masked by, the heterogeneity of the interpatch matrix. Even at the scale of a single mountain, failure to account for variance in local social structure can limit the ability to understand the processes controlling genetic differentiation.

Conclusions

Our results demonstrate a link between a relatively fixed habitat feature measurable by remotely sensed data and processes that affect genetic variation and demography, thereby allowing a more complete assessment of a habitat's conservation value [45]. In cases where financial and/or personnel resources are not available for genetic monitoring or direct management [46], being able to predict how evolutionary forces vary between potential restoration areas may help better to prioritize conservation efforts and to understand how species of concern may respond to habitat degradation [47]. In particular, we can quantify rapidly and remotely: glade size, bedrock clustering within glades, distance between glades, and slopes in the intervening matrix. From these four easily obtained variables we can make detailed predictions about local population size using the regression models found in Table 6.4, predict dispersal using the appropriate Circuitscape model for each gender (Tables 6.8 and 6.10), predict from regression models the number and source of dispersing individuals (Table 6.7) and their preferred destinations (Table 6.8), predict genetic diversity within each glade (Table 6.9), and genetic differentiation among glades (Table 6.10). This allows an efficient and cost-effective way of prioritizing different areas for restoration and preservation of genetic variation in collared lizards in the Ozarks.

Our work shows how many types of networks can be applied to a single biological system to gain insight at multiple biological levels. Starting with social behavioral networks within glades (Fig. 6.4), we were able to demonstrate that the visual openness of a glade has a significant impact on male territoriality and the potential to monopolize part of a glade. We also showed that visual openness influenced the level of genetic diversity and population size of local glades. The degree of visual openness also affected the probability of a male to disperse from a glade, which in turn had an impact on dispersal networks (Fig. 6.5) and observed and predicted gene flow networks (Figs. 6.5 and 6.6; Table 6.10). Thus the impact of one network can influence another network at a different biological level. Although there have been many applications of networks to biology, few studies have investigated the percolating effects of one network upon another. Our study is unique in looking at these percolating effects by integrating studies on behavior, genetics, demography, dispersal, and gene flow.

The increasing availability of classified land use-rasters, high-resolution aerial photographs, and LIDAR coverage will make remote assessment of a variety of relevant habitat resources more tractable for conservation managers in the future. An improved understanding of the processes controlling habitat connectivity using networks will aid conservation managers in the preservation of fragmented populations. Scarce resources can be better prioritized by identifying key subpopulations for which preservation, habitat restoration, or translocation of individuals would most benefit the metapopulation as a whole [48]. Understanding the relative influence of local and landscape variation on patterns of dispersal and gene flow can allow for more accurate modeling of ecological processes, as well as inform movement predictions for organisms that interact across a range of scales [49,50].

Author contributions

AKC designed, executed, and supervised all of the behavioral work, performed most of the data analyses, and wrote the initial drafts of this manuscript. JLN generated all of the genetic data used in this study and helped in the writing. ART participated in some of the field work, performed some of the analyses, and wrote the final draft. All coauthors have read and approved this manuscript.

Acknowledgments

We dedicate this article to Eviatar Nevo in honor of his 90th birthday and the many contributions he has made to evolutionary biology. We greatly appreciate the cooperation and logistical support from the Missouri Department of Conservation, the Department of Natural Resources of the State of Missouri, the National Parks Service, and The Nature Conservancy of Missouri. We are indebted to many graduate and undergraduate students for helping with the fieldwork, and particularly Katie Zelle, Josh Holter, and Paul Lee. This work was supported by National Science Foundation Grant DEB-9610219 to ART and a National Science Foundation Graduate Research Fellowship to AKC. Publication charges are supported by Washington University Department of Biology fund 94037Z to ART.

Ethics approval

Field protocols for handling and toe-clipping lizards were performed under Animal Welfare Assurance Approval No. 20130074. All field work was done under Missouri Department of Conservation Permit No. 14786, a Missouri Department of Natural Resources Permit, and U.S. National Park Service Permit OZAR-2010-SCI-003.

Availability of data and materials

The genetic data that support the findings of this study are openly available in DRYAD entry at http://doi:10.5061/dryad.ps736. The demographic data used in this paper are available in Templeton AR, Brazeal H, Neuwald JL (2011), the transition from isolated patches to a metapopulation in the eastern collared lizard in response to prescribed fires, *Ecology* 92, 1736–1747, and in the associated Ecological Archives E092-148-A1 and Ecological Archives E092-148-S1. An Excel file giving the details of all the individual behavioral observations is available upon request from either Drs. Amy Conley (amy.conley@dec.ny.gov) or Alan Templeton (temple_a@wustl.edu).

References

[1] R. Holderegger, H.H. Wagner, Landscape genetics, BioScience 58 (2008) 199–208 [Online]. Available from: <http://www.ingentaconnect.com/content/aibs/bio/2008/00000058/00000003/art00006> or <https://doi.org/10.1641/B580306>.
[2] A.R. Templeton, Population Genetics and Microevolutionary Theory, first ed., John Wiley & Sons, Inc, Hoboken, NJ, 2006, p. 705.
[3] P. Bertrand, J. Bowman, R.J. Dyer, M. Manseau, P.J. Wilson, Sex-specific graphs: relating group-specific topology to demographic and landscape data, Mol. Ecol. 26 (15) (2017) 3898–3912. Available from: https://doi.org/10.1111/mec.14174.
[4] A. Khimoun, W. Peterman, C. Eraud, B. Faivre, N. Navarro, S. Garnier, Landscape genetic analyses reveal fine-scale effects of forest fragmentation in an insular tropical bird, Mol. Ecol. 26 (19) (2017) 4906–4919. Available from: https://doi.org/10.1111/mec.14233.
[5] C. Gubili, et al., Environmental and anthropogenic drivers of connectivity patterns: a basis for prioritizing conservation efforts for threatened populations, Evol. Appl. 10 (2) (2017) 199–211. Available from: https://doi.org/10.1111/eva.12443.
[6] A.R. Templeton, H. Brazeal, J.L. Neuwald, The transition from isolated patches to a metapopulation in the eastern collared lizard in response to prescribed fires, Ecology 92 (9) (2011) 1736–1747. Available from: https://doi.org/10.1890/10-1994.1. 2011/09/01.
[7] A.R. Templeton, R.J. Robertson, J. Brisson, J. Strasburg, Disrupting evolutionary processes: the effect of habitat fragmentation on collared lizards in the Missouri Ozarks, Proc. Natl. Acad. Sci. U. S. A. 98 (10) (2001) 5426–5432.
[8] A.R. Templeton, The crisis of partial extinction, Nat. Areas J. 2 (3) (1982) 35–38.
[9] C.L. Brewster, S.J. Beaupre, J.D. Willson, Habitat loss and local extinction: linking population declines of eastern collared lizards (*Crotaphytus collaris*) to habitat degradation in Ozark glades, J. Herpetol. 52 (3) (2018) 352–360. Available from: https://doi.org/10.1670/18-002.

[10] Missouri, Species and Communities of Conservation Concern Checklist, Missouri Department of Conservation Publication, 2021. Available from: https://nature.mdc.mo.gov/sites/default/files/downloads/2021_SOCC.pdf.

[11] J.L. Neuwald, A.R. Templeton, Genetic restoration in the eastern collared lizard under prescribed woodland burning, Mol. Ecol. 22 (14) (2013) 3666–3679. Available from: https://doi.org/10.1111/mec.12306.

[12] A.R. Templeton, J.L. Neuwald, H. Brazeal, R.J. Robertson, Restoring demographic processes in translocated populations: the case of collared lizards in the Missouri Ozarks using prescribed forest fires, Isr. J. Ecol. Evol. 53 (2) (2007) 179–196.

[13] S. Lima, P. Zollner, Towards a behavioral ecology of ecological landscapes, Trends Ecol. Evol. 11 (3) (1996) 131–135 [Online]. Available from: <http://linkinghub.elsevier.com/retrieve/pii/0169534796810949>.

[14] L. Fahrig, G. Merriam, Conservation of fragmented populations, Conserv. Biol. 8 (1) (1994) 50–59.

[15] E. Revilla, T. Wiegand, F. Palomares, P. Ferreras, M. Delibes, Effects of matrix heterogeneity on animal dispersal: from individual behavior to metapopulation-level parameters, Am. Nat. 164 (5) (2004) E130–E153 [Online]. Available from: <Go to ISI>://000224700600002.

[16] E. Revilla, T. Wiegand, Individual movement behavior, matrix heterogeneity, and the dynamics of spatially structured populations, Proc. Natl. Acad. Sci. U. S. A. 105 (49) (2008) 19120–19125. Available from: https://doi.org/10.1073/pnas.0801725105.

[17] M.A. Murphy, R. Dezzani, D.S. Pilliod, A. Storfer, Landscape genetics of high mountain frog metapopulations, Mol. Ecol. 19 (17) (2010) 3634–3649. Available from: https://doi.org/10.1111/j.1365-294X.2010.04723.x.

[18] D. Serrano, D. Oro, U. Esperanza, J.L. Tella, Colony size selection determines adult survival and dispersal preferences: Allee effects in a colonial bird, Am. Nat. 166 (2) (2005) E22–E31 [Online]. Available from: <Go to ISI>://000230721200001.

[19] S.L. Zeigler, M.C. Neel, L. Oliveira, B.E. Raboy, W.F. Fagan, Conspecific and heterospecific attraction in assessments of functional connectivity, Biodivers. Conserv. 20 (2011) 2779–2796.

[20] T.A. Baird, M.A. Acree, C.L. Sloan, Age and gender-related differences in the social behavior and mating success of free-living collared lizards, *Crotaphytus collaris*, Copeia 2 (1996) 336–347 [Online]. Available from: <Go to ISI>://A1996UM94900008.

[21] A.M. Schwartz, T.A. Baird, D.K. Timanus, Influence of age and prior experience on territorial behavior and the costs of defense in male collared lizards, Ethology 113 (1) (2007) 9–17. Available from: https://doi.org/10.1111/j.1439-0310.2006.01297.x.

[22] T.A. Baird, D.K. Timanus, Social inhibition of territorial behaviour in yearling male collared lizards, *Crotaphytus collaris*, Anim. Behav. 56 (1998) 989–994 [Online]. Available from: <Go to ISI>://000076755700021.

[23] T.A. Baird, Male collared lizards, *Crotaphytus collaris* (Sauria: Crotaphytidae), signal females by broadcasting visual displays, Biol. J. Linn. Soc. 108 (3) (2013) 636–646. Available from: https://doi.org/10.1111/bij.12003.

[24] J. Husak, Signal use by collared lizards, *Crotaphytus collaris*: the effects of familiarity and threat, Behav. Ecol. Sociobiol. 55 (6) (2004) 602–607. Available from: https://doi.org/10.1007/s00265-003-0748-3.

[25] A.K. Lappin, J.F. Husak, Weapon performance, not size, determines mating success and potential reproductive output in the collared lizard (*Crotaphytus collaris*), Am. Nat. 166 (3) (2005) 426–436 [Online]. Available from: <Go to ISI>://000231314300013.

[26] J.M. Morales, et al., Building the bridge between animal movement and population dynamics, Philos. Trans. R. Soc. B: Biol. Sci. 365 (2010) 2289–2301.

[27] Geospatial modeling environment, v.0.7.2, Spatial Ecology LLC (2012). Available from: http://www.spatialecology.com/gme/.
[28] J.R. Row, G. Blouin-Demers, Kernels are not accurate estimators of home-range size for herpetofauna, Copeia 4 (2006) 797–802 [Online]. Available from: <Go to ISI>://000243120300021.
[29] S.P. Borgatti, M.G. Everett, L.C. Freeman, Ucinet for Windows: Software for Social Network Analysis, Sage Publications, Ltd, London, 2013.
[30] S.P. Borgatti, NetDraw Software for Network Visualization, Analytic Technologies, Lexington, KY, 2002.
[31] V.A. Formica, M.E. Augat, M.E. Barnard, R.E. Butterfield, C.W. Wood, E.D. Brodie, Using home range estimates to construct social networks for species with indirect behavioral interactions, Behav. Ecol. Sociobiol. 64 (7) (2010) 1199–1208. Available from: https://doi.org/10.1007/s00265-010-0957-5.
[32] S.A. Cushman, B.W. Compton, K. McGarigal, (Chapter 20) Habitat Fragmentation Effects Depend on Complex Interactions Between Population Size and Dispersal Ability: Modeling Influences of Roads, Agriculture and Residential Development Across a Range of Life-History Characteristics, Springer Japan, Tokyo, 2010pp. 369–385.
[33] FRAGSTATS v4: Spatial Pattern Analysis Program for Categorical and Continuous Maps. Computer Software Program, University of Massachusetts, Amherst, 2012. Available from: <http://www.umass.edu/landeco/research/fragstats/fragstats.html>.
[34] B.H. McRae, Isolation by resistance, Evolution 60 (8) (2006) 1551–1561 [Online]. Available from: <Go to ISI>://000240530300003.
[35] B. McRae, B. Dickson, T. Keitt, V. Shah, Using circuit theory to model connectivity in ecology, evolution, and conservation, Ecology 89 (10) (2008) 2712–2724 [Online]. Available from: <http://www.esajournals.org/doi/abs/10.1890/07-1861.1>.
[36] B.H. McRae, P. Beier, Circuit theory predicts gene flow in plant and animal populations, Proc. Natl. Acad. Sci. U. S. A. 104 (50) (2007) 19885–19890. Available from: https://doi.org/10.1073/pnas.0706568104.
[37] J.M. Alves, et al., Parallel adaptation of rabbit populations to myxoma virus, Science 363 (6433) (2019) 1319–1326. Available from: https://doi.org/10.1126/science.aau7285.
[38] C.W. Birky Jr., Uniparental inheritance of organelle genes, Curr. Biol. 18 (16) (2008) R692–R695.
[39] B.G. Dickson, et al., Circuit-theory applications to connectivity science and conservation, Conserv. Biol. 33 (2) (2019) 239–249. Available from: https://doi.org/10.1111/cobi.13230.
[40] P.M. Visscher, M.E. Goddard, From R.A. Fisher's 1918 paper to GWAS a century later, Genetics 211 (4) (2019) 1125–1130. Available from: https://doi.org/10.1534/genetics.118.301594.
[41] P.E. Smouse, J.C. Long, R.R. Sokal, Multiple regression and correlation extensions of the Mantel test of matrix correspondence, Syst. Zool. 35 (1986) 627–632.
[42] S.A. Cushman, K.S. McKelvey, J. Hayden, M.K. Schwartz, Gene flow in complex landscapes: testing multiple hypotheses with causal modeling, Am. Nat. 168 (4) (2006) 486–499 [Online]. Available from: <http://www.journals.uchicago.edu/cgi-bin/resolve?AN41553>.
[43] J. Jaquiery, T. Broquet, A.H. Hirzel, J. Yearsley, N. Perrin, Inferring landscape effects on dispersal from genetic distances: how far can we go? Mol. Ecol. 20 (4) (2011) 692–705. Available from: https://doi.org/10.1111/j.1365-294X.2010.04966.x.
[44] P.J. Greenwood, Mating systems, philopatry and dispersal in birds and mammals, Anim. Behav. 28 (4) (1980) 1140–1162 [Online]. Available from: <http://www.sciencedirect.com/science/article/pii/S0003347280801035>.

[45] R.J. Petit, A. El Mousadik, O. Pons, Identifying populations for conservation on the basis of genetic markers, Conserv. Biol. 12 (4) (1998) 844–855. Available from: https://doi.org/10.1111/j.1523-1739.1998.96489.x.
[46] R. Lande, Genetics and demography in biological conservation, Science 241 (1988) 1455–1460.
[47] K.E. Lacy, E.P. Martins, The effect of anthropogenic habitat usage on the social behaviour of a vulnerable species, *Cyclura nubila*, Anim. Conserv. 6 (2003) 3–9. Available from: https://doi.org/10.1017/s1367943003003020.
[48] D. Serrano, J.L. Tella, Dispersal within a spatially structured population of lesser kestrels: the role of spatial isolation and conspecific attraction, J. Anim. Ecol. 72 (2003) 400–410.
[49] R. Biek, L.A. Real, The landscape genetics of infectious disease emergence and spread (in English)Mol. Ecol. 19 (17) (2010) 3515–3531. Available from: https://doi.org/10.1111/j.1365-294X.2010.04679.x.
[50] T.S. Talley, Which spatial heterogeneity framework? Consequences for conclusions about patchy population distributions, Ecology 88 (2007) 1476–1489.

CHAPTER 7

Sensory perception of mole-rats and mole rats: assessment of a complex natural global evolutionary "experiment"

Hynek Burda
Department of Game Management and Wildlife Biology, Faculty of Forestry and Wood Sciences, Czech University of Life Sciences, Praha, Czech Republic

Background

In the late 1970s my interest in sensory perception of subterranean mammals was awoken; particularly in audition and magnetoreception of the Ansell's mole-rats (*Fukomys anselli*, Bathyergidae), formerly assumed to represent the same species as the South-African *Cryptomys hottentotus* and in papers published before 2006 named so. In that time, Eviatar (Eibi) Nevo drew my attention to another subterranean rodent species, the blind mole rat (then designated as *Spalax ehrenbergi*, Spalacidae; note also the difference in spelling between mole-rat and mole rat). Our research of both species in conjunction with our friendship started then. In 1989 Eibi invited me to give a talk at the Fifth International Theriological (today: Mammalogical) Congress in Rome, at the First Symposium on Subterranean Mammals he convened together with Late Osvaldo A. Reig. A year later, successful and since then highly cited proceedings [1] of the symposium were published. The chapter on senses of subterranean mammals [2] summarized the state-of-the-art knowledge on morphology, physiology, ecological relevance, and evolution of sensory systems of subterranean mammals. In fact, it triggered further research interest on the topic. While in 1989, the list of publications dedicated to sensory ecology, a termed coined by Dusenbery [3], of subterranean mammals was still manageable, 30 years later it can be hardly surveyed.

Subterranean mammals, particularly African mole-rats (Bathyergidae) and Eurasian mole rats (Spalacidae) have dug their ways to biomedical laboratories as model animals for studies of i.a. aging, cancer resistance,

New Horizons in Evolution
DOI: https://doi.org/10.1016/B978-0-323-90752-1.00006-7

hypoxia (for the most recent reviews see Refs. [4–7]), and sensory systems. They represent nonrelated lineages of rodents, which independently adapted to life underground under pressure of the same environmental stresses and they constitute an example par excellence of convergent evolution [8]. The underground ecotope is very specific—it is simple, monotonous, safe, stable and predictable, factors, which support structural and functional reduction and degeneration due to lack of use of organs. On the other hand, it is very demanding on orientation in space and time, communication, foraging, and high-energy expenditure due to digging, moreover in a humid, hypoxic and hypercapnic atmosphere of underground burrows. More specifically, since the sensory environment of subterranean rodents appears to be globally rather uniform, either convergent evolution due to exertion of the same selection pressures or stochastic random degeneration due to lack of stimulation is expected.

From the point of view of biomedical research, subterranean mammals are of interest since they represent organisms with certain "handicaps" (e.g., they are naturally microphthalmic, hypothyroid, etc.—see below). Contrary to laboratory rodents, we do not need to make them "sick" to study the response of their organ systems but instead we can study solutions and compensations of those handicaps as designed by nature.

In this present contribution, we will revisit the 30-year-old article [2] and selected examples will be shown regarding research achievements in the fields of sensory biology and ecology of subterranean rodents, and how they changed and extended our knowledge and what the challenges are for further research. In this sense, the present article is a follow-up of that older paper and in most cases, older findings and references cited in Ref. [2] are not repeated. Due to the large amount of literature, the three senses and only a few species of subterranean rodents are focused on. Unless otherwise specified, the generic names *Fukomys* and *Spalax* are used here sensu *F. anselli* and *S. ehrenbergi* superspecies, respectively.

Eye and vision: adaptation, neutral evolution, or side effect or . . .?

Although microphthalmia, that is, reduced eyes, is one of the most conspicuous traits of subterranean mammals, surprisingly, the visual system in subterranean mammals has remained rather understudied for a long time. Living in darkness apparently does not enable vision and, therefore, does not require eyes. Eyes and the whole visual system have been assumed to be regressed, and, for a long time, considered rather uninteresting for researchers. Yet, the question arises why the eyes, as energetically costly organs, have not completely disappeared.

Ecology: what is the optic environment of subterranean rodents like?

Optic environment in subterranean burrows can, no doubt, be designated as (absolutely) dark. Light may surely penetrate into superficial foraging burrows—but how much and at which wavelengths is dependent on geography, season of the year, daytime, vegetation cover, and soil characteristics. We can assume that no light penetrates to nest chambers which, in the case of mole-rats, are usually placed 50 cm under surface and deeper [9]. Although it has been suggested that low acuity vision plays an important role in predator avoidance (see below), only a single study to measure light conditions in artificially opened burrows has been made so far [10]. Only about 0.2%–2.5% of the ambient visible light entered the opened burrow. Light intensity attenuated quickly and reached mesopic light levels (at which both cones and rods contribute to vision) within a few centimeters from the burrow opening; scotopic light levels (at which only rods operate) were estimated to be reached at one to a few meters (of a straight! tunnel) from the opening. Rod-mediated light sensation in straight (!) tunnels seems to be possible over distances much longer than 100 m, implying that it is the burrow architecture (tortuosity and branching) that limits light sensation under natural conditions. As expected, longer wavelength (green) light propagates at longer distances than the short wavelengths (blue) [10].

Morphology: what does it look like?

The eye and optic pathways in *Spalax* have been subject to several studies (for references to older studies see Ref. [2]; for more recent studies see Refs. [11–19]). The findings can be summarized as follows: eyes in *Spalax* are miniscule (0.6 mm), embedded in a hypertrophied Harderian gland and covered by furred skin; they are quantitatively and qualitatively the most rudimentary and regressed of all mammals studied to date. The optic nerve and retinal projections to visual nuclei, although present, are severely atrophied.

The *Spalax* retina contains fully functional rod and long-wavelength-sensitive (LWS) cone photopigments [20], yet it lacks a functional short-wavelength-sensitive (SWS) cone photopigment [21]. *Spalax* can thus be considered a LWS monochromat. Besides that, the retina includes at least three photopigments (rhodopsin, cone opsin, and melanopsin) that may (theoretically) collect the scant light penetrating the soil [22]; overview [8,19,23].

Despite living under the same optic conditions in the darkness of sealed burrows, the visual system of *Fukomys* exhibits rather different properties. *F. anselli* are also microphthalmic; nevertheless, their eyes are superficial and bigger (2 mm) than those of *Spalax*. Qualitatively, they are normally developed, with no signs of degeneration [24,25]. The olivary pretectal nucleus and ventral lateral geniculate nucleus, that is, brain structures involved in brightness discrimination, pupillary light reflex, and orchestrating sleep and circadian responses to light, are relatively well developed, receive robust retinal input, and express c-Fos in response to light exposure [26,27].

The retina is rod-dominated but accommodates a high cone proportion (c. 10%) of cones, which is higher than in nocturnal surface dwellers with 0.5%–3% cones (reviewed in Ref. [28]). The vast majority of the cones are strongly SWS-opsin immunoreactive (the so-called blue cones). In *Fukomys* c. 20% of the cones showed exclusive SWS-opsin label, c. 10% exclusive LWS-opsin label, and c. 70% dual pigment cones expressing mainly SWS-opsin and small amounts of LWS-opsin as determined by immunohistochemistry. The finding of SWS-opsin dominance and low levels of LWS-opsin across the entire retina is unique among mammals [25,28–30]. Thus, in total, approximately 90% of cones showed a predominant SWS-opsin expression.

To summarize, the main differences: the eye in *Spalax* is subcutaneous, and qualitatively degenerated, while the eye in *Fukomys* is superficial and only quantitatively reduced but qualitatively normally developed, that is, not degenerated. *Spalax* possess LWS-opsin but no SWS-opsin, whereas the situation in *Fukomys* is opposite.

Physiology: what is its capacity?

As expected, *Spalax* is blind and does not respond to visual images [31,32]. Morphological findings suggest that *Fukomys* can distinguish light intensities and that light may affect its behavior. In addition, light exposure elicits c-Fos expression in the retrosplenial cortex which suggests that *Fukomys* is even attentive to visual stimuli [26]. Recent morphological and immunocytochemical studies indicate that vision in *Fukomys* might be functional [24–27]. However, small eye size per se imposes severe optical and neural constraints on visual acuity. The very low number of optic nerve fibers and, as a result, retinal ganglion cells indicates poor (if any) visual resolution [27].

In a laboratory experiment, *Fukomys* reacted to white light during nest building and preferably placed their nests in a dark nest box [33]. Behavioral experiments show that *Fukomys* can perceive monochromatic light in the blue and green wavelength range [34].

Evolution: what is it for?

Photoperiod sensation?

Despite its blindness and degenerated eyes, *Spalax* has a circadian clock and light can entrain circadian rhythms of locomotor behavior [20,35,36], and induce gene expression in the suprachiasmatic nucleus [14]. Enucleation abolishes these reactions to light, confirming that they rely upon retinal photoreceptors [20,36]. Indeed, *Spalax* occurs in geographic regions with a photoperiod changing throughout the year, and it is a seasonal breeder. Accordingly, the differential reduction of visual structures, yet retention of the photoperiod-receptive function of the retina in *Spalax* were explained as an adaptive response to the stresses of the underground environment [21]. It should be noted, however, that *Spalax* circadian photo-entrainment has been demonstrated in laboratory conditions with an artificial light/dark cycle, and that although the circannual sensation of photoperiod length would be adaptive, the relevance of light being used as a *zeitgeber* for a circadian clock in constant darkness remains enigmatic [20]. Recent radio-tracking studies [37] showed unimodal daily activity pattern likely related to temperature.

For seasonal entrainment, light in the nearly constant 12:12 LD-rhythm in Zambia would hardly be of use as a *zeitgeber*. Accordingly, Zambian *Fukomys* species are not seasonal breeders [38,39]. Results of early laboratory studies on circadian rhythms in *F. anselli* were inconclusive [40,41]. Radio-tracking of *F. anselli* in the field [42] revealed individually different and flexible activity patterns with an afternoon peak, a situation similar to that in *Spalax* [37]. Later laboratory studies [43] and radio-tracking studies in the field [44] indicated rather nocturnal activity in other *Fukomys* species. Nevertheless, the amount of the individual's activity and its daily patterns are apparently flexible and can be modified in response to actual environmental and social conditions. All recent studies agree that daily activity patterns in *F. anselli* and related species in the field are individually flexible and may be dependent on temperature rather than on the photoperiod changes.

Seeing the light at the end of the tunnel?

Note that although *Fukomys* mole-rats and *Spalax* mole-rats live in the same ecotope and face the same physiological challenges and share the same foraging strategies, they differ in their social behavior and geography and habitats. While *Fukomys* mole-rats are social living in families, *Spalax* mole rats are solitary. *Fukomys* occur in African savannahs, *Spalax* inhabit primarily Mediterranean steppes. Particularly the latter aspect determined predation and evolution of antipredatory strategies.

The risk of a damage of foraging tunnels (running about 10 cm under surface) by large herbivores and subsequent predation by snakes, birds, and carnivores in African savannahs is much higher than in the Mediterranean region. Perhaps significantly, hominine evolution took place parallel and sympatrically with the evolution of African mole-rats; and hominines exerted perhaps the most significant predation pressure upon mole-rats [45–47]. The most frequent method of hunting mole-rats is based on opening the tunnels and waiting for a mole-rat approaching the opening in order to seal the tunnel. We suggested that the ability to perceive (i.e., detect and interpret) light in a dark world could enable mole-rats to "see light at the end of the tunnel" in an accidentally or intentionally opened burrow and to plug it for the sake of safety and stable climatic conditions. This seemed the most plausible explanation for adaptive significance of sight in *Fukomys*, and also an explanation for the prevalence of SWS-cones detecting blue skylight penetrating into an open burrow.

However, it has been demonstrated that blue light propagates less efficiently than green light in natural burrow systems, and light intensities drop quickly to scotopic levels, by which only rods are activated [10]. These findings make it unlikely that the SWS-opsin-rich retina has any adaptive value for life in subterranean burrow systems. Furthermore, some neurobiological findings are not fully consistent with this interpretation. A marked reduction of visual input to the superior collicle suggests that light stimuli are of little importance for *Fukomys*, indicating that these animals are unable to generate spatially appropriate orientation responses, that is, to react behaviorally promptly on light stimuli [27]. Moreover, brain regions involved in visual orientation in three-dimensional environments are severely reduced in bathyergid mole-rats [48] explaining recent behavioral data [49]. These findings suggest that the mole-rats' visual system serves simple light detection rather than high acuity image forming or color discrimination. Thus the mechanisms leading to the unique opsin ratio remained unclear.

To summarize: Mammals usually possess a majority of LWS-opsin cones and a minority of SWS-opsins in the retina, enabling dichromatic vision. Unexpectedly, *Spalax* are LWS monochromats and *Fukomys* are SWS, albeit with no apparent adaptive value.

Short-wavelength-sensitivity as a byproduct of adaptation to low metabolism?

The thyroid hormones (THs) play an important role in the regulation of growth, neural development, thermogenesis, body temperature, and cellular metabolism but are also a major factor influencing cone opsin expression via the production of L-thyroxine (T4) which is converted by deiodination in tissues to metabolically active 3, 5, 30-triiodo-L-thyronine (T3) [50–54]. The mechanism is rather complicated and its description would be beyond the scope of this chapter. Importantly in this connection is the fact that THs suppress SWS-opsin expression and instead activate LWS-opsin expression [51,55,56]. Low serum TH levels, a state referred to as hypothyroid, thus leads to downregulation of LWS-opsin expression with a high plasticity throughout adulthood, at least in mice, rats, and likely also in humans [53,57].

Interestingly, we have previously shown that Ansell's mole-rats have T4 levels which are approximately 10-fold lower compared to other rodents, while T3 levels are in the rodent-typical range [58]. Total TH levels (free and bound fraction) appeared to be downregulated by reduced T4 synthesis and increased protein binding rate. We have interpreted our results in view of the low basal metabolic rate (BMR) of *Fukomys* [59], a typical trait among bathyergid mole-rats [60]. Low BMR is a pivotal adaptation of African mole-rats because in (sealed) underground burrows oxygen availability is low, risk of overheating is high, and food availability is scarce [59,60]. THs regulate metabolic pathways by which the balance between energy expenditure and storage is maintained. Excess THs increase energy expenditure by central and peripheral mechanisms. By the same token, low TH levels reduce energy expenditure [61]. Thus it appears reasonable that natural selection has favored low T4 in *Fukomys* to maintain low BMR. In contrast, as stated above, there is no reason to assume that a similar selective pressure was acting on color vision properties. Instead, the SWS-opsin majority found in *Fukomys* mole-rats is likely to be only a side effect of low T4 levels.

Because THs are pivotal for LWS-opsin expression and metabolic rate regulation, we have, manipulated TH levels in the Ansell's mole-rat using osmotic pumps [62]. We measured gene expression levels in the eye, BMR, and body mass in TH-treated animals. T4 treatment increased both, SWS- and LWS-opsin expression, albeit LWS-opsin expression at a higher degree. However, this plasticity was only given in animals up to approximately 2.5 years. Mass-specific BMR was not affected following T4 treatment, although body mass decreased. Furthermore, the T4 inactivation rate is naturally higher in *Fukomys* compared to laboratory rodents. This is the first experimental evidence that the SWS-opsin majority in Ansell's mole-rats is a side effect of low T4, which is downregulated to keep BMR low. The T4 level is naturally low, likely as an adaptation to the subterranean ecotope, by keeping BMR low.

This scenario is nice yet also not fully satisfactory. *Spalax* has a low BMR and low T4 levels, yet, in contrast to *Fukomys*, there is no a hypothyroidism-like effect upon SWS-opsin expression in the retina of *Spalax* [63]. The large differences in cone opsin arrangements between *Fukomys* and *Spalax* argue for species-specific adaptations to different demands and against convergent adaptation to the common subterranean darkness. Perhaps a complete absence of light does not exert any selective pressure, because large increases in sensitivity would not enable vision [28]. Yet, these differences speak also against a nonadaptive byproduct exerted by THs being a common explanation.

Ear and hearing: degeneration or adaptation?

Ecology: what is the acoustic environment of subterranean rodents like?

Contrary to aboveground, high-frequency sounds (characterized by short wavelengths) that can be localized well by small mammals, and which are employed in echolocation, are quickly dampened underground. In burrows of both *Spalax* and *Fukomys*, airborne sounds of low frequencies (200−800 Hz) consistently showed the lowest attenuation, that is, they propagated better than tested signals of lower and higher frequencies despite different tunnel diameters (7 and 5−6 cm, respectively), soil types, and habitats [64,65]. Moreover, in tunnels there is a tendency for amplification of sound amplitude or accentuation in the given low-frequency range over short distances (up to four times over 1 m) [65].

Low frequencies are characterized by long waves (170–42.5 cm for frequencies 200–800 Hz) and, as such, cannot be localized easily by small animals. Sound in tunnels can reach rodent occupants only from in front or from behind. Accordingly, we can assume that directional hearing is obviously not necessary. Thus far, studies on sound localization in subterranean mammals have been conducted on *Geomys*, *Spalax*, and *Heterocephalus* [66–68]. It has been shown that the animals need noise stimuli of relatively long duration (> 400 ms) to separate 180 degrees (90 degrees to their left and right), and that the threshold of separating sound sources is much larger in the subterranean rodents examined compared to their aboveground living counterparts.

Morphology: what does it look like?

The pinna of the outer ear is missing in both *Spalax* and *Fukomys*. Although general text-book wisdom considers missing auricles to be an adaptation to ease burrowing and preventing dirt shoveling, this explanation cannot withstand comparison with the many fossorial mammals with pinnae which apparently do not interfere with burrowing. As argued earlier [2], pinnae are of no use for locating sounds (their main function) of long waves in two-directional tunnels. Consequently, their reduction or absence in subterranean mammals should be considered to be primarily a regression due to lack of use and represents only a secondary mechanistic adaptation to digging. The outer ear canals in *Spalax* and *Fukomys* are of relatively small lumen and filled with cerumen (pers. obs.). These features may have a protective function in that they prevent dirt collecting during burrowing. However, at the same time, they reduce hearing sensitivity. Taking into account that sounds of most audible frequencies may be enhanced in tunnels (see above), we may consider the features also to be sensory compensatory adaptations.

The middle ears of *Fukomys* [69–71] and *Spalax* [69,72,73] are characterized by many common features: for example, a rather round and relatively large eardrum without a *pars flaccida*, reduced gonial, freely swinging ossicles (i.e., the malleus not firmly connected to the tympanic bone), reduced and straightened transversal part of the malleus, enlarged incus, increased and rather flat incudo-mallear joint, rather parallel position of the mallear manubrium and long incudal process which are fused to form an incudo-mallear complex. The stapedial footplate area is relatively large. Freely swinging ossicles and incus parallel to malleus occur also in the guinea pig [74] but not in the Norway rat [69]. Taking into account comparative data and their functional interpretation (Ref. [75], and references cited above), the middle ears in *Spalax*, *Fukomys*, and the guinea pig can be denoted as low-frequency tuned apparatus. In relation to the skull size and compared to the rat, the eardrum is slightly enlarged and the stapedial footplate is relatively large, resulting in low area ratio. Both, the malleus and incus are relatively large; the enlargement of incus is particularly conspicuous, so the lever ratio is low. The resulting transformer ratio, which is a product of the area ratio and the lever ratio and expresses efficiency of sound transmission (pressure and force amplification, respectively), is markedly low in mole rats compared to their epigeic counterparts. These characteristics might be associated with poor sensitivity to airborne sound.

The cochlea and the cochlear duct of the inner ear in *Spalax* [72,76,77] and *F. anselli* [78–80] show some similarities to each other and also to those of the guinea pig ([81], pers. obs.), yet they exhibit differences to the Norway rat [82,83]. Cochleae in mole-rats and the guinea pig are higher, more coiled, and, compared to skull size, exhibit longer cochlear ducts (basilar membranes) than in the rat. Since hair cell densities in the studied species are similar, the much longer basilar membrane accommodates more hair cells. More hair cells enable better frequency and/or intensity discrimination. The number of hair cells per octave is, on average, higher in mole-rats and the guinea pig than in the rat, suggesting that mean discrimination capacity in these species is also better. It should be noted that, in cochlear regions where the best audible frequencies are received, hair cell density is higher in all the considered species (see references above). Furthermore, in *F. anselli* a clear overrepresentation of the frequencies between 0.6 and 1 kHz (denoted as "acoustic fovea") was revealed: in this frequency range the slope of the place-frequency map amounted to 5.3 mm/octave [79], which actually means that in this species 3000 hair cells are dedicated to process one octave.

As in other mammals, the width of the basilar membrane in mole-rats increases and its thickness decreases from the cochlear base toward the cochlear apex. However, remarkably, there is no, or little, change in the basilar membrane width and thickness between 40% and 85% of the length of the cochlear spiral in *F. anselli* [79], further evidence of cochlear specialization leading to extension of a region devoted to processing certain frequencies and thus resulting in better frequency resolution. It is also noteworthy that many aspects of the cochlear morphology, for example, the low scalae (especially scala tympani), thin spiral ligament, less regular geometric pattern of the reticular lamina of the Corti organ, higher Corti organ and immature outer hair cells, characterizing in other mammals only the most apical, that is, low frequency, regions, are displayed in mole-rats throughout most of the cochlear spiral.

Physiology: what is its capacity?

Hearing characteristics of *Spalax* were studied neurophysiologically [76,84] and behaviorally [67]. For *Fukomys*, neurophysiological [85,86], cochlear [87], and behavioral [88] audiograms are available. These studies show that the hearing of *Spalax* and *Fukomys* is most sensitive in the lower frequency range (0.8–1 kHz). This is quite unusual among small mammals because optimal hearing in related surface dwellers of a comparable body size is usually much higher: about 8–16 kHz or higher. The hearing ranges of *Spalax* and *Fukomys* are very narrow; a high-frequency limit covering the octave of 8–16 kHz, and frequencies of about 16 kHz and higher cannot be perceived. Moreover, hearing thresholds in the mole-rats studied are higher (i.e., hearing sensitivity is lower) than in their epigeic counterparts (reviewed in Ref. [89]). It should be mentioned at this point that behavioral audiograms of *Spalax* and *Fukomys* reveal similar characteristics to audiograms of all other subterranean or fossorial rodents studied thus far: *Geomys bursarius* [66], *Heterocephalus glaber* [68], *Spalacopus cyanus* [90], and *Heliophobius argenteocinereus* [91].

Evolution: what is it for?

It is widely accepted that transmission, frequency tuning and amplification of sound by middle and inner ears follow basic mechanical principles, and that these principles are reflected in parameters of the ear structures. By examining and measuring the ear structures and comparing them with available biomechanical models, we can assess tuning and sensitivity of the ear cf. [92]. As von Békésy [93] stated: "Assuming that during the evolution, the most efficient system always survived. We expect that the physical laws served as guidelines to the evolution of the structures and functions of the middle and inner ear." In other words, we expect that any convergence in hearing parameters will be reflected in (and will be due to) convergence of ear morphologies. On the other hand, any degeneration due to lack of use should lead, in such a morphologically and developmentally complex apparatus as the ear, to stochastic regressions of different components, and the pattern of these regressive changes would probably be different in different phylogenetic lineages.

The similar ossicular morphologies of the mole-rats and guinea pig studied suggest hearing tuned to low frequencies. However, the middle ear of the mole-rats does not reveal any peculiarities, compared to that of the guinea pig, which would explain restriction of hearing range at the high-frequency end. From quantitative and qualitative points of view, none of the components of the middle ear transmitting chain is reduced or degenerated. On the contrary, some parts, particularly the incus and the stapedial footplate, are markedly enlarged, leading to reduced overall sensitivity. This hearing feature can, therefore, be considered an adaptation with a clearly defined morphological substrate. In fact, it has been demonstrated (in the few species for which such comparisons are possible) that the middle ear is, by itself, not responsible for limiting high-frequency hearing. Therefore the tonotopic organization of the cochlea should play a crucial role in setting the frequency limits of cochlear sensitivity and hence in determining the bandwidth of hearing [94]. The cochlea in all its aspects is specialized for sensitive perception and particularly for high resolution of low frequencies.

The frequency parameters of hearing correspond to those characterizing tunnel acoustics. Since hearing is not responsive to higher frequencies (and these are of no biological significance), some scholars explain the restricted hearing of subterranean rodents as being degenerated or vestigial due to lack of use [66–68]. However, the consistent shift of the best hearing range to match the frequencies best propagated in tunnels, and the conservation, or even improvement, of hearing sensitivity in low-frequency range, demonstrate that, at least in the lower frequency range (<16 kHz), hearing has not been degenerated. In fact, compared to the laboratory rat, the hearing range in mole-rats is not restricted—it covers about eight octaves—but is only shifted toward lower frequencies. Hearing range is, however, expanded in the guinea pig.

The question, why (in ultimate terms) sensitivity in best hearing range is much lower than that in other rodents, can be explained by the existence of the "stethoscope effect" proposed by Quilliam [95], and later demonstrated [65]. In tunnels, sound in low-frequency range may be significantly amplified locally so that we assume that, in the course of evolution, subterranean mammals had to reduce their hearing sensitivity in order to compensate for the increase in sound intensity in the environmental channel and thus avoid overstimulation. Hearing in subterranean rodents was studied, however, in laboratory in epigeic open field conditions. We may expect that their best hearing sensitivity in their natural environment, in burrows, is comparable to that of other rodents aboveground. Hence, both frequency tuning and low sensitivity of hearing can be considered convergent adaptations to unique acoustic environment.

Magnetoreception underground: new possibilities for an old sense!

Magnetoreception, the ability to sense the Earth's magnetic field, has been demonstrated for a wide array of animal species including mammals [96–98]. Contrary to general trends in biology, studies on mammals are still underrepresented in this field of research. Among vertebrates, not only birds, but also teleosts, sea turtles and amphibians have attracted more attention from researchers studying magnetoreception than mammals have, which is why birds became researchers' favorite subject, having historical and methodological roots. Homing and navigation abilities of pigeons and migratory bird species have fascinated people for centuries, and research on orientation and navigation of birds was established well before magnetoreception was established as a sensory capability. Although long-distance migrations are known to occur also in mammals (whales, bison, caribou, East African gnus, zebras, etc.) and the homing abilities of dogs, cats, horses and bears have been well (albeit only anecdotally) documented, experimental paradigms, such as use of the Emlen funnel to study the migratory orientation of birds, were not suitable for mammals. Displacement experiments with large mammals and homing experiments with cats, dogs, and horses are, for ethical and technical reasons, also not straightforward. It was only very relatively recently that new methods and insights have emerged that enable researchers to study the distribution and nature of the magnetic sense in mammals on a broader, more ecologically realistic, scale.

Thirty years ago, the study on magnetoreception in subterranean rodents just started and so, in our review article [2] we dedicated this sense only two sentences: "Magnetic-compass orientation in *Cryptomys* was suggested only recently [99]. In the same year, we have published the first convincing experimental evidence for that sense in mole-rats, and indeed in mammals in general [100].” In fact, African mole-rats have become the prime model to study mammalian magnetoreception. They spend their entire life living in total darkness, within complex and up to hundreds to thousands of meters long underground maze of tunnels [101–105]. Currently, they are the only mammals in which a magnetic sense is a widely accepted established fact and for which a reproducible behavioral assay is available. Nevertheless, even after being 30 years on the stage, the chapter devoted to this sense will still be rather fragmentary and will not follow the above-used scheme of chapter divisions.

How do we know?

Digging straight burrows

Subterranean mammals are able to dig long straight tunnels. These tunnels are built in the course of many days or weeks, and in social species, by multiple individuals. How do subterranean mammals manage to keep the course of digging? This question was first raised approximately 70 years ago in 1952 when Eloff mentioned the “remarkable ability (of the South-African *Cryptomys* mole-rats) to follow a direction or find a spot where it previously bored a tunnel” [106]. The African mole-rat’s ability to maintain its course while digging long, straight tunnels [106] gave rise to speculations on possible orientation cues: air currents [107–109] and acoustic cues [85,106,110] have been discussed, as well as internal mechanisms based on kinesthetic and/or vestibular cues [111]. However, none of these mechanisms provide a satisfactory explanation for the highly efficient directional orientation observed in *C. hottentotus* [106].

Inspired by the mentioned enigmatic ability, about 40 years after Eloff, Burda [99] postulated that the mole-rats might use magnetic cues to orientate and navigate. This ability was subsequently explicitly proven in laboratory (see below paragraph nest-building assay, [100]). Following that finding, Lovegrove et al. [112] investigated directional orientation of burrow systems of Damaraland mole-rats (*Fukomys damarensis*) in the field, inferring that the “orientation of the burrow system with respect to a specific compass orientation may represent an intrinsic requirement of successful geomagnetic orientation and direction finding” [112] p. 631. However, no correlation between the Earth’s magnetic field and the burrow system’s orientation was found. Since a much simpler assay for testing magnetosensitivity in rodents was developed and used in the meantime (see below), the question of straight burrows has been forgotten for the next few decades.

Only recently, we opened the question again and analyzed directional orientation of 68 burrow systems in five subterranean rodent species (*F. anselli, Fukomys mechowii, H. argenteocinereus, Spalax galili*, and *Ctenomys talarum*) on the basis of detailed maps of burrow systems [113]. The analysis revealed that the average tunnel orientation in the vast majority of burrow systems showed a significant deviation from a random distribution. In the burrows systems of all the studied species but *S. galili*, roughly N-S orientation prevailed. Burrows of *S. galili* were randomly oriented, a fact, which might be related to high population density of mole rats in the studied area.

Nest-building preferences in rodents

When released into a circular arena and given access to evenly distributed nesting material and food, *Fukomys* mole-rats will construct a nest, in most cases near or against the arena wall. Burda et al. [100] recognized that the nests were not randomly distributed, but rather concentrated in the south and south-eastern part of the arena. Changing the horizontal direction of the magnetic field led to a corresponding shift in nest position.

Magnetic resting direction preferences have further demonstrated a magnetic sense in other African mole-rats, the silvery mole-rat (*Heliophobius argenteus*) and the giant mole-rat (*F. mechowii*) [114]. *Spalax* mole rats also showed a nest-building preference relative to the magnetic field [115,116]. Spontaneous directional preferences for nest building were also found in epigeic rodents such as bank voles, *Myodes glareolus* [117] and wood mice, *Apodemus sylvaticus* [118]. The preferred nest-building preferences were not consistent across species and in some cases not even within the same species tested in different labs. Similar nest-building studies with epigeic rodents have provided evidence for a learned magnetic-compass orientation. Hamsters (*Phodopus sungorus, Phodopus roborowskii*) and C57BL/6J laboratory mice build nests against the wall of a circular arena in a direction relative to the magnetic field corresponding to the dark end of the home "training" cage in which the animals were kept prior to testing [119–123].

In summary, nest-building assays proved to be robust, eliciting consistent responses of various rodent species to the magnetic field [116,118–121] that were not observed using other assays available at the time, for example, [124,125]. Later development of other types of rodent assays confirmed sensitivity to magnetic cues, for example, [126,127], but the nest-building assay has proven most useful in characterizing the biophysical mechanisms underlying rodent magnetoreception (see below), yielding basic insights into neuronal mechanisms [128,129] and extending our knowledge of the taxonomic distribution of magnetosensitivity.

Magnetic novel object assay

Many animals respond to unfamiliar objects in their surroundings with explorative interest or fearful avoidance. Similarly, they may react positively or negatively to new sensory stimuli. While failure to respond to new stimuli or objects does not necessarily mean that the animal does not detect them, a clear reaction provides evidence of detection and/or perception. This general principle has become a basis for the magnetic novel object assay. Kremers et al. [130] showed that captive bottlenose dolphins (*Tursiops truncatus*) approached a strong magnet with shorter latency compared to a control consisting of a demagnetized block that was identical in form and density and therefore indistinguishable with echolocation. Malewski et al. [127] applied this assay as an initial screening method for the existence of magnetoreception and demonstrated consistently longer exploration of the magnet in three different rodent species: Ansell's mole-rat (*F. anselli*), C57BL/6J laboratory mouse, and naked mole-rat (*H. glaber*). Still, this approach provides only an initial screen and indicates possible existence of magnetoreception since inductive nonspecific effects of the strong magnetic gradients associated with permanent magnets on non-magnetosensory tissue cannot be excluded cf. [131,132].

Orientation in a maze

The question, how subterranean rodents can orientate in their complex underground maze, has been repeatedly addressed [99,106,116,133] but only seldom explicitly studied. The challenge of maze studies in at least *Fukomys* is (own obs.) that the animals learn too quickly the maze problem, so that in subsequent runs errors cannot be distinguished from "deliberate" exploration trips, and, especially, that mazes have to be large and these cannot be accommodated in available magnetic coils, in order to manipulate the surrounding magnetic field. Nevertheless, Kimchi and Terkel [116] showed that *Spalax* uses the magnetic sense to accomplish conditioned orientation tasks, enabling the mole rats to return more rapidly and efficiently (using shortcuts) to the starting point in a maze [116,133].

Morphology and physiology: what it looks like and how does it function?

In fact, the magnetic sense is the least understood major sensory system in vertebrates. While the existence of a magnetic sense can no longer be disputed, its seat, transduction mechanisms, neuronal pathways of vertebrate magnetic sense are still unknown. Receptors for the detection of magnetic fields have not yet been conclusively demonstrated in any animal reviewed in [134]. However, findings from behavioral, histological, neuroanatomical, and electrophysiological studies have led to several physically viable theoretical models that might also apply to terrestrial mammals, of which *two mechanisms* are most widely discussed in the literature as putative magnetoreceptors in terrestric animals: the magnetite-based mechanism and the radical-pair mechanism reviewed in Refs. [134–136]. It is important to note that since the magnetic field penetrates bodies, there is no need for a peripheral "antenna-like" sensory receiver. Magnetic sensors may be located anywhere in the body, they do not need to be concentrated in (paired) organs, they can be very tiny, and they do not need to function as their technical counterparts (i.e., a technical compass), just as the nucleus suprachiasmaticus that regulates biological rhythms does not resemble a clock.

It should be noted that the existence of one receptor mechanism does not rule out the involvement of others. In amphibians [137–139], and in birds [140,141], the available evidence indicates that there are both magnetite-based receptors mediated by particles of biogenic magnetite, and photopigment-based receptors involving the radical-pair mechanism. The two mechanisms are thought to be used in different tasks, providing directional (compass) or positional (map) information, respectively [141].

Perhaps the most intuitively appealing mechanism to explain magnetosensitivity in animals is the idea of a small permanent magnet inside the animal that acts like a compass needle [142]. Therefore it is not surprising that after the initial discovery of biogenic magnetic material in the teeth of chitons [143] and the subsequent demonstration of magnetite (Fe_3O_4) chains and their crucial role in the magnetotactic "behavior" of certain bacteria [144,145], investigators began to search for biogenic material in vertebrates. Within 20 years, ferrimagnetic particles were demonstrated in a number of different tissues of a variety of mammals, that is, rodents [146], bats [147,148], marine mammals [149,150], and even humans [151]. It has been suggested that a magnetic sense based on magnetite emerged early in the evolution of animals, and is likely to have been maintained by natural selection in all animal phyla [152].

Mole-rats [153,154] and bats [155], both being microphthalmic mammals living in dark habitats, have been shown to respond to the polarity of the magnetic field, in the absence of light. Moreover, they are affected by brief, high intensity magnetic pulses, but not by weak radiofrequency fields, which is consistent with the involvement of a magnetite-based magnetoreceptor. Neuronal activation studies in mole-rats revealed magnetic field-dependent activity in a layer of the superior colliculus that primarily receives trigeminal input [129,156]. Iron-positive (stained by Prussian blue) particles have been found in the corneal epithelium of mole-rats (*F. anselli*), although it remains to be established whether or not these particles are biogenic magnetite [157]. The corneal epithelium is innervated by the ophthalmic branch of the trigeminal nerve that no longer functions in vision. This nerve has also been suggested to innervate magnetite magnetoreceptors in fish [158] and birds [159–162]. The highly mechano-sensitive innervation of the mammalian cornea makes it a good candidate for the location of sensors that translate the torque exerted by the magnetic field on single-domain (SD) magnetite particles into a change in membrane potential. Further support for the presence of magnetite-based magnetoreceptors in the cornea of mole-rats is discussed below.

When mole-rats were treated with a strong magnetic pulse prior to testing in the nest-building assay, they changed the direction of their nest-building preference by 90-degree relative to control groups, an effect that lasted at least for several weeks [153]. A long-lasting effect of pulse treatment is in accordance with a receptor involving SD particles of magnetite—but since no recovery from the effect was reported in the mole-rat experiment, unspecific effects on the receptors could not be ruled out. However, taken together with the finding that mole-rats respond to the polarity of the magnetic field rather than to its inclination [153,163] the involvement of SD magnetite is likely.

Bats also possess a polarity-sensitive magnetic compass that responds to pulse remagnetization [164]. Thus reliance on a nonlight-dependent magnetite compass (along with the generalized reduction of the visual system [165]) could be an adaptation to aphotic habitats, in contrast to the light-dependent magnetic compasses of the amphibians and birds studied to date cf. [166]. The remarkable finding that magnetic-compass responses in mole-rats and birds are mediated by fundamentally different biophysical mechanisms has been confirmed by experiments showing that low-level radio frequency fields that should interfere with a light-dependent chemical compass, but not a magnetite-based compass, disrupt responses to magnetic cues in migratory birds, but not in mole-rats [154,167].

Ecology and evolution: what is it for?

Magnetoreception has been shown to provide both directional information (i.e., magnetic-compass sense) [168], as well as positional information (magnetic map sense) [169]. A magnetic-compass sense in combination with a reference map would be useful for true navigation in long-distance migrants, but also for spatial orientation in landscapes without distinct landmarks, and in assembling local landmark arrays into register to form a larger map of an individual's familiar space. Indeed, most evidence for the use of a magnetic sense in mammals has been so far collected (and searched for) in the context of local spatial behavior [97,170].

One possibility is that magnetic alignment helps to put the animal into register with a known orientation of a mental (cognitive) map, reducing the complexity of local and long-distance navigation, and reduces the demands on spatial memory [171]. These would be analogous strategies used in human orientation; it is much simpler and intuitive to navigate when the navigators align themselves with a physical map (i.e., the users rotate their body direction to coincide with the alignment of the physical map), rather than to navigate by mentally rotating the physical map to align with the user's orientation. Therefore we suggest that the mental map in animals is fixed in alignment with respect to the geomagnetic field indeed important component(s) of the cognitive map may be derived from the magnetic field (see below), and spontaneous magnetic alignment behavior puts the animal into register with this map. This relatively simple alignment strategy would help animals to reliably and accurately "read" their cognitive map and/or extend the range of their maps when exploring unfamiliar environments [170,172]. It can be assumed that this is the oldest, ubiqitous usage of magnetic sense in animals, which later was utilized in long-distance navigation.

Spatial cells (head direction, grid, and place cells) form the basis for quantitative spatiotemporal representation of places, routes, and associated experiences during behavior and in memory, that is, the "cognitive map" reviewed in i.a. [173]. In rats and mice the firing fields of the spatial cells are dependent on the visual cues reviewed in i.a. [174]. This is, however, not an option for subterranean rodents. In the extensive maze of their underground burrow systems, where there are no visual landmarks, magnetic field may provide reference to spatial cells.

It is unknown why the animals spontaneously prefer to build their nests in an unknown circular arena in a specific magnetic direction, and therefore difficult to explain whether the differences are due to, for example, the various testing locations, tested populations, or holding conditions. Nevertheless, in light of the above arguments regarding cognitive mapping, it is possible that the nest under given conditions serves as a spatial reference for (a landmark) for the map.

A magnetic-compass sense alone (i.e., without a map) could also be very helpful in a variety of contexts. It could be used, for instance as a heading/direction indicator, to keep the course of digging in subterranean mammals, maintain the course of swimming in aquatic mammals and, more generally, provide a compass reference that increases the range and accuracy of path integration systems [175,176].

Keeping a digging direction, for example, during foraging [99] or during natal or mating dispersal [8,31,177] would be advantageous, as digging curvy tunnels would imply increased or even devastating energetic costs; note that costs for digging are between 360 and up to 3400 times higher than if the animal moves the same distance aboveground [178]. We may assume that straight tunnels longer than, for example, one m are not the product of a single digging bout, and in the case of social species, not even of a single individual [179]. This fact points out the necessity of a heading indicator to keep the course of digging. While diverse heading indicators (visual or olfactory landmarks, sun position, wind direction) can be used to keep the course of locomotion aboveground, availability of such cues is restricted underground.

Knowledge will be unearthed from this neuroethological gold mine and can be used to shed light on a fundamental problem in biology: orientation in space and time in the dark.

The magnetic orientation may play a role also during sporadic aboveground activity of subterranean mammals—for example, during foraging or searching for partner and subsequent homing. Magnetic orientation may be useful also for connecting (damaged) tunnels of a burrow system or bypassing obstacles. Magnetoreception may be used in synergy with the kinesthetic sense to provide a compass reference that increases the range and accuracy of path integration strategies [133,175,176].

Magnetoreception could theoretically also be utilized in chronobiology. The natural daily variation of the magnetic field follows a regular temporal pattern and might provide a *zeitgeber* to animals like mole-rats that live in a monotonous, stable, uniform sensory environment deprived of light cues (i.e., day—night variation in light intensity) that help to time daily activities, and synchronize the behavior of group members in social species.

Existence of magnetic sense per se cannot be considered an adaptation to life underground—since it was most probably a property already of epigeic ancestors of subterranean mammals. However, we can assume that of the two putative magnetoreceptor systems (magnetite and radical-pair mechanism), only the magnetite-based magnetoreception was preserved in the darkness of underground burrows, while the light-dependent radical-pair mechanism has emancipated from the selection pressure and degraded.

The studies of magnetoreception in subterranean mammals have thus a large heuristic potential to unearth one of the fundamental problems in biomedical sciences magnetoreception and influence electromagnetic fields upon physiology.

Declarations

Competing interests

None

Funding

I acknowledge support by the grant "EVA4.0," No. CZ.02.1.01/0.0/0.0/16 019/0000803 financed by OP RDE.

Acknowledgments

I dedicate this chapter in gratitude and esteem to my mentor of more than 30 years, Eviatar (Eibi) Nevo. I am very much grateful to Solomon Wasser for inviting me to take part at the Symposium celebrating Eibi's jubilee and achievements, and to contribute to this volume. I thank my friends and (former) colleagues who shared with me their fascination for subterranean rodents and their senses and contributions to the findings reported here as well as the many inspiring discussions that helped me gain insight into the fascinating underground sensory environment. They are, among others (in alphabetical order): Sabine Begall, Simone Lange, Sandra Malewski, Erich Pascal Malkemper, Stephan Marhold, Regina Moritz, Marcus Müller, Pavel Němec, Helmut Oelschläger, Leo Peichl, Radim Šumbera, Roswitha and Wolfgang Wiltschko. I would not have started and would not have been able to study mole-rats if it were not for Volkmar Bruns, Mathias

Kawalika, Wolfgang Poduschka, and Jürgen Winckler. My largest thanks are due to my wife Jana for her understanding, help, and support—she has accompanied me in the field as well as in the lab and she too has indeed adopted the mole-rats.

References

[1] E. Nevo, O.A. Reig (Eds.), Evolution of Subterranean Mammals at the Organismal and Molecular Levels, Alan R. Liss, New York, 1990.

[2] H. Burda, V. Bruns, M. Müller, Sensory adaptations in subterranean mammals, in: E. Nevo, O.A. Reig (Eds.), Evolution of Subterranean Mammals at the Organismal and Molecular Levels, Alan R Liss, New York, 1990, pp. 269–293.

[3] D.B. Dusenbery, Sensory Ecology: How Organisms Acquire and Respond to Information, W H Freeman, New York, 1992.

[4] P. Dammann, A. Scherag, N. Zak, K. Szafranski, S. Holtze, S. Begall, et al., Comment on 'Naked mole-rat mortality rates defy Gompertzian laws by not increasing with age', eLife 8 (2019) e45415.

[5] M.E. Pamenter, Y. Dzal, W.A. Thompson, W.K. Milsom, Do naked mole rats accumulate a metabolic acidosis or an oxygen debt in severe hypoxia? J. Exp. Biol. 222 (2019) jeb191197.

[6] H. Schmidt, A. Bicker, G. Poetzsch, A. Avivi, I. Shams, T. Hankeln, Hypoxia tolerance, longevity and cancer-resistance in the mole rat *Spalax* – a liver transcriptomics approach, Sci. Rep. 7 (2017) 14348.

[7] K.R. Taylor, Ń.A. Milone, C.E. Rodriguez, Four cases of spontaneous neoplasia in the naked mole-rat (*Heterocephalus glaber*), a putative cancer-resistant species, J. Gerontol. A Biol. Sci. Med. Sci 72 (2017) 38–43.

[8] E. Nevo, Mosaic Evolution of Subterranean Mammals: Regression, Progression and Global Convergence, Oxford University Press, Oxford, 1999.

[9] H. Burda, R. Šumbera, S. Begall, Microclimate in burrows of subterranean rodents—revisited, in: S. Begall, H. Burda, C. Schleich (Eds.), Subterranean Rodents—News From Underground, Springer, Heidelberg, 2007, pp. 21–33.

[10] O. Kott, R.E. Moritz, R. Šumbera, H. Burda, P. Němec, Light propagation in burrows of subterranean rodents: tunnel system architecture but not photoreceptor sensitivity limits light sensation range, J. Zool. 294 (2014) 67–75.

[11] S. Sanyal, H.G. Jansen, W.J. DeGrip, E. Nevo, W.W. de Jong, The eye of the blind mole rat, *Spalax ehrenbergi*: rudiment with hidden function? Invest. Ophtalmol. Vis. Sci. 31 (1990) 1398–1404.

[12] H.M. Cooper, M. Herbin, E. Nevo, Ocular regression conceals adaptive progression of the visual system in a blind subterranean mammal, Nature 61 (1993) 156–159.

[13] H.M. Cooper, M. Herbin, E. Nevo, Visual system of a naturally microphthalmic mammal: the blind mole rat, *Spalax ehrenbergi.*, J. Comp. Neurol. 328 (1993) 313–350.

[14] P. Vuillez, M. Herbin, H.M. Cooper, E. Nevo, P. Pevet, Photic induction of Fos immunoreactivity in the suprachiasmatic nuclei of the blind mole rat (*Spalax ehrenbergi*), Brain Res. 654 (1994) 81–84.

[15] J. Negroni, E. Nevo, H.M. Cooper, Neuropeptidergic organization of the suprachiasmatic nucleus in the blind mole rat (*Spalax ehrenbergi*), Brain Res. Bull. 44 (1997) 633–639.

[16] I. Tobler, M. Herrmann, H.M. Cooper, J. Negroni, E. Nevo, P. Achermann, Rest-activity rhythm of the blind mole rat *Spalax ehrenbergi* under different lighting conditions, Behav. Brain Res. 96 (1998) 173–183.

[17] A. Avivi, H. Oster, A. Joel, A. Beiles, U. Albrecht, E. Nevo, Circadian genes in a blind subterranean mammal II: Conservation and uniqueness of the three period homologs in the blind subterranean mole rat, *Spalax ehrenbergi* superspecies, Proc. Natl. Acad. Sci. U.S.A. 99 (2002) 11718–11723.

[18] R. Cernuda-Cernuda, W.J. DeGrip, H.M. Cooper, E. Nevo, J.M. Garcia-Fernandez, The retina of *Spalax ehrenbergi*: novel histologic features supportive of a modified photosensory role, Invest. Ophthalmol. Vis. Sci. 43 (2002) 2374–2383.

[19] J. Hannibal, P. Hindersson, E. Nevo, J. Fahrenkrug, The circadian photopigment melanopsin is expressed in the blind subterranean mole rat, *Spalax*, NeuroReport 13 (2002) 1411–1414.

[20] Z.K. David-Gray, J.W.H. Janssen, W.J. DeGrip, E. Nevo, R.G. Foster, Light detection in a 'blind' mammal, Nat. Neurosci. 1 (1998) 655–656.

[21] Z.K. David-Gray, J. Bellingham, M. Munoz, A. Avivi, E. Nevo, R.G. Foster, Adaptive loss of ultraviolet-sensitive/violet-sensitive (UVS/VS) cone opsin in the blind mole rat (*Spalax ehrenbergi*), Eur. J. Neurosci. 16 (2002) 1186–1194.

[22] Z.K. David-Gray, H.M. Cooper, J.W.H. Jannsen, E. Nevo, R.G. Foster, Spectral tuning of a circadian photopigment in a subterranean "blind" mammals (*Spalax ehrenbergi*), FEBS Lett. 461 (1999) 343–347.

[23] J.W.H. Janssen, P.H.M. Bovee-Geurts, Z.P.A. Peeters, J.K. Bowmaker, H.M. Cooper, Z.K. David-Gray, et al., A fully functional rod visual pigment in a blind mammal: a case for adaptive functional reorganization? J. Biol. Chem. 275 (2000) 38674–38679.

[24] R. Cernuda-Cernuda, J.M. Garcia-Fernandez, M.C.M. Gordijn, P.H.M. Bovee-Geurts, W.J. DeGrip, The eye of the African mole-rat *Cryptomys anselli*: to see or not to see? Eur. J. Neurosci. 17 (2003) 709–720.

[25] L. Peichl, P. Němec, H. Burda, Unusual cone and rod properties in subterranean African mole-rats (Rodentia, Bathyergidae), Eur. J. Neurosci. 19 (2004) 1545–1558.

[26] H.A. Oelschläger, M. Nakamura, M. Herzog, H. Burda, Visual system labeled by c-Fos immunohistochemistry after light exposure in the "blind" subterranean Zambian mole-rat (*Cryptomys anselli*), Brain Behav. Evol. 55 (2000) 209–220.

[27] P. Němec, H. Burda, L. Peichl, Subcortical visual system of the African mole-rat *Cryptomys anselli*, Eur. J. Neurosci. 20 (2004) 757–768.

[28] L. Peichl, Diversity of mammalian photoreceptor properties: adaptations to habitat and lifestyle? Anat. Rec. A 287 (2005) 1001–1012.

[29] D.M. Hunt, L. Peichl, S cones: evolution, retinal distribution, development, and spectral sensitivity, Vis. Neurosci. 31 (2014) 115–138.

[30] P.K. Ahnelt, H. Kolb, The mammalian photoreceptor mosaic-adaptive design, Prog. Retin. Eye Res. 19 (2000) 711–777.

[31] R. Rado, G. Bronchti, Z. Wollberg, J. Terkel, Sensitivity to light of the blind mole rat: Behavioral and neuroanatomical study, Isr. J. Zool. 38 (1992) 323–331.

[32] M. Herbin, J.P. Rio, J. Reperant, H.M. Cooper, E. Nevo, M. Lemire, Ultrastructural study of the optic nerve in blind mole-rats (Spalacidae, *Spalax*), Vis. Neurosci. 12 (1994) 253–261.

[33] R.E. Wegner, S. Begall, H. Burda, Light perception in "blind" subterranean Zambian mole-rats: proof by spontaneous heliophobia or scotophilia, Anim. Behav. 71 (2006) 1021–1024.

[34] O. Kott, R. Šumbera, P. Němec, Light perception in two strictly subterranean rodents: life in the dark or blue? PLoS One 5 (2010) e11810.

[35] R. Rado, J. Terkel, Circadian activity of the blind mole rat, *Spalax ehrenbergi*, monitored by radiotelemetry, in seminatural and natural conditions, in: E. Spanier, Y. Steinberger, M. Luria (Eds.), Environmental Quality and Ecosystem Stability, vol. IV-B, ISEEQS, Jerusalem, 1989, pp. 391–400.

[36] R. Rado, H. Gev, B.D. Goldman, J. Terkel, Light and circadian activity in the blind mole-rat, in: E. Riklis (Ed.), Photobiology, Plenum Press, New York, 1991, pp. 581–589.
[37] J. Šklíba, M. Lövy, S.C.W. Koeppen, L. Pleštilová, M. Vítámvás, E. Nevo, et al., Activity of free-living subterranean blind mole rats *Spalax galili* (Rodentia: Spalacidae) in an area of supposed sympatric speciation, Biol. J. Linn. Soc. 118 (2016) 280–291.
[38] H. Burda, Reproductive biology (behaviour, breeding, and postnatal development) in subterranean mole-rats, *Cryptomys hottentotus* (Bathyergidae), Z. Säugetierk 54 (1989) 360–376.
[39] H. Burda, Constraints of pregnancy and evolution of sociality in mole-rats. With special reference to reproductive and social patterns in *Cryptomys hottentotus* (Bathyergidae, Rodentia), Z. Zool. Syst. Evolut-forsch 28 (1990) 26–39.
[40] P. Fritzsche, G. Miethe, R. Gattermann, Circadiane Motorik und Synchronisation durch unterschiedliche Zeitgeber beim subterranen Graumull *Cryptomys hottentotus* Lesson, 1826, Z. Säugetierk Suppl. (1998) 62.
[41] R. Gattermann, H. Burda, Soziale isolation und 24-h-Rhythmik beim eusozialen Graumull, Verh. Dtsch. Zool. Ges. 240 (1994) 39.
[42] J. Šklíba, M. Lövy, M. Hrouzková, O. Kott, J. Okrouhlík, R. Šumbera, Social and environmental influences on daily activity pattern in free-living subterranean rodents: the case of a eusocial bathyergid, J. Biol. Rhythms 29 (2014) 203–214.
[43] M.K. Oosthuizen, N.C. Bennett, The effect of ambient temperature on locomotor activity patterns in reproductive and non-reproductive female Damaraland mole-rats, J. Zool. 297 (2015) 1–8.
[44] M. Lövy, J. Šklíba, R. Šumbera, Spatial and temporal activity patterns of the free-living giant mole-rat (*Fukomys mechowii*), the largest social Bathyergid, PLoS One 8 (2013) e55357.
[45] H. Burda, M. Kawalika, Evolution of eusociality in the Bathyergidae: the case of the giant mole-rat (*Cryptomys mechowi*), Naturwiss 80 (1993) 235–237.
[46] J.D. Yeakel, N.C. Bennett, P.L. Koch, N.J. Domiony, The isotopic ecology of African mole rats informs hypotheses on the evolution of human diet, Proc. R. Soc. B 274 (2007) 1723–1730.
[47] G. Ossendorf, A.R. Groos, T. Bromm, M.G. Tekelemariam, B. Glaser, J. Lesur, et al., Middle Stone Age foragers resided in high elevations of the glaciated Bale Mountains, Ethiopia, Science 365 (2019) 583–587.
[48] P. Němec, P. Cveková, O. Benada, E. Wielkopolska, S. Olkowicz, K. Turlejski, et al., The visual system in subterranean African mole-rats (Rodentia, Bathyergidae): Retina, subcortical visual nuclei and primary visual cortex, Brain Res. Bull. 75 (2008) 356–364.
[49] O. Kott, P. Němec, A. Fremlová, V. Mazoch, R. Šumbera, Behavioural tests reveal severe visual deficits in the strictly subterranean African mole-rats (Bathyergidae) but efficient vision in the fossorial rodent coruro (*Spalacopus cyanus*, Octodontidae), Ethology 122 (2016) 1–13.
[50] L. Ng, J.B. Hurley, B. Dierks, M. Srinivas, C. Saltó, B. Vennström, et al., A thyroid hormone receptor that is required for the development of green cone photoreceptors, Nat. Genet. 27 (2001) 94–98.
[51] M.R. Roberts, M. Srinivas, D. Forrest, G. Morreale de Escobar, T.A. Reh, Making the gradient: thyroid hormone regulates cone opsin expression in the developing mouse retina, Proc. Natl. Acad. Sci. U.S.A. 103 (2006) 6218–6223.
[52] M.L. Applebury, F. Farhangfar, M. Glösmann, K. Hashimoto, K. Kage, J.T. Robbins, et al., Transient expression of thyroid hormone nuclear receptor TRbeta2 sets s opsin patterning during cone photoreceptor genesis, Dev. Dyn. 236 (2007) 1203–1212.

[53] A. Glaschke, J. Weiland, D. Del Turco, M. Steiner, L. Peichl, M. Glösmann, Thyroid hormone controls cone opsin expression in the retina of adult rodents, J. Neurosci. 31 (2011) 4844–4851.
[54] D. Forrest, A. Swaroop, Mini review: the role of nuclear receptors in photoreceptor differentiation and disease, Mol. Endocrinol. 26 (2012) 905–915.
[55] A. Glaschke, M. Glösmann, L. Peichl, Developmental changes of cone opsin expression but not retinal morphology in the hypothyroid *Pax8* knockout mouse, Invest. Ophthalmol. Vis. Sci. 51 (2010) 1719–1727.
[56] M.R. Roberts, A. Hendrickson, C.R. McGuire, T.A. Reh, Retinoid x receptor γ is necessary to establish the s-opsin gradient in cone photoreceptors of the developing mouse retina, Invest. Ophthalmol. Vis. Sci. 46 (2005) 2897–2904.
[57] M. Cakir, O.B. Turgut, E. Turan, G. Gonulalan, I. Polat, K. Gunduz, The effect of hypothyroidism on color contrast sensitivity: a prospective study, Eur. Thyroid J. 4 (2015) 43–47.
[58] Y. Henning, C. Vole, S. Begall, M. Bens, M. Broecker-Preuss, A. Sahm, et al., Unusual ratio between free thyroxine and free triiodothyronine in a long-lived mole-rat species with bimodal ageing, PLoS One 9 (2014) e113698.
[59] C.K.M. Schielke, H. Burda, Y. Henning, J. Okrouhlík, S. Begall, Higher resting metabolic rate in long-lived breeding Ansell's mole-rats (*Fukomys anselli*), Front. Zool. 14 (2017) 45.
[60] J. Zelová, R. Šumbera, F. Sedláček, H. Burda, Energetics in a solitary subterranean rodent, the silvery mole-rat, *Heliophobius argenteocinereus*, and allometry of RMR in African mole-rats (Bathyergidae), Comp. Biochem. Physiol. 14 (2007) 412–419.
[61] R. Mullur, Y.-Y. Liu, G.A. Brent, Thyroid hormone regulation of metabolism, Physiol. Rev. 94 (2014) 355–382.
[62] Y. Henning, N. Mladěnková, H. Burda, K. Szafranski, S. Begall, Retinal S-opsin dominance in Ansell's mole-rats (*Fukomys anselli*) is a consequence of naturally low serum thyroxine, Sci. Rep. 8 (2018) 4337.
[63] A. Avivi, E. Nevo, K. Cohen, N. Sotnichenko, A. Heckbergs, M. Band, et al., They live in the land down under: thyroid function and basal metabolic rate in the blind mole rat, Spalax, Endocr. Res. 39 (2014) 85.
[64] G. Heth, E. Frankenberg, E. Nevo, Adaptive optimal sound for vocal communication in tunnels of a subterranean mammal (*Spalax ehrenbergi*), Experientia 42 (1986) 1287–1289.
[65] S. Lange, R. Wegner, P. Dammann, M. Kawalika, S. Begall, H. Burda, Living in a "stethoscope" acoustics in underground tunnels promotes auditory adaptation in subterranean mole-rats (*Cryptomys* sp.), Naturwiss. 94 (2007) 134–138.
[66] R.S. Heffner, H.E. Heffner, Vestigial hearing in a fossorial mammal, the pocket gopher (*Geomys bursarius*), Hear. Res. 46 (1990) 239–252.
[67] R.S. Heffner, H.E. Heffner, Hearing and sound locatization in blind mole-rats (*Spalax ehrenbergi*), Hear. Res. 62 (1992) 206–216.
[68] R.S. Heffner, H.E. Heffner, Degenerate hearing and sound localization in naked mole rats (*Heterocephalus glaber*), with an overview of central auditory structures, J. Comp. Neurol. 331 (1993) 418–433.
[69] H. Burda, V. Bruns, G.C. Hickman, The ear in subterranean Insectivora and Rodentia in comparison with ground-dwelling representatives. I. Sound conducting system of the middle ear, J. Morphol. 214 (1992) 49–61.
[70] S. Lange, H. Burda, Comparative and functional morphology of the middle ear in Zambezian mole-rats (Coetomys *Cryptomys*, Bathyergidae), Belg. J. Zool. 135 (Suppl. 1) (2005) 5–10.
[71] S. Lange, H. Burda, N.C. Bennett, P. Němec, Middle ear ossicles as a taxonomic trait in African mole-rats (Rodentia: Bathyergidae), in: B.A. Huber, B.J. Sinclair,

K.H. Lampe (Eds.), African Biodiversity: Molecules, Organisms, Ecosystems, Springer, Berlin, 2005, pp. 329–337.

[72] H. Burda, V. Bruns, E. Nevo, Middle ear and cochlear receptors in the subterranean mole-rat, *Spalax ehrenbergi*, Hear. Res. 39 (1989) 225–230.

[73] R. Rado, M. Himelfarb, B. Arensburg, J. Terkel, Z. Wollberg, Are seismic communication signals transmitted by bone conduction in the blind mole rat? Hear. Res. 41 (1989) 23–30.

[74] M.J. Mason, Middle ear structures in fossorial mammals: a comparison with non-fossorial species, J. Zool. 255 (2001) 467–486.

[75] G. Fleischer, Studien am Skelett des Gehörorgans der Säugetiere, einschließlich des Menschen, Säugetierk Mitt. 131 (1973) 131–239.

[76] V. Bruns, M. Müller, W. Hofer, G. Heth, E. Nevo, Inner ear structure and electrophysiological audiograms of the subterranean mole rat, *Spalax ehrenbergi*, Hear. Res. 33 (1988) 1–10.

[77] Y. Raphael, M. Lenoir, R. Wroblewski, R. Pujol, The sensory epithelium and its innervation in the mole rat cochlea, J. Comp. Neurol. 314 (1991) 367–382.

[78] M. Müller, H. Burda, V. Bruns, Structure and function of the inner ear in subterranean mammals, in: J. Erber, R. Menzel, H.J. Pflüger, D. Todt (Eds.), Neural Mechanisms of Behavior, Stuttgart: G Thieme, 1989, p. 144.

[79] M. Müller, B. Laube, H. Burda, V. Bruns, Structure and function of the peripheral auditory system in the African mole rat (*Cryptomys hottentotus*): evidence for a low frequency acoustic fovea, J. Comp. Physiol. A 171 (1992) 469–476.

[80] S. Lange, Sinnesökologie afrikanischer Sandgräber (Bathyergidae) am Beispiel von Hör- und Geruchssinn (Ph.D. thesis), Univ Duisburg-Essen, Essen, 2006.

[81] H. Burda, Guinea pig cochlear hair cell density; its relation to frequency discrimination, Hear. Res. 14 (1984) 315–317.

[82] H. Burda, L. Voldrich, Correlation between the hair cell density and the auditory thresholds in the white rat, Hear. Res. 2 (1980) 91–93.

[83] H. Burda, L. Ballast, V. Bruns, Cochlea in the Old World mice and rats (Muridae), J. Morphol. 198 (1988) 269–287.

[84] G. Bronchti, P. Heil, H. Scheich, Z. Wollberg, Auditory pathway and auditory activation of primary visual targets in the blind mole rat (*Spalax ehrenbergi*): I. 2-Deoxyglucose study of subcortical centers, J. Comp. Neurol. A 284 (1989) 253–274.

[85] M. Müller, H. Burda, Restricted hearing range in a subterranean rodent, *Cryptomys hottentotus* (Bathyergidae), Naturwiss 76 (1989) 134–135.

[86] P. Gerhardt, Y. Henning, S. Begall, E.P. Malkemper, Audiograms of three subterranean rodent species (genus *Fukomys*) determined by auditory brainstem potentials reveal extremely poor high-frequency cut-offs, J. Exp. Biol. 220 (2018) 4377–4382.

[87] M. Kössl, G. Frank, H. Burda, M. Müller, Acoustic distortion products from the cochlea of the blind African mole rat, *Cryptomys* spec, J. Comp. Physiol. A 178 (1996) 427–434.

[88] G. Brückmann, H. Burda, Hearing in blind subterranean Zambian common mole-rats (*Cryptomys* sp., Bathyergidae, Rodentia), J. Comp. Physiol. A 181 (1997) 83–88.

[89] H. Burda, Ear and eye in subterranean mole-rats, *Fukomys anselli* (Bathyergidae) and *Spalax ehrenbergi* (Spalacidae): progressive specialization or regressive degeneration? Anim. Biol. 56 (2006) 475–486.

[90] S. Begall, H. Burda, B. Schneider, Hearing in coruros (*Spalacopus cyanus*): special audiogram features of a subterranean rodent, J. Comp. Physiol. A 190 (2004) 963–969.

[91] E. Knotková, R. Šumbera, F. Sedláček, Vocalization and behavioral hearing range in silvery mole-rat (*Heliophobius argenteocinereus*), in: Abstract of the Ninth International Mammalogical Congress, Sapporo, Japan, 2005, p. 222.

[92] E.M. Relkin, Introduction to the analysis of middle-ear function, in: A.F. Jahn, J. Santos-Sacchi (Eds.), Physiology of the Ear, Raven Press, New York, 1988, pp. 103–123.
[93] G. von Békésy, Introduction, in: W.D. Keidel, W.D. Neff (Eds.), Handbook of Sensory Physiology. Vol V/I, Auditory System, Anatomy, Physiology (Ear), Springer, Berlin, 1974, pp. 1–8.
[94] M.A. Ruggero, A.N. Temchin, The roles of the external, middle, and inner ears in determining the bandwidth of hearing, Proc. Natl. Acad. Sci. U.S.A. 99 (2002) 13206–13210.
[95] T.A. Quilliam, The mole's sensory apparatus, J. Zool. Lond. 149 (1966) 76–88.
[96] R. Wiltschko, W. Wiltschko, Magnetic Orientation in Animals, Springer, Heidelberg, 1995.
[97] S. Begall, H. Burda, E.P. Malkemper, Magnetoreception in mammals, Adv. Study Behav. 46 (2014) 45–88.
[98] H. Mouritsen, Long-distance navigation and magnetoreception in migratory animals, Nature 558 (2018) 50–59.
[99] H. Burda, Magnetische Navigation bei den Graumullen, *Cryptomys hottentotus* (Bathyergidae)? Z. Säugetierk 61 (1987) 12. Suppl.
[100] H. Burda, S. Marhold, T. Westenberger, R. Wiltschko, W. Wiltschko, Evidence for magnetic compass orientation in the subterranean rodent *Cryptomys hottentotus* (Bathyergidae), Experientia 46 (1990) 528–530.
[101] R.A. Brett, The ecology of naked mole-rat colonies: burrowing, food, and limiting factors, in: P.W. Sherman, J.U.M. Jarvis, R.D. Alexander (Eds.), The Biology of the Naked Mole-Rat, Princeton University Press, Princeton, NJ, 1991, pp. 137–184.
[102] J. Šklíba, V. Mazoch, H. Patzenhauerová, E. Hrouzková, M. Lövy, O. Kott, et al., A maze-lover's dream: burrow architecture, natural history and habitat characteristics of Ansell's mole-rat (*Fukomys anselli*), Mammal. Biol. 77 (2012) 420–427.
[103] R. Šumbera, H. Burda, W.N. Chitaukali, J. Kubová, Silvery mole-rats (*Heliophobius argenteocinereus*, Bathyergidae) change their burrow architecture seasonally, Naturwiss 90 (2003) 370–373.
[104] R. Šumbera, J. Šklíba, M. Elichová, W.N. Chitaukali, H. Burda, Natural history and burrow system architecture of the silvery mole-rat (*Heliophobius argenteocinereus*) from Brachystegia woodland, J. Zool. Lond. 274 (2008) 77–84.
[105] R. Šumbera, V. Mazoch, H. Patzenhauerová, M. Lövy, J. Šklíba, J. Bryja, et al., Burrow architecture, family composition and habitat characteristics of the largest social African mole-rat: the giant mole-rat constructs really giant burrow systems, Acta Theriol. 57 (2012) 121–130.
[106] G. Eloff, Orientation in the mole-rat *Cryptomys*, Br. J. Psychol. 41 (1952) 134–145.
[107] G. De Graaff, On the mole-rat (*Cryptomys hottentotus damarensis*) (Rodentia) in the Kalahari Gemsbok National Park, Koedoe 15 (1972) 25–35.
[108] G. Eloff, The functional and structural degeneration of the eye of the South African rodent moles, *Cryptomys bigalkei* and *Bathyergus maritimus*, S Afr. J. Sci. 54 (1958) 293–302.
[109] W. Poduschka, A. der Mullratte, *Cryptomys hottentottus* (Lesson, 1826), Säugetierk Mitt. 26 (1978) 260–274.
[110] D.R. Rosevear, The Rodents of West Africa, British Museum of Natural History, London, 1969, p. 604.
[111] E. Teroni, V. Portenier, R. Maurer, A.S. Etienne, The control of spatial orientation by self-generated and visual cues in the golden hamster, Sci. Technol. Anim. Lab. 13 (1988) 99–102.

[112] B.G. Lovegrove, G. Körtner, G. Körtner, The magnetic compass orientation of the burrows of the Damara mole-rat *Cryptomys damarensis* (Bathyergidae), J. Zool. 226 (1992) 631–633.

[113] S. Malewski, S. Begall, C.E. Schleich, C.D. Antenucci, H. Burda, Do subterranean mammals use the Earth's magnetic field as a heading indicator to dig straight tunnels? PeerJ 6 (2018) 5819.

[114] L. Oliveriusová, P. Němec, Z. Králová, F. Sedláček, Magnetic compass orientation in two strictly subterranean rodents: learned or species-specific innate directional preference? J. Exp. Biol. 215 (2012) 3649–3654.

[115] S. Marhold, A. Beiles, H. Burda, E. Nevo, Spontaneous directional preference in a subterranean rodent, the blind mole-rat, *Spalax ehrenbergi*, Folia Zool. 49 (2000) 7–18.

[116] T. Kimchi, J. Terkel, Magnetic compass orientation in the blind mole rat *Spalax ehrenbergi*, J. Exp. Biol. 204 (2001) 751–758.

[117] L. Oliveriusová, P. Němec, Z. Pavelková, F. Sedláček, Spontaneous expression of magnetic compass orientation in an epigeic rodent: the bank vole, *Clethrionomys glareolus*, Naturwiss 101 (2014) 557–563.

[118] E.P. Malkemper, S.H.K. Eder, S. Begall, J.B. Phillips, M. Winklhofer, V. Hart, et al., Magnetoreception in the wood mouse (*Apodemus sylvaticus*): influence of weak frequency-modulated radio frequency fields, Sci. Rep. 5 (2015) 9917.

[119] M.E. Deutschlander, M.J. Freake, S.C. Borland, J.B. Phillips, R.C. Madden, L.E. Andersson, et al., Learned magnetic compass orientation by the Siberian hamster, *Phodopus sungorus*, Anim. Behav. 65 (2003) 779–786.

[120] R. Muheim, N.M. Edgar, K.A. Sloan, J.B. Phillips, Magnetic compass orientation in C57BL/6J mice, Learn. Behav. 34 (2006) 366–373.

[121] M.S. Painter, M. Davis, S. Ganesh, E. Rak, K. Brumet, H. Bayne, et al., Evidence for plasticity in magnetic nest-building orientation in laboratory mice, Anim. Behav. 138 (2018) 93e100.

[122] S. Malewski, S. Begall, H. Burda, Light-independent magnetoreception in the Djungarian hamster (*Phodopus sungorus*), Mammal. Biol. 91 (2018) 91–94.

[123] S. Malewski, S. Begall, H. Burda, Learned and spontaneous magnetosensitive behaviour in the Roborovski hamster (*Phodopus roborovskii*), Ethology 124 (2018) 12744.

[124] R.C. Madden, J.B. Phillips, An attempt to demonstrate a magnetic compass orientation in two species of mammals, Anim. Learn. Behav. 15 (1987) 130–134.

[125] C.E. Schleich, C.D. Antinuchi, Testing magnetic orientation in a solitary subterranean rodent *Ctenomys talarum* (Rodentia: Octodontidae), Ethology 110 (2004) 485–495.

[126] J.B. Phillips, P.W. Youmans, R. Muheim, K.A. Sloan, L. Landler, M.S. Painter, et al., Rapid learning of magnetic compass direction by C57BL/6 mice in a 4-armed 'plus' water maze, PLoS One 8 (2013) e73112.

[127] S. Malewski, H. Burda, F. Sedláček, R. Šumbera, K. Caspar, E.P. Malkemper, et al., Attracted by a magnet: exploration behaviour of rodents in the presence of magnetic objects, Behav. Proc. 151 (2018) 11–15.

[128] T. Burger, M. Lucová, R.E. Moritz, H.H.A. Oelschläger, R. Druga, H. Burda, et al., Changing and shielded magnetic fields suppress c-Fos expression in the navigation circuit: input from the magnetosensory system contributes to the internal representation of space in a subterranean rodent, J. R. Soc. Interface 7 (2010) 1275–1292.

[129] P. Němec, J. Altmann, S. Marhold, H. Burda, H.H.A. Oelschläger, Neuroanatomy of magnetoreception: the superior colliculus involved in magnetic orientation in a mammal, Science 294 (2001) 366–368.

[130] D. Kremers, Jl Marulanda, M. Hausberger, A. Lemasson, Behavioural evidence of magnetoreception in dolphins: detection of experimental magnetic fields, Naturwiss. 101 (2014) 907–911.
[131] V. Zablotskii, T. Polyakova, A. Dejneka, Cells in the non-uniform magnetic world: how cells respond to high-gradient magnetic fields, Bioessays 40 (2018) e1800017.
[132] V. Zablotskii, T. Polyakova, O. Lunov, A. Dejneka, How a high-gradient magnetic field could affect cell life, Sci. Rep. 6 (2016) 37407.
[133] T. Kimchi, A.S. Etienne, J. Terkel, A subterranean mammal uses the magnetic compass for path integration, Proc. Natl. Acad. Sci. U.S.A. 101 (2004) 1105–1109.
[134] G.C. Nordmann, T. Hochstoeger, D.A. Keays, Magnetoreception—a sense without a receptor, PLoS Biol. 15 (2017) e2003234.
[135] H. Mouritsen, Sensory biology: search for the compass needles, Nature 484 (2012) 320–321.
[136] S. Johnsen, Open questions: we don't really know anything, do we? Open questions in sensory biology, BMC Biol. 15 (2017) 43.
[137] J.B. Phillips, Magnetic compass orientation in the eastern red-spotted newt (*Notophthalmus viridescens*), J. Comp. Physiol. A 158 (1986) 103–109.
[138] J.B. Phillips, S.C. Borland, Behavioral evidence for the use of a light-dependent magnetoreception mechanism by a vertebrate, Nature 359 (1992) 142–144.
[139] J.B. Phillips, S.C. Borland, M.J. Freake, J. Brassart, J.L. Kirschvink, Fixed-axis' magnetic orientation by an amphibian: non-shoreward-directed compass orientation, misdirected homing or positioning a magnetite-based map detector in a consistent alignment relative to the magnetic field? J. Exp. Biol. 205 (2002) 3903–3914.
[140] U. Munro, J.A. Munro, J.B. Phillips, R. Wiltschko, W. Wiltschko, Evidence for a magnetite-based navigational map in birds, Naturwiss. 84 (1997) 26–28.
[141] R. Wiltschko, W. Wiltschko, The magnetite-based receptors in the beak of birds and their role in avian navigation, J. Comp. Physiol. A 199 (2012) 89–98.
[142] E.D. Yorke, A possible magnetic transducer in birds, J. Theor. Biol. 77 (1979) 101–105.
[143] H.A. Lowenstam, Magnetite in denticle capping in recent chitons (Polyplacophora), Geol. Soc. Am. Bull. 73 (1962) 435–438.
[144] R. Blakemore, Magnetotactic bacteria, Science 190 (1975) 377–379.
[145] R.B. Frankel, R.P. Blakemore, R.S. Wolfe, Magnetite in freshwater magnetotactic bacteria, Science 203 (1979) 1355–1356.
[146] J.G. Mather, Magnetoreception and the search for magnetic material in rodents, in: J.L. Kirschvink, D.S. Jones, B.J. MacFadden (Eds.), Magnetite Biomineralization and Magnetoreception in Organisms: A New Biomagnetism, Plenum Press, New York, 1996, pp. 509–533.
[147] E.R. Buchler, P.J. Wasilewski, Bats have magnets, EOS Trans. Am. Geophys. Union 63 (1982) 156.
[148] E.R. Buchler, P.J. Wasilewski, Magnetic remanence in bats, in: J.L. Kirschvink, D. S. Jones, B.J. MacFadden (Eds.), Magnetite Biomineralization and Magnetoreception in Organisms: A New Biomagnetism, Plenum Press, New York, 1985, pp. 483–487.
[149] J. Zoeger, J.R. Dunn, M. Fuller, Magnetic material in the head of the common Pacific dolphin, Science 213 (1981) 892–894.
[150] G.B. Bauer, M. Fuller, A. Perry, J.R. Dunn, J. Zoeger, Magnetoreception and biomineralization of magnetite in cetaceans, in: J.L. Kirschvink, D.S. Jones, B.J. MacFadden (Eds.), Magnetite Biomineralization and Magnetoreception in Organisms: A New Biomagnetism, Plenum Press, New York, 1985, pp. 489–507.
[151] J.L. Kirschvink, A. Kobayashi-Kirschvink, B.J. Woodford, Magnetite biomineralization in the human brain, Proc. Natl. Acad. Sci. U.S.A. 89 (1992) 7683–7687.

[152] J.L. Kirschvink, M.M. Walker, C.E. Diebel, Magnetite-based magnetoreception, Curr. Opin. Neurobiol. 11 (2001) 462–467.
[153] S. Marhold, H. Burda, I. Kreilos, W. Wiltschko, Magnetic Orientation in Common Mole-Rats From Zambia, Royal Institute of Navigation, London, 1997.
[154] P. Thalau, T. Ritz, H. Burda, R.E. Wegner, R. Wiltschko, The magnetic compass mechanisms of birds and rodents are based on different physical principles, J. R. Soc. Interface 3 (2006) 583–587.
[155] Y. Wang, Y. Pan, S. Parsons, M. Walker, S. Zhang, Bats respond to polarity of a magnetic field, Proc. R. Soc. B 274 (2007) 2901–2905.
[156] P.J. May, The mammalian superior colliculus: laminar structure and connections, Progr. Brain Res. 151 (2006) 321–378.
[157] R.E. Wegner, S. Begall, H. Burda, Magnetic compass in the cornea: local anaesthesia impairs orientation in a mammal, J. Exp. Biol. 209 (2006) 4747–4750.
[158] M.M. Walker, C.E. Diebel, C.V. Haugh, P.M. Pankhurst, J.C. Montgomery, C.R. Green, Structure and function of the vertebrate magnetic sense, Nature 390 (1997) 371–376.
[159] G. Fleissner, E. Holtkamp-Rötzel, M. Hanzlik, M. Winklhofer, G. Fleissner, N. Petersen, et al., Ultrastructural analysis of a putative magnetoreceptor in the beak of homing pigeons, J. Comp. Neurol. 458 (2003) 350–360.
[160] M. Hanzlik, C. Heunemann, E. Holtkamp-Rötzler, M. Winklhofer, N. Petersen, G. Fleissner, Superparamagnetic magnetite in the upper beak tissue of homing pigeons, Biometals 13 (2000) 325–331.
[161] M.N. Williams, J.M. Wild, Trigeminally innervated iron-containing structures in the beak of homing pigeons, and other birds, Brain Res. 889 (2001) 243–246.
[162] M. Winklhofer, E. Holtkamp-Rotzler, M. Hanzlik, G. Fleissner, N. Petersen, Clusters of superparamagnetic magnetite particles in the upper-beak skin of homing pigeons: evidence of a magnetoreceptor? Eur. J. Miner. 13 (2001) 659–669.
[163] S. Marhold, W. Wiltschko, H. Burda, A magnetic polarity compass for direction finding in a subterranean mammal, Naturwiss. 84 (1997) 421–423.
[164] R.A. Holland, J.L. Kirschvink, T.G. Doak, M. Wikelski, Bats use magnetite to detect the Earth's magnetic field, PLoS One 3 (2008) e1676.
[165] P. Němec, P. Cveková, H. Burda, L. Peichl, Visual systems and the role of vision in subterranean rodents: diversity of retinal properties and visual system designs, in: S. Begall, H. Burda, C. Schleich (Eds.), Subterranean Rodents—News From Underground, Springer, Heidelberg, 2007, pp. 129–160.
[166] R.E. Moritz, H. Burda, S. Begall, P. Němec, Magnetic compass: a useful tool underground, in: S. Begall, H. Burda, C.E. Schleich (Eds.), Subterranean Rodents—News From Underground, Springer, Heidelberg, 2007, pp. 161–174.
[167] P. Thalau, T. Ritz, K. Stapput, R. Wiltschko, W. Wiltschko, Magnetic compass orientation of migratory birds in the presence of a 1.315 MHz oscillating field, Naturwiss 92 (2005) 86–90.
[168] W. Wiltschko, R. Wiltschko, Magnetic orientation and magnetoreception in birds and other animals, J. Comp. Physiol. A 191 (2005) 675–693.
[169] M.J. Freake, R. Muheim, J.B. Phillips, Magnetic maps in animals: a theory comes out of age? Quart. Rev. Biol. 81 (2006) 4.
[170] H. Burda, S. Begall, V. Hart, E.P. Malkemper, M.S. Painter, J.B. Phillips, Magnetoreception in mammals, in: B. Fritzsch, B. (Ed.), H. Bleckmann (Volume Ed.), The Senses: A Comprehensive Reference, vol. 7, Elsevier, Academic Press, 2020, pp. 421–444.
[171] J.B. Phillips, R. Muheim, P.E. Jorge, A behavioral perspective on the biophysics of the light-dependent magnetic compass: a link between directional and spatial perception? J. Exp. Biol. 213 (2010) 3247–3255.

[172] V. Hart, P. Nováková, E.P. Malkemper, S. Begall, V. Hanzal, M. Ježek, et al., Dogs are sensitive to small variations of the Earth's magnetic field, Front. Zool. 10 (2013) 80.
[173] E.I. Moser, E. Kropff, M.-B. Moser, Place cells, grid cells, and the brain's spatial representation system, Annu. Rev. Neurosci. 31 (2008) 69–89.
[174] J. O'Keefe, Spatial cells in the hippocampal formation, in: Nobel Lecture, 7 December 2014. <https://www.nobelprize.org/prizes/medicine/2014/okeefe/lecture/>.
[175] A. Cheung, S.W. Zhang, C. Stricker, M.V. Srinivasan, Animal navigation: the difficulty of moving in a straight line, Biol. Cybern. 97 (2007) 47–61.
[176] A. Cheung, S.W. Zhang, C. Stricker, M.V. Srinivasan, Animal navigation: general properties of directed walks, Biol. Cybern. 99 (2008) 197–217.
[177] G.C. Hickman, Adaptiveness of tunnel systems features in subterranean mammal burrows, in: E. Nevo, O.A. Reig (Eds.), Evolution of Subterranean Mammals at the Organismal and Molecular Levels, Wiley-Liss, New York, 1990, pp. 185–210.
[178] D. Vleck, The energy costs of burrowing by the pocket gopher *Thommomys bottae*, Physiol. Zool. 52 (1979) 122–136.
[179] J.U.M. Jarvis, M.J. O'Riain, N.C. Bennett, P.W. Sherman, Mammalian eusociality—a family affair, Trends Ecol. Evol. 9 (1994) 47–51.

CHAPTER 8

Evolutionary agriculture domestication of wild emmer wheat

Junhua Peng[1], Zhiyong Liu[2], Xionglun Liu[3], Jun Yan[4], Dongfa Sun[5] and Eviatar Nevo[6]

[1]Center of Crop Germplasm Enhancement and Utilization, Huazhi Bio-Tech Co., Ltd., Changsha, P.R. China
[2]State Key Laboratory of Plant Cell and Chromosome Engineering, Institute of Genetics and Developmental Biology, The Innovative Academy of Seed Design, Chinese Academy of Sciences, Beijing, P.R. China
[3]Southern Regional Collaborative Innovation Center for Grain and Oil Crops in China, Hunan Agricultural University, Changsha, P.R. China
[4]Key Laboratory of Coarse Cereal Processing, Ministry of Agriculture Rural Affairs, School of Pharmacy and Bioengineering, Chengdu University, Chengdu, P.R. China
[5]College of Plant Science and Technology, Huazhong Agricultural University, Wuhan, P.R. China
[6]Institute of Evolution, University of Haifa, Haifa, Israel

Introduction

Wheat is the universal cereal of Old World agriculture and the most important crop plant in the world [1–7], followed by rice and maize. Wheat was among the earliest domesticated crop plants, dating back ~10,000 years ago in the prepottery Neolithic Near East Fertile Crescent [8]. Modern wheat cultivars usually refer to two species, hexaploid bread wheat, *Triticum aestivum* ($2n = 6x = 42$, AABBDD), and tetraploid hard or durum-type wheat, *Triticum turgidum durum* ($2n = 4x = 28$, AABB) used for macaroni and low-rising bread. Bread wheat accounts for ~95% of world wheat production, and the other 5% is durum wheat. Nowadays, wheat ranks the first in the world's grain production and accounts for more than 20% of the total human food calories. Wheat is now extensively grown on 17% of all crop areas, in the temperate, Mediterranean-type, and subtropical parts of both world hemispheres from 67°N to 45°S. It is the major cereal crop of temperate regions and is the staple food for 40% of the world's population (faostat.fao.org; www.croptrust.org). The world's main wheat-producing regions are in temperate and southern Russia, the central plains of the United States, southern Canada, Mediterranean basin, north-central China, India, Argentina, and southwestern Australia.

New Horizons in Evolution
DOI: https://doi.org/10.1016/B978-0-323-90752-1.00007-9

Human history is tightly correlated with development of wheat, barley, and possibly rye because they belong to the Neolithic founder crops from which the western agriculture was built [9]. Wheat is also a superb model organism for the evolutionary theory of allopolyploid speciation, adaptation, and domestication in plants. The domestication, primarily in modern breeding practices, led to its genetic erosion and increasing susceptibility and vulnerability to environmental stresses, pests, and diseases [10–12]. Thus the future genetic improvement as a high-quality nutritional food is paramount for feeding the everincreasing human population. The best strategy for wheat improvement is to utilize the adaptive genetic resources of the wild progenitors, such as wild emmer wheat (*Triticum dicoccoides*), etc. [13,14]. Recently, a comprehensive study showed that historic gene flow from wild relatives made a substantial contribution to the adaptive diversity of modern bread wheat [15].

Because of the high self-pollination, genetic diversity of wheat is represented in the wild by numerous clones and in cultivation by some 25,000 different cultivars. Cultivated primitive forms have hulled grains, whereas advanced forms are free-threshing. But *T. diccocoides* has brittle rachis that make spikes disarticulate at maturity into individual spikelets. Each spikelet constitutes an arrow-like device that inserts the seed into the ground [16]. All the cultivated forms of wheat have nonbrittle spikes that stay intact after maturation for harvesting, threshing and sowing by humans [17]. The free-threshing of cultivated wheat is controlled by the Q locus [18], located on chromosome 5A. This Q locus may have arisen from the *q* gene of the hulled varieties by a series of mutations [2].

There is no wild type (WT) of cultivated common wheat found in nature. But wild emmer wheat, *T. dicoccoides* is proved to be the direct progenitor for both the tetrapoid and hexaploid cultivated wheats, occupies a pivotal position in wheat domestication process, and harbors rich genetic resources for wheat improvement [5,6]. Prof. Eviatar Nevo from the Institute of Evolution at the University of Haifa initiated large-scale germplasm collection and performed extensive studies on wild cereals, mainly wild barley and *T. dicoccoides* in 1970s. To date, he has established the largest gene bank of wild cereals in the world, and has about 430 cereal publications, over 120 of which are for *T. dicoccoides* (http://evolution.haifa.ac.il/images/cereal_list_sep_2019.pdf). As one piece of celebration events for Prof. Nevo's 90th birthday, we present this review paper to summarize extensively studies on domestication evolution and gene discovery in *T. dicoccoides*, and also briefly the breeding application of the wild emmer germplasm in China.

Evolutionary domestication of *Triticum dicoccides*

Triticum dicoccides, the wild emmer wheat ($2n = 4x = 28$, genome AABB) is a tetraploid predominantly self-pollinated plant species. It originated from a spontaneous hybridization between wild diploid einkorn wheat, *Triticum urartu* ($2n = 2x = 14$, genome AA), and a close relative of the goat grass *Aegilops speltoides* ($2n = 2x = 14$, genome SS, where S is closely related to B) [19,20]. Wild emmer wheat presumably originated in and adaptively diversified from north-eastern Israel and the Golan Height into the Near East Fertile Crescent, across a variety of ecological conditions [21]. The wide range of ecological conditions, such as temperature [17,22], soil [17,23], water availability [17,21], light intensity [17,22], humidity [11,14,17,24–26], etc., may exert diverse selection pressures, thus determine the evolutionary course while shaping its genetic structure.

Triticum dicoccoides is of great importance in wheat domestication and breeding

Wheat domestication increased food production, expanded sedentism and human population, and promoted development of early human civilization. Wheat cultivars are superior to most other cereals in their nutritive value. They contain 60%–80% starch and 8%–15% protein, which rise in elite wild genotypes of *T. dicoccoides* up to 13.9%–28.9% [27–33]. Wild emmer wheat is also extremely rich in high-molecular-weight glutenins [34], thus an important source of elite baking quality. Wheat is the staple food for billions of people, and wild emmer wheat has unique bread-baking qualities. In the still exploding world population (approaching 10 billion in 2050), wheat will continue to serve as the major food ingredient through bread production [10,11].

Wheat evolution studies showed that hexaploid bread wheat is derived from a spontaneous hybridization between tetraploid wheat and the diploid D genome donor, *Aegilops tauchii*, and does not have a wild hexaploid progenitor [35–40]. Thus the wheat domestication process mainly occurred in tetraploid *T. dicoccoides* containing A^uA^uBB genomes. The A^u genome in bread and durum wheat is different from A^m in *Triticum boeoticum* but the same as in *T. dicoccoides*. Therefore the *T. dicoccoides* is actually the core in wheat domestication evolution.

The earliest evidence for wheat utilization is from Ohalo, a site near the Lake of Galilee where a 19,000-year-old *T. dicoccoides* sample with brittle rachis was found, permitting sedentism and cereal agriculture [41]. However, wild emmer was first cultivated in the southern Levant in the Pre-Pottery Neolithic A (PPNA) 10,300–9500 BP. Domesticated emmer (with a *nonbrittle* rachis) appeared several hundred years later in the late PPNB (9500–9000 BP), which was grown mixed with wild emmer in many Levantine sites. Types with naked free-threshing grains emerged in the late PPNB (9000–7500 BP) [42]. Mutations affecting spike traits including shattering, also called brittle rachis (controlled by genes *Br1* and *Br2*), tough glume (controlled by genes *Tg* and *Sog*), and speltoid spike (*q*, nonfree-threshing) were largely responsible for wheat domestication [43].

Because of *T. dicoccoides*' full interfertility with domesticated emmer wheat, this wild species can serve as one of the most important genetic resources to improve durum as well as bread wheat [44]. *T. dicoccoides* possesses important beneficial traits, stripe (yellow) and stem rust resistance, powdery mildew resistance, soil born wheat mosaic virus, amino acid composition, grain protein content (GPC) and storage protein genes (HMW glutenins), high photosynthetic yield, salt and drought tolerance, herbicide resistance, amylases and alpha-amylase inhibitors, micronutrients such as Zn and Fe [45,46], and genotypic variation for diverse traits such as germination, biomass, earliness, nitrogen content, and yield, short stature, and high tillering capacity [17]. However, *T. dicoccoides* also shows agriculturally deleterious features such as brittle rachis, no-free-threshing characteristic, few, small, and light spikes, and small grains. Nevertheless, among the 75 domestication QTL effects for 11 traits, wild QTL alleles of *T. dicoccoides* for 18 (24%) effects were agriculturally beneficial, for example, contributing to short plant, early HD, more spike number/plant, higher spike weight/plant, more kernel number/spikelet (KNL), higher GWH, and higher yield [47]. Thus this large portion of cryptic beneficial alleles together with genes for resistance or tolerance to biotic and abiotic stresses and high protein content [17,47] could substantially advance the utilization of *T. dicoccoides* for wheat improvement [3,6,21,48,49]. As of today much of the adaptive vast potential genetic resources existing in wild emmer remain to be tapped and exploited for wheat improvement.

Triticum dicoccides has played a central role in wheat evolutionary domestication

The family Poaceae (grasses) evolved 50–70 million years ago (Mya) [50,51] and the subfamily Pooideae including wheat, barley, and oats has diverged around 20 Mya [52]. Wild diploid wheat (*T. urartu*, $2n = 2x = 14$, genome AA) hybridized with goat grass (*Ae. speltoides*, $2n = 2x = 14$, genome BB) 300,000–500,000 BP [51,53] to produce *T. dicoccoides* ($2n = 4x = 28$, genome AABB). The earliest evidence that man collected and used these cereals is from Ohalo II, a permanent site of epipaleolithic (19,000 BP) hunter-gatherers on the southwestern shore of the Sea of Galilee, Israel [42]. Here, Kislev et al. [41] found grains of wild barley and wild emmer, and presented evidence for grain processing and baking of flour. About 10,000 BP hunter-gatherers began to cultivate wild emmer. Subconscious plant selection slowly created a cultivated emmer (*Triticum dicoccum*, $2n = 4x = 28$, genome AABB) that spontaneously hybridized with another goat grass [*Ae. tauschii* ($2n = 2x = 14$, genome DD)] around 9000 BP to produce an early spelt (*T. spelta*, $2n = 6x = 42$, genome AABBDD). About 8500 BP, natural mutation changed the ears of both emmer and spelt to a more easily threshed type that later evolved into the free-threshing ears of durum wheat (*T. durum*, $2n = 4x = 28$, genome AABB) and bread wheat (*T. aestivum*, $2n = 6x = 42$, genome AABBDD).

It is accepted that *T. aestivum* originated from a cross between domesticated hulled tetraploid emmer *T. dicoccum* (or the free-threshing hard wheat *T. durum*, or the free-threshing *Triticum parvicoccum*) and the goat grass *Ae. tauschii* (DD) [35–40]. This cross should have taken place after emmer wheat cultivation spread east from the Fertile Crescent into the natural distribution area of *Ae. tauschii*. The cross occurred most probably in south or west of the Caspian Sea about 9000 years ago [54–56]. History of wheat evolution clearly shows that *T. dicoccoides* is located in the center of wheat domestication evolution.

Where was *Triticum dicoccides* domesticated?

Based on the ploidy level described above, wheat species can be actually divided into three groups: (1) diploid $2n = 2x = 14$ einkorn wheat, (2) tetraploid $2n = 4x = 28$ = emmer wheat, and (3) hexaploid $2n = 6x = 42$ = common wheat or bread wheat [57–59]. There are two wild diploid *Triticum* species recognized as *T. boeoticum* (A^bA^b) and *T. urartu* (A^uA^u). The former is the ancestor of einkorn wheat *Triticum monococcum* but has been proved to be unrelated with cultivated tetraploid and hexaploid wheats [39,54,60–67]. The latter, *T. urartu*, was never domesticated but played a critical role in wheat evolution and donated the A^u genome to all tetraploid and hexaploid wheats [1,20].

The economically most important wheat is *T. aestivum* or bread wheat (A^uA^uBBDD). However, no wild hexaploid wheat has ever been found in nature, and only a semiwild weedy form of hulled and brittle hexaploid wheat, *Triricum tibetanum*, has been discovered as a weed in barley and wheat fields [12,35–40,54–56,68–70].

Wheat domestication occurred mainly in the wild tetraploid wheats. There are two wild tetraploid wheat species known as *T. dicoccoides* and *Triticum araraticum*. They are similar in morphology, but different in their genomic constitution: *T. dicoccoides* has the genomic formula A^uA^uBB and *T. araraticum* A^uA^uGG [1]. *T. dicoccoides* naturally grows in all over the Fertile Crescent. *T. dicoccoides* was rediscovered in nature by Aaron Aaronsohn [71]. The first isolated spikelet of *T. dicoccoides* were collected in 1855 by T. Kotschy but these were recognized as wild wheat only in 1873 by Kornicke who published his first note on it in 1889 [72]. Aahronson found in 1906 an isolated specimen of *T. dicoccoides* near Rosh Pinna, eastern Galilee [71]. The domesticated form of *T. dicoccoides* is known as *T. dicoccum* (emmer, A^uA^uBB).

The wheat was believed to be probably domesticated in southeast Turkey [73–76]. A reconsideration of the geography of domestication of tetraploid emmer wheats has been proposed by Özkan et al. [74] and by Luo et al. [76]. Phylogenetic analysis indicates that two different races of *T. dicoccoides* exist, the western one, colonizing Israel, Syria, Lebanon and Jordan; and the central-eastern one, which has been frequently sampled in Turkey and rarely in Iraq and Iran. It is the central-eastern race that has played the role as progenitor of the domesticated germplasm [73,75,76], which indicates that the Turkish Karacadag population has a tree topology consistent with that of the progenitor of domesticated genotypes.

However, we believe more in the multisite model of domestication of wild emmer and not in single site in southeast Turkey. Nevo and Beiles [14] studied *T. dicoccoides* and found no evidence for two races of *T. dicoccoides*. Review of archeological findings from the PPNA (10,300–9500 BP) indicates that wild emmer was first cultivated in the southern Levant (the western part of the Fertile Crescent). A recent archeological study showed that the *T. dicoccoides* and cultivated emmer coexisted about 9800 BP at Chogha Golan in the eastern wing of the Fertile Crescent/Central Zagros, west Iran [77]. Domesticated emmer (with a nonbrittle spike) appeared several hundred years later in the early PPNB (9500–9000 BP), and for a millennium or more was grown in a mixture with *T. dicoccoides* in many Levantine sites. After the appearance of domesticated emmer, types with naked, free-threshing grains emerged in the late PPNB (9000–7500 BP). These archeological findings of *T. dicoccoides* cultivation and domestication do not support the idea of domestication within a small core area, but rather indicate the polycentric origin of agriculture in the

Levant [41,42]. We strongly support the model of multiple-site independent domestication of *T. dicoccoides* across the Levant. According to this model, the genes for nonbrittleness were transferred to numerous wild emmer genotypes through numerous spontaneous hybridizations, followed by human selection. Consequently, domesticated emmer wheat evolved as polymorphic populations rather than as single genotypes [42].

A recent study [78] confirmed that *T. dicoccoides* evolved in the southern Levant, and the *T. dicoccoides* populations in south-eastern Turkey and the Zagros Mountains are relatively recent reticulate descendants of a subset of the Levantine wild populations. It is proposed that during a predomestication period, diverse wild populations were collected from a large area in west of the Euphrates and cultivated in mixed stands. Within these cultivated stands, spontaneous hybridization gave rise to lineages displaying reticulated genealogical relationships with their ancestral populations. Gradual movement of early farmers out of the Levant introduced the predomesticated reticulated lineages to the northern and eastern parts of the Fertile Crescent, giving rise to the local wild populations but also facilitating fixation of domestication traits. Throughout the domestication process, humans played an intuitive role [78].

Several cultivated tetraploid A^uA^uBB wheats were derived later from the domesticated emmer, *Triticum carthlicum* (Persian wheat), *Triticum polonicum* (Polish wheat), *Triticum ispahanicum*, *Triticum turanicum* (Khurasan wheat), and *T. turgidum* (English or pollard wheat). *T. dicoccum* was the favored crop for bread-making in ancient Egypt. Emmer wheat cultivation has significantly declined and can be found only in some traditional farming communities, mainly in Russia and Ethiopia. *T. durum* (macaroni or hard wheat) also originated from *T. dicoccum* somewhat later [79] and possibly independently [55,74]. This free-threshing naked wheat is widely cultivated today for pasta production. The wild tetraploid wheats including both *T. dicoccoides* and *T. araraticum* are distributed over the same area in the eastern part of the Fertile Crescent, Turkey, Iran, and Iraq [1]. These two species are morphologically indistinguishable [80] and can be distinguished only by crossing or molecular tests. While *T. dicoccoides* crosses easily with cultivated tetraploid wheats, *T. araraticum* does not cross with *T. dicoccoides*, most probably due to relevant differences in the genome, likely the existence of several translocations between B and G chromosomes [81]. *T. araraticum* was also domesticated but its cultivated form, *Triticum timopheevii* (A^uA^uGG; Timopheev's wheat), has been found in West Georgia together with the hexaploid wheat *Triticum zhukovskyi* ($A^mA^mA^uA^uGG$; Zhukovskyi's wheat) [63]. It is speculated that when emmer cultivation spread to Transcaucasia, local populations of *T. araraticum* were colonizing as a weed in the fields of emmer crops and, by being incorporated into the agricultural cycle of harvest and sowing, became domesticated [54].

Additionally, Mori et al. [82] found by using chloroplast SSR markers that *T. dicoccoides* is obviously more diverse that *T. araraticum*, and domesticated timopheevii wheat (*T. timopheevii*) is monophyletically originated from *T. araraticum*. The plastotypes revealed clear differences between the chloroplast DNA of timopheevii wheat and wild emmer wheat, and thus supported the hypothesis that these two wheat species originated independently. None of the *T. araraticum* plastotypes collected in Transcaucasia were closely related to the *T. timopheevii* plastotype. But the plastotypes found in northern Syria and southern Turkey showed closer relationships with *T. timopheevii*. Therefore the domestication of timopheevii wheat might have occurred also in the Fertile Crescent region including southern Turkey and northern Syria other than in Transcaucasia [82].

How fast is the domestication process of *Triticum dicoccides*?

The earliest cereal gathering or wheat domestication occurred in the Near East 19,000 years before the present (years B.P.) [41,42,83]. Conventionally, wheat domestication studies have been focusing on a few quality traits (brittle rachis, tough glume, and free-threshing) controlled by single major genes (*Br/br*, *Tg/tg*, and *Q/q*). If ancient wheat breeders or farmers only selected the nonshattering or indehiscent, soft glume and free-threshing mutants in the wild wheat populations, the wheat plant would have been domesticated in a very short period, or the domestication should have been a rapid event. Hillman and Davies [84] performed natural selection of barley, einkorn and emmer wheat under primitive farming and concluded that perhaps only 20–30 years would be enough to completely domesticate these plants. Honne and Heun [85] believe this conclusion is appropriate. The fact that the archaeobotanical record shows that remains of wild and domesticated forms of the same plant overlap for a long time (up to 3000 years) appears inconsistent with rapid domestication [83,86,87].

The earliest indehiscent domestic wheat has been recognized in archeological levels dated to ~9250 years B.P. How long was *T. dicoccoides* cultivated before this date? Estimates vary from less than 200 to at least several hundred years [84]. A recent archeological study conducted in northern Syria and south-eastern Turkey indicated that indehiscence took over one millennium to become established events [83]. This means that early farmers did not only focus on indehiscence, but also on other important traits, spike size, heading date (HD)/growth duration, plant height, grain size, etc., in the harvest process of wild wheat. Measurements taken from ancient grains demonstrate that the size of wheat and barley grains remained essentially the same between 9500 and 6500 years B.P. [88]. Therefore selection for large cereal grains is slow because grain size is controlled by polygenes [47] and thus depends more on the position on the ear and the coupling of environmental conditions and genetic diversity than solely on genetic diversity.

If early farmers harvested spikes after the ears began to shatter, indehiscent mutants would be rapidly adopted. But farmers probably harvested before the spikelets fell to avoid loss and paid close attention to important agronomic and economic traits (yield and yield components, plant height and HD, etc.), thus indehiscence was not advantageous. Furthermore, when crops failed, farmers would have had to gather spikes from the wild. These two practices lowered the probability of the rare indehiscent mutant being selected. Domestication was a series of events occurring at different places over thousands of years, during which wild emmer wheat persisted in cultivated fields. Therefore the process of wheat domestication was slow, spanned over one thousand years, occurred in multiple sites of the Fertile Crescent, and fitted a gradualist and multisite model [41,83]. This multiplace and long period of domestication seems much more realistic than the fast domestication. Furthermore, domesticated grasses, changes in grain size and shape evolved prior to nonshattering ears or panicles. Initial grain size increases may have evolved during the first centuries of cultivation, within perhaps 500–1000 years. Nonshattering infructescence was much slower, becoming fixed about 1000–2000 years later [86,87].

Wheat traits subjected to domestication selection

There exist significant differences between the domesticated cereal crops and their wild relatives. Many of these differences are obviously due to the intentional selections of humans. The key wheat traits affected by domestication were the brittle rachis, tough glume, free-threshing state, and a set of quantitative traits [4,5].

Brittle rachis

The breakage of rachis sheds seeds at maturity of any wild forms of wheat. This trait is agriculturally deleterious, and thus transformation of brittle rachis (Br) to nonBr is perhaps the first symbol of domestication in wheat [47]. Loss of seed shattering was a key event in the domestication of major cereals [89]. The modification of the brittle rachis trait has been critical for the origin of agriculture and sedentary societies. In nature, the spikelets of the wild ears fall apart at ripening through fragmentation of the rachis (by shattering or disarticulation). This mechanism is necessary for seed dispersal and self-planting. In a tough, nonbrittle rachis the formation of fracture zones at the rachis is suppressed until mature spikes are harvested by man. It is thought that the spikes of nonbrittle mutated plants were consciously selected by early farmers and that their frequency increased constantly in cultivated fields. But this process was slow and establishment of the nonbrittle ancient cultivar took over one millennium [83,86,87]. The brittle rachis was dominant to the tough rachis, and was controlled by a single gene. In the cross of semiwild wheat with *T. aestivum spp. spelta*, three genes interact to control three types of rachis fragility, that is, semiwild wheat-type, spelta-type, and the tough rachis of common wheat. Semiwild wheat differs from common wheat in rachis fragility. This wheat also differs from other wheats with fragile rachis (*T. aestivum ssp. spelta, macha*, and *vavilovii*) in the pattern and degree of rachis disarticulation [90].

The brittle rachis character is mapped to the homeologous group 3 chromosomes in wheats [55,91–93]. In einkorn, this trait is under the control of two genes that segregate 15 brittle to 1 tough rachis in the F_2 progeny of wild × domesticated crosses [94]. Cao et al. [90] identified a single dominant gene, *Br1*, responsible for rachis fragility in a feral form of *T. aestivum* from Tibet. The gene was later localized on chromosome 3DS [95], as supported by Rong et al. [96] in a cross of *T. dicoccoides* × *T. aestivum*. Other dominant genes are *Br2* and *Br3* on chromosomes 3A and 3B, respectively [90,95,97]. The mature spike rachis of wild emmer disarticulates spontaneously between each spikelet leading to the dispersion of wedge-type diaspores. By contrast, the spike rachis of domesticated emmer fails to disarticulate and remains intact until it is harvested. This major distinguishing character between wild and domesticated emmer wheat is controlled by two major genes, *br2* and *br3*, on the short arms of chromosomes 3A and 3B, respectively [98].

Recently, Avni et al. [99] revealed genomic regions regulating the BR phenotype, including two major loci on *T. dicoccoides* chromosomes 3A and 3B (15.5 and 32.5 Mb, respectively) containing homologies to the *Btr1* and *Btr2* genes controlling BR in barley. In a complex of gene duplication clusters (three or four copies for each gene in the A and B subgenomes), the orthologous wheat genes (chromosome-3A, *TtBtr1-A* and *TtBtr2-A*; chromosome-3B, *TtBtr1-B* and *TtBtr2-B*) were identified [99].

The previous studies point to (1) multiple genetic pathways controlling the trait(s), and (2) different genetic origins of loci controlling shattering in polyploids [55]. These considerations, combined with the mapping of QTL for shattering, allow the analyses of microsyntenous relationships of these traits in the Triticeae and other grasses. *Br* in *T. dicoccoides* functions as an abscission layer in millet, seed dispersal in sorghum and maize, and seed shedding in rice [47].

Glume tenacity

Glume tenacity is another key trait closely related to free-threshing habit and is modified by the domestication process in wheat. The wild emmer wheat floret is wrapped by tough glumes that make spikes difficult to thresh, whereas cultivated wheats have relatively soft glumes and are free-threshing. Major and minor mutations were involved in the evolution of the free-threshing habit in hexaploid wheat (*T. aestivum*). The nonfree-threshing habit of semiwild wheat (*T. tibetanum*) was dominant to the free-threshing habit of common wheat, and glume tenacity of semiwild wheat was controlled by a single gene in the cross of semiwild wheat with the wheat cultivar Columbus. In the cross of semiwild wheat with *T. aestivum spp. spelta*, the F_2 and F_3 population did not segregate for glume tenacity. Semiwild wheat differs from common wheat in glume tenacity [90].

The *Tg1* locus on chromosome 2D confers the free-threshing habit in hexaploid wheat [100]. Genetic analysis showed that at least two genes controlled the free-threshing trait in crosses involving synthetic wheats [101]. Jantasuriyarat et al. [102] detected several QTL on chromosomes 2A, 2B, 2D, 5A, 6A, 6D, and 7B that significantly affect the free-threshing characteristic. However, the free-threshing habit was predominantly affected by a QTL on chromosome arm 2DS (corresponding to the *Tg1* gene) and to a lesser extent by a QTL on chromosome arm 5AL (corresponding to the *Q* factor). Recently *Tg1* was mapped to a more precise location on the 2DS [103].

The soft glume (*sog*) gene in a diploid wheat relative, *T. monococcum*, was found to be close to the centromere on the chromosome arm 2AS. But in common wheat the tenacious glume (*Tg*) gene was located in the most distal region on the chromosome arm 2DS. The different positions suggest that the threshability mutations have independent evolutionary origins [104]. The *Tg* controls the speltoid phenotype and inhibits the expression of Q. The suppression of the free-threshing character was thought to be due to a partially dominant *Tg* allele on chromosome 2D, derived from *A. tauschii* and thus leading to tenacious glumes. The conclusion is that free-threshing hexaploids have the genotype *tgtg*, QQ [100,101]. Faris et al. [105] demonstrated that the gene inhibiting threshability in *T. dicoccoides* was homoeologous to *Tg-D*1 and therefore designated *Tg-B1*. It is known that *T. dicoccoides* possesses the *Tg-B1* allele for tenacious glumes and the *q-A1* allele. It is also possible that *T. dicoccoides* has a *Tg-A1* allele as suggested in the studies of Faris et al. [105], which would increase glume tenacity and further inhibit threshability. Therefore the genotype *Tg-A1TgA1/Tg-B1TgB1/q-A1q-A1* likely confers the phenotype of seed threshability in *T. dicoccoides*.

Free-threshing

The early wheat varieties were characterized by hulled seeds that required drying to be liberated from the chaff. When species characterized by a low degree of glume tenacity and by fragile rachis and free-threshing habit were selected by the farmers, harvesting grains became efficient. Free-threshing wheats have thinner glumes and paleas that allow an early release of naked kernels. After threshing, free grains are winnowed and stored ready for milling. Free-threshing varieties, like tetraploid hard wheat (*T. durum*), represent the final steps of wheat domestication.

Obviously, threshing efficiency is a significant incentive for rapid domestication of emmer wheat. Tzarfati et al. [106] found that the transition from a brittle hulled *T. dicoccoides* phenotype (wild emmer) to nonbrittle hulled *T. dicoccocum* phenotype (cultivated emmer) reduced ~30% in threshing time, whereas the transition from *T. dicoccocum* to nonbrittle free-threshing *T. durum* cultivars further reduced 85% in threshing time. Therefore both nonbrittle rachis and free-threshing are labor-saving traits that increase the efficiency of postharvest processing in tetraploid hexploid wheats, which could have been an incentive for rapid domestication of the Near Eastern cereals.

Major and minor mutations have been proposed to explain the evolution of the free-threshing habit in wheat [102,107]. A major gene Q located on the chromosome arm 5AL inhibits speltoidy but also has pleiotropic effects on rachis fragility and glume tenacity. All nonfree-threshing wild wheats carry the recessive *q* allele and all free-threshing tetraploid and hexaploid wheats carry the dominant Q allele. In *T. aestivum*, the

Q allele supports the formation of square-headed ears with good threshability, besides inducing softening of the glumes, reduction of ear length, more spikelets per ear, and toughness of the rachis [108–115]. Disruption of the Q gene generates a *q* mutant phenotype, known as speltoid type because *q* mutants have tenacious glumes similar to that of spelt (*T. spelta; qq* genotype). Bread wheat lines harboring both Q and *q* alleles have intermediate phenotypes. Muramatsu [116] also showed that the *q* allele is active by creating genotypes with 1–5 doses of either Q or *q* alleles. He showed that a square-headed hexaploid ear derives from either two doses of Q or five doses of *q*.

In a population of recombinant inbred lines (RIL) derived from a cross between the durum variety Rusty and the cultivated emmer accession PI 193883, Sharma et al. [117] demonstrated that QTL associated with spike length, spikelets per spike, and spike compactness were primarily associated with known genes such as the pleiotropic domestication gene Q. But rachis fragility was not associated with the Q locus, and can be influenced by the genetic background. Threshability QTL were identified on chromosome arms 2AS, 2BS, and 5AL corresponding to the tenacious glume genes *Tg2A* and *Tg2B* as well as the Q gene, respectively. Effects of these three genes are mostly additive, with Q having the most profound effects on threshability, and that free-threshing alleles are necessary at all three loci to attain a completely free-threshing phenotype [117].

In hexaploid wheat, the polygenic component controlling free-threshing is scattered throughout all three genomes. In tetraploid wheats, QTL studies identified four putative loci [118], located on chromosomes 2B, 5A, and 6A. Two of these QTL correspond in position to the Q and *Tg* loci. A recent mapping effort led to the identification of two QTL affecting both glume adherence and threshability [98], suggesting that threshability is a function of glume adherence [103].

Seed size

The evolution from small-seeded wild plants with natural seed dispersal to larger-seeded nonshattering plants is evident. In domesticated grasses, initial grain size increases may have evolved during the first centuries of cultivation, within perhaps 500–1000 years [119]. Seed size was strongly selected in all domesticated cereals, wheat, barley, oats, and rye in the Near East, maize in America, rice in Asia, and sorghum and millet in Africa [47]. Seed size, and thus grain yield, was positively selected during domestication. The genetic control of seed size in domesticated versus wild tetraploid wheats was analyzed by using *T. dicoccoides* substitution lines in *T. durum* background [120]. Kernel size is under a complex polygenic control, and genes with alleles contributing to increase and decrease in kernel size have been mapped to chromosomes 1A, 2A, 3A, 4A, 7A, 5B, and 7B. In an experiment with a cross of *T. dicoccoides* × *T. durum*, we mapped eight QTL for grain weight/grain size on chromosomes 1B, 2A, 4A, 5A, 5B, 6B, 7A, and 7B. Major grain weight QTL were located on chromosomes 2A, 4A, and 5B. These three major seed-size QTL correspond closely in sorghum, rice and maize, and another five QTL correspond between two of these genera when the taxa are compared in a pairwise fashion. Parallel synteny existing between wheat and rice chromosomes indicates that all the detected seed-size QTL in *T. dicoccoides* correspond to their rice counterparts [47].

Developmental timing

Flowering time was also selected in the major cereals. Short-day flowering wild grasses were transformed into domesticates in which flowering time was unaffected by day-length [121]. HD/flowering time is an important criterion for regional adaptation and yield in all cereals. The control of HD is critical for reproductive success and has a major impact on grain yield in Triticeae. Wild progenitors of domesticated cereals are well adapted to the prevailing environmental conditions in the Fertile Crescent. The first cereals domesticated in this region presumably showed the photoperiodic and vernalization phenotypes of their progenitors. However, during the domestication process and the spread of agriculture from the Fertile Crescent, novel adaptive traits suited for the new environments were selected. One key event was the selection of spring types that can be sown after winter. These spring types lack the vernalization requirement and show different response to long days. Reduced photoperiod response is important in Europe and North America, where growing seasons are long [122]. In our study, the wild parent, *T. dicoccoides*, was sensitive to day-length and flowering was later than in the cultivar Langdon. Four HD QTL were mapped on chromosomes 2A, 4B, 5A, and 6B [47]. The wild allele for the QTL on 5A will increase the value of HD and so is responsible for the late flowering of *T. dicoccoides*, whereas the wild HD alleles on chromosomes 2A, 4B, and 6B can accelerate the flowering date. These "earliness" alleles, plus the early genes from the *T. durum* cultivar, might explain the significant transgressive segregation (the majority of the individuals were earlier than the early parent

Langdon) for HD in the mapping population [47]. In the long period of observation, we found that there is immense genetic variation in flowering time in *T. dicoccoides*. Wild emmer from Mt. Hermon in north Israel, for example, accession H52, flower late in April and ripen in May whereas those from Gitit in the Samaria steppes in central Israel flower in February–March and ripen in April. Thus there is a widespread range in flowering of *T. dicoccoides* from cold (late) to warm (early) localities.

In temperate cereals, *Vrn* and *Ppd* have been involved in domestication and adaptation to local environments. The evolution of spring types from a predominantly winter ancestral state is a key event in the postdomestication spread of temperate cereals [123]. On the basis of the map positions, it can be postulated that HD QTL on 5A may be similar to the *VRN1* gene mapped on chromosome 5A in *T. momococcum*. This gene is similar to the Arabidopsis MADS-box transcription factor Apetala 1 (AP1), which initiates the transition from the vegetative to the reproductive state of the apical meristem [124]. The HD QTL is located in a collinear position with the photoperiod response (*Ppd*) genes on the short arm of the group 2 chromosomes in wheat and barley. In common wheat, the allelic series of *Ppd* loci has decreasing potency from *Ppd-D1* to *Ppd-B1* to *Ppd-A1* [125]. Further major photoperiod related genes/gene families appear to be conserved between barley and Arabidopsis, involving the GIGANTEA (GI), CONSTANS (CO), and FLOWERING LOCUS T (FT) genes in Arabidopsis and their orthologs in barley HvGI, HvCO, and HvFT [126–129]. Nevertheless, none of the grass QTL associated with flowering time cosegregate with orthologous Arabidopsis "flowering" genes, that is, different major determinants of photoperiod have been selected in the Triticeae [126,130].

Grain yield

Primary domestication targets were likely the genes that facilitated harvesting and enabled colonization of new environments. Yield must have soon assumed priority, minimizing labor input and land needs. Generally, wild wheat, *T. dicoccoides*, has poor yielding potential. In a mapping population derived from *T. dicoccoides* × *T. durum*, the wild parent had a very poor yield of 0.5 g/plant which characterizes marginal, steppic populations of *dicoccoides*, whereas the domesticated parent had a much higher yield of 8.2 g/plant. Eight yield QTL were mapped on chromosomes 1B, 2A, 3A, 5A, and 5B, and overlapped with QTL for other traits on chromosomes 1B, 2A, 3A, and 5A [47]. QTL conferring *T. aestivum* yield traits were also mapped to chromosomes 3A, 4A, and 5A [131–134]. In a recent association mapping analysis, using simple sequence repeat (SSR) markers and a collection of bread wheat cultivars, QTL for kernel size were detected on chromosomes 2D and 5A/5B [135].

Other quantitative traits modified through domestication

During the domestication process involving the abovedescribed qualitative and quantitative traits, many other quantitative traits were also subjected to selections of ancient farmers via hitch-hiking effects. These traits include plant height (HT), spike number/plant (SNP), spike weight/plant (SWP), single spike weight (SSW), kernel number/plant (KNP), kernel number/spike (KNS), KNL, and spikelet number/spike (SLS). We detected over 50 QTL effects for these eight traits in a wild emmer × durum wheat mapping population [47].

Plant height is an extremely important target trait in modern wheat breeding since the "green revolution" in cereals was achieved by reducing plant height, thus the lodging susceptibility and increase in grain yield [136]. Modern wheats are short because they respond abnormally to gibberellin. The *Rht-1* gene in wheat encodes a repressor of GA signaling orthologous to *Arabidopsis GAI* (gibberellic acid "insensitive"), maize dwarf8 (*d8*), and barley Slender1 (*Sln1*) [136–139]. Pleiotropic effects are not surprising for genes controlling hormone action and may be a common occurrence for the traits targeted by domestication and breeding [55,140]. *Rht-B1b* and *Rht-D1b* genes on wheat chromosomes 4B and 4D are semidominant mutant alleles of the *Rht-1* gene conferring dwarfism [136]. In addition, genes were identified that reduce plant height without affecting early growth, or coleoptile length and vigor. These genes were mapped to different wheat chromosomes, thus widening their exploitation in plant breeding [141]. However, dwarf wheat cultivars were used only in commercial production after the 1960s, and most of the wheat landraces are tall. Therefore ancient farmers did not select the dwarf but selected the tall mutants that had higher biomass and yielding potential during the domestication. Two pairs of linked QTL for plant height were detected in chromosomes 5A and 7B, respectively, in our study. One of *T. dicoccoides* alleles on chromosome 5A could reduce plant height by 9.6–15.2 cm [47].

Spike number is one of the most important yield components and greatly correlates with the tillering capacity in wheat. It must have undergone selection during the domestication. The grassy wild wheat, for example, *T. dicoccoides*, usually has strong tillering ability and can be used as a source to increase the tillering capacity or spike number of wheat cultivars. In the *T. dicoccoides* × *T. durum* cross, seven QTL effects for spike number were detected in five chromosomes 1B, 2A, 2B, 5A, and 7A, among which, 1B and 7A were the most significant and each carried a pair of linked QTL, respectively [47]. The genetic variation for tillering capacity was assessed for the wheat gene pool, that is, low tillering genotypes frequently have a uniculm phenotype, enlarged spike, and modified leaf morphology [142]. In wheat, a single recessive gene (*tin*) located on chromosome 1AS was found to control tiller number [143]. This gene is perhaps a homoeologous allele of the striking spike number QTL on chromosome 1B of *T. dicoccoides* [47]. Comparative genomics analyses revealed that *tin*, rice-reduced tillering mutations, and the barley *uniculm2* mutant were located in nonsyntenic chromosomes [144]. Recently, a tiller inhibition gene *tin3* was identified and mapped to the long arm of *T. monococcum* chromosome $3A^m$ that is syntenic to a 324-kb region of rice chromosome arm 1L [145,146].

SWP and SSW are significantly correlated with each other, and also with grain weight/size and yield [47]. Therefore they were also subjected to selection during domestication. In the *T. dicoccoides* × *T. durum* cross, 10 QTL effects were detected for SWP in six chromosomes with linked QTL in chromosomes 1B, 2A, 5A, and 7A, and five QTL effects were detected for SSW in four chromosomes with linked QTL in chromosome 5A. Among these chromosomes, 5A and 2A for both SWP and SSW, and 1B for SWP are extremely important [47]. Thus highly significant domestication selection was applied to these chromosomes or chromosome regions.

KNP, KNS, KNL, and SLS are highly correlated with each other and also with yield [47]. They are important yield components, thus should have been also subjected to selection during the domestication process. In our domestication QTL mapping effort, nine QTL effects for KNP were detected in six chromosomes; seven QTL effects for KNS were identified in five chromosomes; seven QTL effects for KNL were found in six chromosomes; and six QTL effects for SLS were detected in four chromosomes [47]. Among the relevant chromosomes, 5A is extremely important for all these four traits, 2A is greatly important for KNP, KNS, and SLS, and 1B is highly important for KNP. Therefore chromosomes 5A and 2A played a key role in domestication modification of these four spike-related traits. Interestingly, the abovediscussed free-threshing gene Q is located in chromosomes 5A [18]. It is thus highly possible that the key domestication gene, Q, has pleiotropic effects on KNP, KNS, KNL, and SLS.

Recently Hebelstrup [147] suggests that an evolution towards bigger grains with higher glycemic loads in the form of higher relative starch content has resulted in lower relative protein, fiber and mineral contents. It is hypothesized that in addition to the simple explanation that larger grains emerged under domestication, complex and indirect effects such as increased glycemic index and sweet taste should also be taken into consideration [147].

Domestication syndrome factors

Domesticated species differ from their wild ancestors and relatives for a set of traits that is known as the domestication syndrome. The most important syndrome traits include growth habit, flowering time, seed dispersal, and gigantism [148]. In an effort to map quantitatively inherited domestication traits in wild emmer wheat, we found that most of the significant QTL effects are clustered mainly in a limited number of intervals in chromosomes 1B, 2A, 3A, and 5A. Consequently, the total number of intervals carrying domestication QTL was only 16 though as many as 70 QTL effects were detected. The chromosomal regions harboring a cluster of domestication QTL are referred to as domestication syndrome factors (DSFs). Only seven DSFs, each involving a pleiotropic QTL or cluster of QTL affecting 5–11 traits, were found in four chromosomes in wild emmer wheat [47]. Although most domestication traits are quantitatively inherited, the dramatic morphological changes that accompanied domestication may be due to relatively few genes [148].

A general transition from small-seeded plants with natural seed dispersal to larger-seeded nonshattering plants until harvest applies to all seed crops. Domestication genes have been functionally conserved over thousands of years and have similar, though not identical, effects in various species. These parallels transcend the deepest divisions within the angiosperms, with both monocot and dicot crops developing a similar adaptive domestication syndrome to human cultivation over the last 10,000 years [149]. The seven DSFs, in four of 14 chromosomes in tetraploid wheat, contained 80.4% of the 56 strong-to-moderate QTL effects underlying the differences between wild *T. dicoccoides* and cultivated *T. durum* for 11 traits [47]. Independent domestication of sorghum, rice, and maize involved convergent selection for large seeds, nonshattering spikes, and day-length insensitive flowering. These similar phenotypes are largely determined by a small number of QTL that closely resemble each other in the three taxa [150]. Thus the limited number of DSFs of wheat corroborates the results in other cereal crops showing that the domestication syndrome is under a relatively simple and rapidly evolving genetic control [150].

Gene distribution in *Triticeae* chromosomes is highly *nonrandom*, with a few gene-rich regions alternating with gene-poor regions, as in other eukaryotes. Gene-rich regions correspond to hot spots of recombination [151–155]. The map positions of all seven wheat DSFs appeared to overlap with gene-rich regions, and the key domestication gene, Q. Therefore the high pleiotropy and or tight linkage of most wheat domestication QTL suggest an important role of recombination in either consolidation of positive mutations within the DSF clusters [156] and in reducing the antagonism between artificial and background (purifying) selection [157]. The presumed coincidence between DSFs and gene-rich regions could facilitate component dissection of these factors, their further fine mapping, and finally map-based cloning.

Gene discovery in *Triticum dicoccoides*

Domestication evolution of crop plants from wild progenitors resulted in massive erosion of genetic diversity, and the vulnerability and susceptibility of modern genotypes or cultivars to abiotic and biotic environmental stresses [4,5]. Wild relatives have been proved to be the best hope for crop improvement because of their adaptive complexes to abiotic and biotic stresses [11,13,15,158]. Crop improvement can be dramatically advanced by effective utilization of the immense abiotic and biotic genes unraveled in natural populations of crop progenitors. *T. dicoccoides*, wild emmer wheat is the direct progenitor of cultivated wheats, has the same genome formula as durum wheat and has contributed two genomes to bread wheat, and has played a pivotal role in wheat domestication evolution as discussed above [4,5,11,13,158].

Wild emmer wheat distributes across the new Fertile Crescent region including Israel, Syria, Jordan, Lebanon, south-east Turkey, northern Iraq, and western Iran [17]. This wild species has adapted to a broad range of environments and is rich in genetic resources such as drought and salt tolerances [21,159], herbicide tolerances [17,160], Zn and Fe contents [45,46], biotic (viral, bacterial, and fungal) tolerances [17,48], high-quantity and high-quality storage proteins [46,161], and many others. Therefore gene discovery studies on wild emmer wheat are of paramount importance for wheat improvement.

Natural populations of *T. dicoccoides* possess wide genotypic variations in many traits [17]. These traits mainly belong to the following categories: yield-related traits including plant height, flowering time, seed size and yield [11,15,162]; amino acid composition [163]; protein content and quantity including high grain protein, novel gliadins, and glutenins [34,46,161,164–168]; micronutrient mineral contents [45,169]; abiotic stress tolerances including salt, drought, and heat [159,170–173]; herbicide resistances [174–176]; and biotic stress resistances including powdery mildew [177,178], Fusarium head blight (FHB) [179–181], leaf rust, stem rust, stripe rust [15,31,182,183], wheat soil-borne mosaic virus [184], tan spot (*Pyrenophora tritici-repentis*) [185–187], *Stagonospora nodorum* leaf blotch [185,187], and insects (alpha-amylase inhibitor) [188]. Although there are wide genetic variations in *T. dicoccoides*, to date, only a limited number of genes in wild emmer have been identified and mapped. Most mapping efforts were focused on disease resistance and genes controlling high protein content and some other quantitative traits.

Gene loci for quantitatively inherited agronomic traits

The history of wheat evolution clearly shows that *T. dicoccoides* is located in the center of the wheat domestication process [5]. Many agronomically important quantitative traits have been and will continue to be modified during domestication and subsequent breeding processes. These traits include grain yield, seed size, plant height, tillering capacity, and flowering time [4,5].

Grain yield

Yield is always a key trait in all crops and must have high priority long ago in order to minimize labor input and land needs. Generally, *T. dicoccoides* has poor yielding potential. But through analysis of chromosome arm substitution lines, Millet et al. [189] confirmed that several chromosome arms of *T. dicoccoides* could improve grain yield and protein percentage in common wheat. In a mapping population derived from *T. dicoccoides* × *T. durum*, the wild parent had a very poor yield of 0.5 g/plant, which characterizes marginal, steppic populations of *T. dicoccoides*, whereas the domesticated parent had a much higher yield of 8.2 g/plant. In this population, eight significant yield QTL were mapped on chromosomes 1B, 2A, 3A, 5A, and 5B [47]. Merchuk-Ovnat et al. [190] showed that the ancestral QTL alleles from *T. dicoccoides* improved grain yield, biomass and photosynthesis across environments in modern wheat. The effects of *T. dicoccoides* chromosome arms or genes will be presumably enhanced when combination of genes from several "wild" arms are introgressed into a single "domesticated" genotype, and the interaction between the genes and those in the recipient common wheat must be accounted for when higher yield or protein content is desired [189].

Seed size

Elias et al. [120] analyzed the genetic control of seed size using *T. dicoccoides* substitution lines in *T. durum* background and mapped genes with alleles affecting kernel size to chromosomes 1A, 2A, 3A, 4A, 7A, 5B, and 7B. Through QTL analysis, we mapped eight QTL for grain weight/grain size on chromosomes 1B, 2A, 4A, 5A, 5B, 6B, 7A, and 7B. The major QTL were located on chromosomes 2A, 4A, and 5B. These three major seed-size QTL correspond in sorghum, rice and maize, and another five QTL correspond between two of these genera when the taxa are compared in a pairwise fashion. Parallel synteny existing between wheat and rice chromosomes indicates that all detected seed-size QTL in *T. dicoccoides* correspond to their rice counterparts [47].

Flowering time

This trait is closely related with crop adaptiveness and has been subjected to selection in the major cereals. HD/flowering time is an important criterion for regional adaptation and yield in all cereals. The control of HD is critical for reproductive success and has a major impact on grain yield in Triticeae. In our study, *T. dicoccoides* was sensitive to day-length and flowering was later than in the cultivar Langdon. Four HD QTL were mapped on chromosomes 2A, 4B, 5A, and 6B [47]. The wild allele for the QTL on 5A will increase the value of HD and so is responsible for the late flowering of *T. dicoccoides*, whereas the wild HD alleles on chromosomes 2A, 4B, and 6B can accelerate the flowering date. These "earliness" alleles, plus the early genes from the *T. durum* cultivar, might explain the significant transgressive segregation for HD in the mapping population [47]. In the long period of observation, we found that there is immense genetic variation for flowering time in *T. dicoccoides*. For example, *T. dicoccoides* from Mt. Hermon in north Israel flower late in April and ripen in May, whereas those from Gitit in the Samaria steppes in central Israel flower in February–March and ripen in April. Therefore there is a wide range for flowering in *T. dicoccoides* from cold (late) to warm (early) localities.

Based on the map positions, it can be postulated that HD QTL on 5A may be similar to the *VRN1* gene mapped on chromosome 5A in

T. momococcum. This gene is similar to the *Arabidopsis* MADS-box transcription factor *Apetala 1* (*AP1*), which initiates the transition from the vegetative to the reproductive state of the apical meristem [124]. The HD QTL is located in a collinear position with the photoperiod response (*Ppd*) genes on the short arm of group 2 chromosomes in wheat and barley. In common wheat, the allelic series of *Ppd* loci had decreasing potency from *Ppd-D1* to *Ppd-B1* to *Ppd-A1* [125]. Further major photoperiod related genes/gene families appear to be conserved between barley and *Arabidopsis*, involving the GIGANTEA (GI), CONSTANS (CO), and FLOWERING LOCUS T (FT) genes in *Arabidopsis* and their orthologs in barley HvGI, HvCO, and HvFT [126–129]. Nevertheless, none of the grass QTL for flowering time cosegregate with orthologous *Arabidopsis* "flowering" genes, that is, different major determinants of photoperiod have been selected in the Triticeae [126,130].

Chhuneja et al. [191] analyzed 45 accessions of *T. dicoccoides* for the allelic composition of the vernalization genes at *VRN-A1* and *VRN-B1* loci. Thirty-five *T. dicoccoides* accessions had only winter type allele *vrn-A1* at *VRN-A1* locus. One accession each had *VRNA1a* and *Vrn-A1d* allele, and two accessions each showed the presence of *VRN-A1b* and *Vrn-A1c*, potent spring type alleles. Three accessions, however, had multiple alleles at *VRN-A1* locus. At the *VRN-B1* locus, 35 accessions had winter type allele *vrn-B1*, and 10 accessions had dominant *Vrn-B1* allele. These *T. dicoccoides* accessions constitute a very useful resource for incorporation of alternative vernalization alleles in the cultivated wheat gene pool for developing better adaptive wheat varieties.

In a recent mapping effort, Zhou et al. [192] showed that photoperiod-sensitivity gene *Ppd-B1* from *T. dicoccoides* was located in chromosome arm 2BS, and was tightly linked to a major QTL governing late heading of CASL2BS. There is a significant dominance by additive effect of *Ppd-B1* with the LUX gene located on 3AL. Increased copy number could enhance the expression of *Ppd-1*, and result in an earlier HD than CASL2BS in long days.

Plant height

Plant height is an extremely important target trait in wheat breeding programs since the "green revolution" in cereals was realized by reducing plant height, resulting in the lodging resistance and increase in grain yield [136]. Two pairs of linked QTL for plant height were detected in chromosomes

5A and 7B, respectively, in our study. One of the *T. dicoccoides* alleles on chromosome 5A could reduce plant height by 9.6–15.2 cm [47].

Spike number

Spike number is one of the most important yield components and significantly correlates with the tillering capacity in wheat. The grassy wild wheat *T. dicoccoides* usually has strong tillering ability and can be used as a source to increase the tillering capacity or spike number of wheat cultivars. In the *T. dicoccoides* × *T. durum* cross, seven QTL effects for spike number were detected in 1B, 2A, 2B, 5A, and 7A chromosome, among which, 1B and 7A were the most significant and each carried a pair of linked QTL, respectively [47]. Chromosome substitution analysis indicated that the number of spikelets per spike and spike length were controlled by linked, but different, loci on the long arm of *T. dicoccoides* 2A, and comparative mapping revealed that the QTL for number of spikelets per spike overlapped with a previously mapped QTL for FHB susceptibility [193]. In wheat, a single recessive gene (*tin*) located on 1AS chromosome arm was found to control tiller number [143]. This gene is perhaps a homoeologous allele of the striking spike number QTL on chromosome 1B of *T. dicoccoides* [47]. Comparative genomics analyses located *tin*, rice-reduced tillering mutations, and the barley *uniculm2* mutant to nonsyntenic chromosomes [144].

Spike compactness

From both theoretical and practical stand points, genes for spike morphology are of great interest for scientists and breeders since development and morphology of the wheat spike is important because the spike is where reproduction occurs and it holds the grains until harvest [193]. In a 2A chromosomal substitution line of durum wheat with *T. dicoccoides*, it was found that *T. dicoccoides* chromosome 2A confers a short, compact spike with fewer spikelets per spike. A major QTL for spike compactness coincided with the QTL for spike length. The genes for spike length and compactness were not orthologous to either *sog* or *C* genes conferring compact spikes in diploid and hexaploid wheat, respectively. Further analysis showed that the gene for spike length and compactness derived from *T. dicoccoides* could be an ortholog of the barley *Cly1/Zeo* gene, an *AP2*-like gene pleiotropically affecting cleistogamy, flowering time, and rachis internode length. This knowledge of new genetic loci and associated markers may be useful for manipulation of spike morphology in durum wheat [193].

Spike weight

SWP and SSW are significantly correlated with each other, and also with grain weight/size and yield [47]. In the *T. dicoccoides* × *T. durum* cross, 10 QTL effects were detected for SWP in six chromosomes, and five QTL effects were detected for SSW in four chromosomes. Among these chromosomes, 5A and 2A for both SWP and SSW, and 1B for SWP are extremely important [47].

Kernel number

KNP, KNS, KNL, and SLS are highly correlated with each other and also with yield [47]. In our QTL mapping effort, nine QTL effects for KNP were detected in six chromosomes, seven QTL effects for KNS were identified in five chromosomes, seven QTL effects for KNL were found in six chromosomes, and six QTL effects for SLS were detected in four chromosomes [47]. Among the relevant chromosomes, 5A is extremely important for all these four traits, 2A is greatly important for KNP, KNS and SLS, and 1B is highly important for KNP. Therefore chromosomes 5A and 2A played a key role in breeding modification of these four spike-related traits. Interestingly, the free-threshing gene Q is located in chromosomes 5A [18]. It is thus highly possible that the key domestication gene, Q, has pleiotropic effects on KNP, KNS, KNL, and SLS.

Genes for disease resistance

Genes for rust resistance

There are three *species* of wheat rust pathogens, *Puccinia graminis* Pers. *f. sp. tritici* (the pathogen of wheat stem rust), *Puccinia recondita* Rob. ex Desm. *f. sp. tritici* (leaf rust), and *Puccinia striiformis* West. *f. sp. tritici* (stripe, or yellow rust). The rust diseases caused by these pathogens are significantly affecting wheat production worldwide. While rust diseases are readily controlled by resistant cultivars, virulent pathotypes are a constant threat [194]. During the past several decades, great progress has been made in breeding for resistance to rust diseases. However, the rust diseases are still severe and threatening the world's wheat production because of the continuous and rapid virulence evolution in these pathogens [195].

The stripe rust disease, caused by *P. striiformis* West. *f. sp. tritici*, continues to be a problem in cooler and humid wheat-growing areas at high elevation or higher latitudes [194]. This disease usually attacks young wheat plants during the winter and early spring and makes the plants stunted and weakened. In extreme situations, stripe rust can result in up to 100% yield losses [195]. *T. dicoccoides*, the wild emmer wheat has considerable genetic diversity for rust resistance and is a promising source of new rust resistance genes for cultivated wheats. Gerechter-Amitai and Stubbs [196] and Nevo et al. [31] found that *T. dicoccoides* populations indigenous to Israel are valuable sources of stripe rust resistance. Several novel resistance genes to stripe rust in wild emmer wheat were identified [197–200]. But only a few of them, designated *Yr15*, *YrH52*, *Yr36*, *YrG303*, and *YrTZ2*, have been characterized in detail and mapped or cloned [197,198,201–212].

Gerechter-Amitai and Stubbs [196] reported that accession G-25 from Rosh Pinna, Israel, was resistant to many races of *P. striiformis* from different geographical origins and the responsible gene was later identified as *Yr15* [197,198]. *Yr15* was mapped on 1BS by using cytogenetic [202] and molecular marker analyses [201,203,205,210,213]. Peng et al. [205] developed 14 molecular markers linked to *Yr15*. These markers have been applied to marker-assisted selection (MAS) in US wheat breeding programs (https://maswheat.ucdavis.edu/protocols/Yr15). Meanwhile, Abdollahi Mandoulakani et al. [214] developed six REMAP- and IRAP-derived SCAR markers for MAS of *Yr15* gene in wheat breeding programs.

An interesting fact is that most of the tested accessions from Mt. Hermon, Israel are highly resistant to stripe rust at both the seedling and adult stages, and a large number of the highly resistant accessions to stripe rust in Israeli wild emmer wheat are from the Mt. Hermon population. Nevo et al. [31] tested 113 wild emmer accessions from 11 populations using rust race 2EO, and found that all the accessions from the Mt. Hermon population were entirely immune to stripe rust at seedling and adult stages, and accounted for 55% and 67% of the total adult and seedling highly resistant accessions, respectively. The et al. [215] tested 541 accessions from 23 Israeli wild emmer populations using rust race 110 E143 A + , and found that all accessions from the Mt. Hermon population were highly resistant to stripe rust, and account for 74% of the total highly resistant accessions. van Silfhout et al. [200] tested about 850 wild emmer accessions collected from 31 locations using 28 stripe rust isolates from 19 countries and found that 19 (28%) of the 68 promising resistant accessions were derived from Mt. Hermon. In the effort of mapping stripe rust resistance genes harbored in Mt. Hermon population, Peng et al. [204,205] identified and mapped a stripe rust resistance gene, *YrH52*, also on 1BS. Forty-six molecular markers were linked to the *YrH52* gene. Half of these marker-loci (23) were relatively close to the stripe rust resistance gene *YrH52* with linkage distance less than 10 cm. Fifteen markers were closely linked to *YrH52*, with linkage distance <5 cm. These markers will be significantly helpful for the transfer of *YrH52* gene to cultivated wheats.

High-density molecular mapping of the chromosome region harboring *YrH52* and *Yr15* suggested close linkage between these two genes [205]. However, further investigation is required to determine whether they are truly different. The new isolate with *Yr15*-virulence, identified in Denmark [216], could be helpful for discrimination between these two genes. Using genome-wide association mapping approach, Sela et al. [217] uncovered four significant associations on chromosome arms 1BS, 1BL and 3AL. The locus on 1BS was located in a region known to contain stripe rust resistance genes [204–206]. Recently, a temporarily designated *YrTZ2* gene resistant against the Chinese stripe rust race CYR34 was identified in a *T. dicccoides* accession TZ-2 originally collected from Mt. Hermon. This gene was also mapped to chromosome arm 1BS and flanked by *Xwmc230* and *Xgwm413* with genetic distance of 0.8 cm (distal) and 0.3 cm (proximal), respectively [211]. *YrTZ2* gene actually locates in the same chromosomal interval and links with the same set of molecular markers as the previously reported *Yr15* and *YrH52* genes [204–206].

Recently, the abovementioned *Yr15* gene was eventually cloned [218] after 30 years of continuous endeavors [197,198,201–206]. This broad-spectrum *R*-gene derived from *T. dicoccoides* encodes a putative kinase-pseudokinase protein (wheat tandem kinase 1), contains a unique *R*-gene structure, and belongs to a distinct family of plant proteins, the tandem kinase-pseudokinases (TKPs) [218]. Most interestingly, it was found that the three previously characterized stripe rust resistance genes, *Yr15*, *YrG303*, and *YrH52*, cosegregated in fine-mapping populations and shared identical full-length genomic sequence of functional *Wtk1*. Using EMS mutagenisis approach, the artificially generated susceptible *yrG303* and *yrH52* lines carried single nucleotide mutations in *Wtk1*. Sequence comparison among *yr15*, *yrG303*, and *yrH52* genes showed that while key conserved residues were intact, other conserved regions in critical kinase subdomains were frequently affected. Therefore *Yr15*-, *YrG303*-, and *YrH52*-mediated resistances to stripe rust are encoded by a single locus *Wtk1*, and introgression of *Wtk1* into multiple genetic backgrounds resulted in variable phenotypic responses [212]. Sequence comparison revealed that *YrTZ2* has identical *WTK1* gene as *Yr15* and *YrH52*, indicating they are same allele at the same gene locus (Unpublished result).

The stripe rust resistance gene *Yr35* and leaf rust resistance gene *Lr53*, transferred to common wheat from *T. dicoccoides*, were reported previously to be completely linked on chromosome arm 6BS [219]. A recent linkage analysis demonstrated a recombination of 3% between the genes. Microsatellite markers located on the short arm of chromosome 6B were used to map the genes, with the markers *Xcfd1* and *Xgwm508* being mapped approximately 1.1 and 4.5 cm, respectively, proximal to *Lr53* [220].

A *T. dicoccoides*-derived high-temperature adult-plant resistance gene, *Yr36*, confers resistance to a broad spectrum of stripe rust races at relatively high temperatures (25°C−35°C). This gene was also mapped on chromosome 6BS and was cloned using the map-based cloning approach [207,208]. *Yr36* (*WKS1*) gene includes a kinase and a putative START lipid-binding domain, both of which are necessary to confer resistance. Further analysis has shown that this *T. dicoccoides* gene, *Yr36* (*WKS1*) and its paralogue *WKS2* are present only in the southern distribution range of *T. dicoccoides* in the Fertile Crescent [221]. *Yr36* is present in *T. dicoccoides* but is absent in modern pasta and bread wheat varieties, and therefore can be used to improve resistance to stripe rust in a broad set of varieties [208].

Moseman et al. [182] screened 687 *T. dicoccoides* accessions for seedling responses to culture PRTUS6 of *Puccinia triticina*. Ninety-eight accessions (14%) were at least moderately resistant. Evaluations of seedling and adult plant responses of 742 accessions by Anikster et al. [183] led to the identification of some highly to moderately resistant accessions, and that adult resistance was more common than seedling resistance. Six accessions of *T. dicoccoides* of diverse origin were tested with 10 races of leaf rust. The infection type patterns were all different from those of lines carrying the *Lr* genes on the A or B chromosomes with the same races [222]. The unique reaction patterns are probably controlled by genes for leaf rust resistance that have not been previously identified. It was shown that resistance to leaf rust race 15 in each of the six *T. dicoccoides* accessions is conferred by a single dominant or partially dominant gene. With one exception, the accessions carry different resistance genes: CI7181 and PI 197483 carry a common gene for resistance to leaf rust race 15 [222]. However, only one leaf rust resistance gene *Lr53* derived from *T. dicoccoides* has been mapped to chromosome arm 6BS using monosomic, telosomic, C-banding, and RFLP analyses [219]. This gene is closely linked to *Yr35*, but independent with *Yr36*, which is also located in 6BS [220].

Some polymorphism in response to stem rust was detected in populations of *T. dicoccoides* [183,223], but no stem rust resistance gene has been identified and mapped. Six accessions of *T. dicoccoides* of diverse origin were tested with 10 stem rust races. Their infection type patterns were all different from those of lines carrying the *Sr* genes on the A or B chromosomes with the same races [222]. The unique reaction patterns are probably controlled by genes for stem rust resistance that have not been previously identified. It was shown that resistance to stem rust race 15B-1 in each of the six *T. dicoccoides* accessions is conferred by a single dominant or partially dominant gene [222].

Genes for powdery mildew resistance

Powdery mildew, caused by *Blumeria graminis* f. sp. *tritici*, is a devastating wheat disease worldwide [17,224]. *T. dicoccoides* is a promising genetic source for powdery mildew resistance [14,177,225,226]. Moseman et al. [177] tested 233 *T. dicoccoides* accessions, collected from 10 sites in Israel and elsewhere, and identified frequent resistance and predicted the likelihood of relatively large numbers of resistance genes that were different from the cataloged genes in cultivated wheats. Interestingly, Yin et al. [227] found that all *T. dicoccoides* accessions growing on the south-facing or African slope were susceptible to a composite of *B. graminis*, while those growing on the north-facing or European slope in the Evolution Canyon were highly resistant to *B. graminis* at both seedling and adult stages. Thus natural selection causes adaptive genetic resistance in wild emmer wheat against powdery mildew.

Several resistance genes (*Pm16*, *Pm26*, *Pm30*, *Pm36*, *Pm64*, *MlZec1*, *PmG16*, *PmG3M*, *PmG25*, and *MlIW30*) derived from *T. dicoccoides* were identified and mapped using molecular markers [96,209,226,228–235]. *Pm16*, originally mapped on chromosome 4A by monosomic analysis [226], was later mapped on 5BS using SSR markers [230]. *Pm31*, previously reported as derived from wild emmer [236], was subsequently identified as *Pm21* because of a pedigree error [237].

Recently, two novel powdery mildew resistance genes (temporarily designated *PmG3M* and *PmG16*) were mapped on chromosomes 6BL and 7AL, respectively, using SSR and resistance gene analog markers [209,237]. *PmG3M* from accession G-305-3M conferred resistance to 41 powdery mildew isolates in Israel. *PmG16* harbored in accession 18–16 conferred resistance to 29 of 42 Israeli isolates. *PmG16* was located on the distal region of chromosome arm 7AL. Thirty-two SSR, sequence tagged site (STS, DArT, and CAPS) markers were included in the genetic map of this gene. Colinearity was established between the genomic region spanning the *PmG16* locus within the distal region of chromosome arm 7AL and the genomic regions on rice chromosome 6 and *Brachypodium* Bd1 using four DNA markers [209,238]. The identified *PmG16* may facilitate the use of wild alleles for improvement of powdery mildew resistance in elite wheat cultivars via MAS.

Since *T. dicoccoides*-derived *Pm30* was resistant only to 19 of 42 Israeli isolates [237], but resistant to all 20 tested isolates in China [230], it is expected that *PmG16* might provide good resistance to powdery mildew outside of Israel, and the usefulness of *PmG16* needs to be further investigated [237]. Accession 18−16 might provide both drought and powdery mildew resistance for wheat improvement.

Israeli *T. dicoccoides* accession IW72 is resistant to powdery mildew at both the seedling and adult stages. The resistance was controlled by a single dominant gene, temporarily designated *MlIW72*, which is linked with SSR markers, *Xgwm344*, *Xcfa2040*, *Xcfa2240*, *Xcfa2257*, and *Xwmc525* on the chromosome arm 7AL [239]. In this same chromosome arm 7AL (bin7AL-16-0.86-0.90), Ouyang et al. [240] mapped a dominant powdery mildew resistance gene *MlIW172* harbored in the *T. dicoccoides* accession IW172. The chromosome location and genetic mapping results suggested that the powdery mildew resistance genes *MlIW72* and *MlIW172* might be new alleles at the *Pm1* locus or a new locus closely linked to *Pm1* [239,240].

Powdery mildew resistance in wild emmer accession MG29896 is controlled by a single dominant gene. Molecular and deletion mapping indicated that the powdery mildew resistance gene was located on chromosome arm 5BL and was designated *Pm36*. The 244 bp allele of eSSR marker BJ261635 can be used for MAS in the wheat resistance breeding programs for facilitating gene pyramiding [232]. The temporarily designated gene *PmG25* derived from *T. dicoccoides* accession G25 was proposed to be allelic or closely linked to *Pm36* [233].

Wild emmer accession IW2 collected from Mount Hermon, Israel, is highly resistant to powdery mildew at the seedling and adult plant stages. A single dominant gene was responsible for the resistance in IW2. This gene was mapped on chromosome 3BL using molecular and deletion mapping strategy, and was designated *Pm41* [241]. This gene was successfully cloned and belongs to the type of CC-NBS-LRR (CNL). *Pm41* is detected only in the southern *T. dicoccoides* populations with very low frequency (1.81%) (mainly in Mt. Hermon and Gerizim of Israel and Bekaa of Lebanon), and is completely absent in northern WEW population, domesticated emmer, cultivated durum, and hexaploid wheat, suggesting that *Pm41* is restricted to the place of origin and not involved in the domestication of tetraploid wheat and polyploidization of hexaploid wheat [241].

A powdery mildew resistance gene was transferred from wild emmer accession G-303-1M to susceptible common wheat and resulted in an inbred line P63. The powdery mildew resistance in P63 was controlled by a single recessive gene, *pm42*. This gene was physically mapped on chromosome 2BS using deletion mapping strategy. Nine genomic SSR markers (*Xbarc7*, *Xbarc55*, *Xgwm148*, *Xgwm257*, *Xwmc35*, *Xwmc154*, *Xwmc257*, *Xwmc382*, and *Xwmc477*), five AFLP-derived SCAR markers (*XcauG3, XcauG6, XcauG10, XcauG20*, and *XcauG22*), three EST–STS markers (BQ160080, BQ160588, and BF146221) and one RFLP-derived STS marker (*Xcau516*) were linked to the resistance gene. The large number of closely linked molecular markers will enable the rapid transfer of *pm42* to wheat breeding populations thus improving the genetic diversity [242].

A hexaploid wheat line 3D232 contained powdery mildew resistance gene, originating from wild emmer wheat accessions collected from Israel. This line is resistant to all the 21 Chinese powdery mildew isolates tested, and the resistance is conferred by a single dominant gene, *Ml3D232*. This gene was mapped in chromosome bin 5BL0.59–0.76. Using comparative genetic analyses, eight EST markers, including a NBS-LRR analog cosegregated with *Ml3D232*, provide a target site for fine genetic mapping, chromosome landing, and map-based cloning of the powdery mildew resistance gene [243]. A comparative genomics study showed that three genes, *MlWE4*, *Pm36*, and *Ml3D232*, were cosegregated with markers *XBD37670* and *XBD37680*, indicating they are likely the same gene or alleles in the same locus. The cosegregated markers provide a starting point for chromosome landing and map-based cloning of *MlWE4*, *Pm36*, and *Ml3D232* [244]. Recent study indicated that *Pm36* and wild emmer interogression lines 3D232 and WE4 were resistant to all the tested *B. graminis* isolates in China, providing broad-spectrum resistance germplasm and valuable resource for wheat breeding programs.

In a recent mapping effort, Geng et al. [234] demonstrated that powdery mildew resistance in a bread wheat lines 2L6 derived from wild emmer wheat accession IW30, was controlled by a single dominant gene, temporarily designated *MlIW30*. This gene was found to be linked with 19 SSR and two STS markers, and finally localized to the long arm of chromosome 4A and flanked by SSR markers *XB1g2000.2* and *XB1g2020.2* within an interval of 0.1 cm. *MLIW30* might be a novel *Pm* gene since no powdery mildew resistance gene in/or derived from *T. dicoccoide* has been mapped in chromosome 4A [234].

Powdery mildew resistance in bread wheat line N0324 was derived from *T. dicoccoides* accession 5055, and controlled by a single recessive gene designated temporarily as *pm5055*. This gene was mapped to chromosome 2B by SSR markers, and nullitetrasomic and ditelosomic lines, and might be closely related to *MlIW170* or *pm42*. The *pm5055* gene can be used to diversify powdery mildew resistance sources in wheat breeding programs [245].

A new powdery mildew resistance gene *Pm64* was identified from common wheat–wild emmer introgression line WE35. *Pm64* conferred an intermediate level of resistance to *B. graminis* f. sp. *tritici* isolate E09 at the seedling stage and a high level of resistance at the adult plant stage. *Pm64* was located in a 0.55 cm genetic interval between markers *WGGBH1364* and *WGGBH612* on 2BL bin 2BL4-0.50-0.89, corresponding to a 15 Mb genomic region on Chinese Spring and Zavitan 2BL, respectively. *Pm64* was linked in repulsion with stripe rust resistance gene *Yr5* and a few recombinant lines with both *Pm64* and *Yr5* genes were identified from a larger segregating population and can be served as a valuable resource in breeding for resistance to powdery mildew and stripe rust [235].

Genes for Fusarium head blight resistance

FHB is a destructive disease of wheat, and is caused mainly by *Fusarium graminearum* in North America and *Fusarium culmorum* in many parts of Western Europe. Host resistance is the most effective approach controlling FHB in wheat [246,247] and resistance to the various *Fusarium* species is assumed to have the same genetic basis. The resistances available in hexaploid wheat have not been transferred to tetraploid durum wheat [180]. Screening *T. dicoccoides* germplasm for resistance revealed some variation in response to FHB [179–181,248]. Test of a set of *T. durum*–*T. dicoccoides* chromosome substitution lines showed that LDN (DIC-3A) is consistently less susceptible than the other substitution lines [180]. Otto et al. [249] found that QTL *Qfhs.ndsu-3AS* linked with microsatellite marker *Xgwm2*, and explained 37% of the phenotypic and 55% of the genotypic variances in FHB variation. Another QTL, *Qfhs.fcu-7AL*, harbored in accession PI478742, was identified and mapped on chromosome 7AL [250].

Chromosome 2A of an Israeli *T. dicoccoides* genotype (Israel A) increases FHB severity when present in durum wheat (*T. turgidum* var.

durum) cv. Langdon (LDN). Depending on the evaluation, the single best SSR marker in a region on 2AL accounted for 21%–26% of the variation for FHB resistance, with the Israel A marker alleles associated with increased FHB susceptibility. The single best markers from each evaluation reside within an interval of approximately 22 cm. This study identifies one or more new QTL on chromosome 2A in tetraploid wheat that can contribute to significant variation in FHB resistance [251]. These sources of FHB resistance or susceptibility in *T. dicoccoides* could be used directly for breeding durum wheat and could also be transferred to hexaploid wheat through MAS or maybe helpful for studying FHP resistance mechanism in wheat.

Genes for grain protein content and flour quality

GPC is of great nutritional value and determines the quality of bread and other wheat products [252,253]. Gerechter-Amitai and Grama [254] reported high GPCs of 20%–24% in *T. dicoccoides*. This content is considerably higher than that in cultivated wheats. Avivi [27,28], Nevo et al. [164], Levy [32,33], and Levy and Feldman [255] presented similar results. Nevo et al. [164] detected high correlations between protein content, kernel weight, protein weight, and two climatic variables, humidity and tropical days. Some allozyme genotypes, for example, *Adh-lBaa*, *Pgi-Acc*, *Lap-cc*, *Pept-2aa*, and *Est-lAdcO*, were significantly associated with high protein levels [164]. A high GPC gene was detected in the *T. dicoccoides* accession FA-15-3 (Israel A) [27]. Substitutions of the chromosomes of accession FA-15-3 in the cultivated durum wheat cultivar Langdon showed that a gene for high protein content was located on chromosome 6B [256], and a QTL for GPC on chromosome 6BS [257]. This QTL accounted for 66% of the GPC variation in the cross Langdon (DIC 6B) substitution line x Langdon. The effect of the gene was about 14 g/kg of grain in both tetraploid and hexaploid lines [257–259]. Olmos et al. [260] mapped this QTL as a simple Mendelian locus, *Gpc-B1*, located within a 0.3-cM interval [261], and molecular markers *Xuhw89* and *Xucw71* within this region flanked a 245-kb physical contig, including *Gpc-B1* [262]. This gene was cloned and annotated as a NAC transcription factor regulating multiple processes including nutrient (protein, zinc, and iron) remobilization to the developing grain during leaf senescence [46,161]. The cloning of *Gpc-B1* provides a better understanding of nutrient remobilization and accumulation. Gene-specific and closely linked markers also provide good tools for transferring *Gpc-B1* to cultivated wheat by MAS. In addition to the *Gpc-B1* locus on chromosome 6BS, three QTL affecting GPC and one yield QTL were identified and mapped on chromosome 5B [263]. Blanco et al. [264] detected three GPC QTL that were located on chromosome arms 2AS, 6AS, and 7BL in *T. dicoccoides* accession MG29896.

Combining different strategies, Distelfeld and Fahima [265] delimited the *Gpc-B1* QTL to a 7.4-kb chromosome region containing only one gene that encodes a NAC transcription factor (TtNAM-B1). Modern wheat varieties carry a nonfunctional TtNAM-B1 allele attributed to a frame shift mutation caused by an insertion of thymine, while *T. dicoccoides* carries a functional TtNAM-B1 allele. The absence of the functional TtNAM-B1 allele in modern germplasm suggests a broad potential impact of the functional *T. dicoccoides* allele in breeding of cultivated durum and bread wheat varieties.

Through testing 23 durum Langdon−*T. dicoccoides* (LDN−DIC) substitution lines based on *T. dicoccoides* accessions PI 481521 and PI 478742, Klindworth et al. [266] identified eight lines including LDN742-6B, LDN521-7B, LDN521-5B, LDN742-7A, LDN742-5B, LDN521-2A, LDN742-7B, and DN521-1A, which had significantly higher GPC than LDN. The results suggested that chromosomes 1A, 2A, 5B, and 7B of PI 481521 and 7A, 5B, 6B, and 7B of PI 478742 may carry high GPC genes. Using allele specific marker *Xuhw89*, LDN742-6B was shown to carry the same *Gpc-B1* allele as in Israel A. The remaining six lines with high GPC are potential sources of new high GPC genes for durum wheat. Also in observing LDN−DIC substitution lines, Salmanowicz et al. [267] found that chromosome DIC-6B primarily is stable source of an enhanced GPC and advantageous dough rheological properties, and similar features seem to be shown by the substitutions with the DIC-1A, DIC-2A, and DIC-6A, Therefore substitution lines, particularly those with DIC-6B and DIC-6A and to a lesser extent DIC-1A and DIC-2A, were distinguished by advantageous grain quality traits, mixing properties and dough functionality in wheat breeding.

Qin et al. [268] characterized four low-molecular-weight-isoleucine (LMW-i) type and one novel chimeric (between LMW-i and LMWmethionine (m) types) low-molecular-weight glutenin subunit (LMW-GS) genes from *T. dicoccides*. These genes are designated as emmer-1 to emmer-5 and possesses the same primary structure shared by other published LMW-GSs. The three genes *emmer-1*, *emmer-3*, and *emmer-5* are similar, with the exception that *emmer-3* and *emmer-5* lost a few repeat motifs compared to *emmer-1*. The *emmer-4* is a chimeric gene generated by illegitimate recombination between LMW-i and LMW-mtype.

Recently, Zhang et al. [269] evaluated dough rheological properties of 27 *T. dicccoides* accessions using the micro-doughlab. The four dough rheological parameters, mean values of water absorption (WA), development time, stability time (ST), and degree of softening, for all the samples were 77.2 %, 1.73, 1.68 minutes, and 116.6 BU, respectively. Based on these four dough rheological parameters, the cluster analysis revealed good flour quality in some accessions. Excitingly, the main quality parameters of TD-129 reach the normal level or even the strong gluten quality in absence of wheat D genome, with the values of WA, ST, Mid-line peak time, angle of descent, and gluten index reaching 74.55%, 9.55, 2.63 minutes, 12°, and 46.5%, respectively [269]. In TD-256 accession, the mobilities of *Glu-A1* and *Glu-B1* subunits are consistent with those of bread wheat, whereas the dough rheological properties reveal its poor flour quality partly due to the protein structure of the HMW-GS [270]. Therefore *T. dicoccoides* might contain valuable genes for high-quality endosperm storage proteins, and thus could be used in bread wheat breeding programs for good baking quality.

Genes for micronutrient mineral content

Genetic variation in micronutrient mineral content in *T. dicoccoides* becomes attractive for wheat scientists in recent years. Cakmak et al. [45] screened 825 accessions from various countries and found that the highest concentrations of zinc and iron in *T. dicoccoides* exceeded those of cultivated wheat. Substitution lines with *T. dicoccoides* chromosomes 6A, 6B, and 5B had higher Zn and Fe concentrations than Langdon and other substitution lines [45]. Recombinant chromosome substitution lines (RSLs) carrying the *Gpc-B1* allele had a 12% higher concentration of Zn, an 18% higher concentration of Fe, a 29% higher concentration of Mn, and a 38% higher concentration of protein than lines with the alternative durum allele. The high concentrations of Zn, Fe, and Mn in grain were consistently expressed over five different environments [271]. Therefore *T. dicoccoides* is an important genetic resource for increasing the concentrations and contents of essential minerals in cultivated wheat.

Peleg et al. [272] found wide genetic variation among RIL derived from a cross between durum wheat and *T. dicoccoides* for all grain minerals with a considerable transgressive effect. A total of 82 QTL were detected for 10 minerals. Most QTL were in favor of the wild allele (50 QTL). Fourteen pairs of QTL for the same trait were mapped to seemingly homoeologous positions reflecting synteny between A and B genomes. A significant positive correlation was found among GPC, Zn, Fe, and Cu, which was supported by significant overlap between the respective QTL, suggesting common physiological and/or genetic factors controlling the concentrations of these mineral nutrients. Several genomic regions on chromosomes 2A, 5A, 6B, and 7A were found to harbor clusters of QTL for GPC and other nutrients. These identified QTL may facilitate the use of wild alleles for improving grain nutritional quality of elite wheat cultivars, especially in terms of protein, Zn, and Fe.

Chatzav et al. [273] reported wide genetic diversity in *T. dicoccoides* accessions for all grain nutrients. The concentrations of grain zinc, iron, and protein in wild accessions were about twofold higher than in the domesticated genotypes. Concentrations of these compounds were positively correlated with one another, with no clear association with plant productivity, suggesting that all three nutrients can be improved concurrently with no yield penalty. Therefore *T. dicoccoides* germplasm offers unique opportunities to exploit favorable alleles for grain nutrient properties that were excluded from the domesticated wheat gene pool.

Most recently, Fatiukha et al. [274] pointed out that most of the elements essential for plants are metals stored in seeds as chelate complexes with phytic acid or sulfur-containing compounds. The involvement of phosphorus and sulfur in metal chelation is the reason for strong phenotypic correlations within ionome. Adjustment of element concentrations for the effect of variation in phosphorus and sulfur seed content thus resulted in drastic change of phenotypic correlations between the elements. The genetic architecture of wheat grain ionome was characterized by QTL analysis in a cross between durum and *T. dicoccoides*. They identified 105 QTL and 617 QTL effects for 11 elements, revealed some potential functional associations between QTL and corresponding genes. Therefore accounting for variation in phosphorus and sulfur is crucial for understanding of the physiological and genetic regulation of mineral composition of wheat grain ionome [274].

Genes for tolerance to abiotic stresses

Drought and salinity are the major abiotic stresses that dramatically threaten the world's food supply. *T. dicoccoides* is distributed over the

Middle East areas having dry and possibly saline soils [171]. Populations of *T. dicoccoides* from Israel did have lower rates of Na + uptake than the durum cultivar Langdon [170], when tested by applying ^{22}Na + to 7-day-old seedlings for 2 days in 1 mM NaCl. In addition, some accessions could grow in 175 or 250 mM NaCl until maturity. The most tolerant accession (Gilboa) had average dry weights per plant at 175 mM NaCl that were 63% of the control [171]. This shows tolerance higher than in cultivated durum wheat, which is typically reduced to about 10% of control when grown to maturity at 150 mM NaCl [275].

In a comprehensive evaluation of 30 *T. dicoccoides* and 14 durum wheat accessions for salt tolerance, Feng et al. [276] identified five *T. dicoccoides* accessions showing high salt tolerance. By screening wild emmer genotypes, we identified a promising salt-tolerant line from Gitit (18–35) in the eastern Samaria steppes. We further investigated the physiological difference of *T. dicoccoides* and cultivated wheats in response to salt stress, and found that salt stress resulted in an increase in lipid peroxidation (malondialdehyde) content and electrolyte leakage, to a greater extent in cultivated wheat genotype Zheng 9023 than in salt-resistant *T. dicoccoides* genotype 18–35, but the latter had higher relative dry weight. Differential expression analysis showed that higher transcript induction folds of genes encoding transcription factor were detected in the resistant plants (*T. dicoccoides*) than in sensitive plants (cultivated wheat) after salt treatment. *T. dicoccoides* thus demonstrated better tolerance to salt stress than cultivated wheat, and the higher tolerance of *T. dicoccoides* is because of high expression of stress-responsive genes encoding transcription factor, including NAC2F, NAC8, DREB3A, MYB3R, and MYB2A. Therefore *T. dicoccoides* is an important germplasm for salt tolerance improvement in cultivated wheats [277].

In *T. dicoccoide*, we cloned and characterize a CBL gene, designated *TdCBL6*, which shares high sequence homology with rice *OsCBL6* [278]. *TdCBL6* transcription was induced by NaCl, polyethylene glycol and abscisic acid. *TdCBL6* expression was much higher in the salt-tolerant line than in the salt-sensitive line when they were subjected to salt treatment. Transgenic Arabidopsis ectopic expression of the *TdCBL6* gene displayed higher levels of photosynthetic efficiency (Fv/Fm) and lower ion leakage (EL) than WT plants under NaCl stress conditions. The *TdCBL6*-overexpressing lines showed low-K^+ (LK)-sensitive phenotypes compared with WT plants. The ectopic expression of *TdCBL6* resulted in reduction of H_2O_2 content, and affected expression of K^+-responsive/H_2O_2-regulated genes under LK stress. Therefore heterologous expression of

TdCBL6 in *Arabidopsis* confers salt tolerance by reducing membrane injury and improving photosynthetic efficiency, and that the *TdCBL6* gene may be involved in response to LK stress by regulating the reactive oxygen species-mediated LK signaling pathway [278].

Among abiotic stressors, drought is a major factor responsible for dramatic yield loss in agriculture. *T. dicoccoides* is capable of engaging known drought stress-responsive mechanisms [279]. In a mapping population consisting of 152 RIL derived from a cross between durum wheat and *T. dicoccoides*, 22 QTL were detected by Peleg et al. [280] for drought susceptibility index. Major genomic regions controlling productivity and related traits were identified on chromosomes 2B, 4A, 5A, and 7B. QTL for productivity were associated with QTL for drought-adaptive traits, suggesting the involvement of several strategies in wheat adaptation to drought stress. Merchuk-Ovnat et al. [281] reported that a near-isogenic line (NIL-7A-B-2), introgressed with a QTL on chromosome 7AS from *T. dicoccoides* into the background of bread wheat *cv.* BarNir, and the introgression of 7AS QTL induced a deeper root system under progressive water stress. The identified QTL may facilitate the use of wild alleles for improvement of drought resistance in elite wheat cultivars.

In transcriptome analysis on terminal drought response in *T. dicoccoides*, Krugman et al. [49] identified a total of 5892 differentially regulated transcripts between drought and well-watered control and/or between drought-resistant (R) and drought-susceptible (S) genotypes. Out of these transcripts, 221 were uniquely expressed or highly abundant transcripts in the R genotype and thus potential candidates for drought resistance genes. Annotation of the 221 genes revealed that 26% of them are involved in multilevel regulation including transcriptional regulation, RNA binding, kinase activity, and calcium and abscisic acid signaling implicated in stomatal closure. These results demonstrate the potential of *T. dicoccoides* as a source for candidate genes for improving drought resistance.

The gene for a *T. dicoccoides* DRE-binding protein, TdicDRF1, was cloned and shown to be drought-responsive with orthologs in other plants [282]. Analysis of the AP2/ERF DNA-binding domain of TdicDRF1 as a GST-fusion protein and its binding to DRE by electrophoretic mobility shift assay indicate functional differences between wheat DREBs and those characterized in *Arabidopsis thaliana*. DREB expression increased in drought-stressed roots, correlating with the RT-PCR results, but not in leaf, showing that tissue-specific regulation occurs at the protein level. Hence, the DREB–DRE interaction undergoes subtle multilevel regulation.

Lucas et al. [283] further cloned and characterized a gene for a drought stress-inducible putative membrane protein from root tissue of *T. dicoccoides*. Sequence analysis indicated that the protein is a member of the widespread but hitherto uncharacterized TMPIT (transmembrane protein inducible by TNF-α) family, so is labeled TdicTMPIT1. Real-time RT-PCR showed that the TdicTMPIT1 gene is upregulated on drought stress in drought-tolerant *T. dicoccoides*, but not in a drought-sensitive accession or in cultivated durum wheat. The TdicTMPIT1 product was predicted to be a membrane protein with four transmembrane helices. The confocal laser microscopy demonstrated that the TdicTMPIT1 tagged with GFP was localized in a membranous compartment. TdicTMPIT1 is a membrane protein associated with the drought-stress response in *T. dicoccoides*, and so it may be useful for the improvement of modern wheat genotypes. Members of this protein family in other organisms are also proposed to be involved in stress responses.

Akpinar et al. [284] performed root differential transcriptome analysis between *T. durum* and *T. dicoccoides* under control and drought conditions. A total of 66 miRNAs were identified from all species, across all conditions, of which 46 and 38 of the miRNAs identified from *T. durum* and *T. dicoccoides*, respectively, had not been previously reported. Genotype and/or stress-specific miRNAs provide insights into our understanding of the complex drought response. Particularly, miR1435, miR5024, and miR7714, identified only from drought-stress roots of drought-tolerant *T. dicoccoides* genotype TR39477, can be candidates for future studies to explore and exploit the drought response to develop tolerant varieties.

Merchuk-Ovnat et al. [190] compared a wheat near isogenic line NIL-7A-B-2 containing a drought-related QTL from *T. dicoccoides* on chromosome 7A and its recurrent bread wheat parent BarNir for drought responses. NIL-7A-B-2 exhibited an advantage over BarNir in grain yield and biomass production under most environments. Physiological analyses suggested that enhanced photosynthetic capacity and photochemistry combined with higher flag leaf area are among the factors underlying the improved productivity of NIL-7A-B-2. These were coupled with improved sink capacity in NIL-7A-B-2, manifested by greater yield components than its recurrent parental line. Therefore ancestral *T. dicoccoides* QTL alleles for drought resistance could improve grain yield, biomass and photosynthesis across environments in modern common wheat [190].

Drought- and salt-tolerant genes and QTL identified in *T. dicoccoides* have great potential in wheat improvement. Advanced backcross QTL analysis, the introgression libraries based on wild wheat as donors, and positional cloning of natural QTL will play prevailing roles in elucidating the molecular control of drought and salt tolerance. Combining tolerant genes and QTL in crop breeding programs aimed at improving tolerance to drought and salinity will be achieved within a multidisciplinary context. Wild genetic resistances to drought and salinity will be shifted in the future from lab/field experiments to the farmers [21].

Breeding application of *Triticum dicoccoides* germplasm in China

With the support of Prof. Eviatar Nevo in the Institute of Evolution at the University of Haifa, Prof. T. Yang at the former Beijing Agricultural University (now the China Agricultural University) initiated the wheat breeding program exploiting *T. dicoccides* germplasm in early 1990s. Especially, great efforts were made mainly in gene discovery and breeding application of powdery mildew resistance from *T. dicoccides* [227,228,230,233,234,239,240,242–245,285]. So far, the scientists in China have discovered the largest number that accounts for over 60% of the genes for powdery mildew resistance derived from *T. dicoccides* in the world. It should be highly noted that *Pm41* is the first powdery mildew resistance gene coloned from wild emmer using map-based cloning approach [241].

A series of bread wheat lines containing *T. dicccoides*-derived genes for powdery mildew resistance have been developed in China. The common wheat inbred line P63 contains the recessive powdery mildew resistance gene *pm42* [242]. In bread wheat line N0324, the powdery mildew resistance is controlled also by a single recessive gene *pm5055*, closely related to *MlIW170* or *pm42* [245]. The bread wheat line 2L6 harbors the novel powdery mildew resistance gene *MlIW30* [234]. N0308 is a common wheat line containing powdery mildew resistance gene *PmG25*, possibly allelic or closely linked to *Pm36* and *Ml3D232* [233]. These common wheat lines and the large number of closely linked molecular markers will accelerate transfer of *T. dicccoides*-derived resistance genes to bread wheat cultivars, and thus improve the genetic diversity of common wheat in China, one of the largest wheat producer in the world.

It is known that *T. dicoccoides* generally has poor processing quality, but still some of the accessions possesses good flour quality traits [269,270]. Recently, a lab at the Sichuan Agricultural University tried to improve common wheat quality through wide hybridization between *T. dicoccoides* and bread wheat. Two hybrids, BAd7–209 and BAd7–210, obtained by this approach, showed enriched 1Ax alleles and significantly enhanced gluten strength [286]. Using the same approach, the active *Glu-1Ay* allele of *T. dioccoides* was integrated into common wheat background of Chuannong 16. The resulted common wheat line TaAy7−40 successfully expressed the wild emmer allele and showed improved flour quality [287]. Therefore *T. dicoccoides* germplasm could be utilized extensively in breeding programs to effectively improve common wheat flour quality in shouthwest China.

Wild emmer has been used for breaking bottleneck of yield improvement in common wheat breeding program of China. For example, Prof. Tsomin Yang of China Agricultural University developed a wheat introgression line WE74 using wild emmer accession G-748-M (kindly provided by Dr. Z.K. Gerechter-Amitai, Agricultural Research Organization, the Volcani Center, Israel) crossed with a Chinese wheat landrace Yanda 1817 and followed by several generations of backcrossing with an elite cultivar Nongda 015. At Prof. Zhiyong Liu's lab, WE74 was selected to make another cross with a commercial wheat near isogenic line D2774-10/6*Shi 4185 to develop a high-yielding wheat line ZK331 with more tillering capacity, disease resistance, stress tolerance and high general combining ability (GCA) (Fig. 8.1). In addition, γ-ray radiation treatment was applied to a F_6 line 13H631−1521 to enhance the genomic recombinations between *T. dicoccoides* and the common wheat.

The newly developed line ZK331 was further used to cross with a series of adapted common wheat lines or cultivars, for example, GY

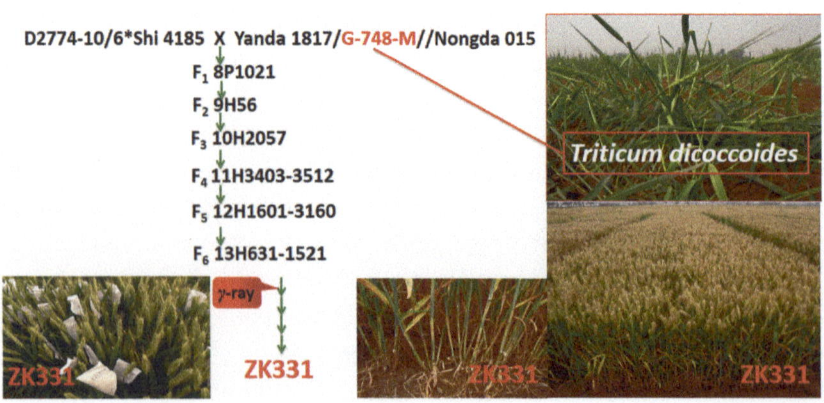

Figure 8.1 Breaking bottleneck of yield improvement in common wheat using *T. dicoccoides* germplasm.

11069, Henong 7069, Zhengfeng 00530, Liangxing 99, and Zhongyu 04zhong36, to develop super cultivars that should adapt to the Huang-Huai winter wheat region, the main wheat production area of China. A large number of super lines of common wheat were developed, and evaluated for yield performance in the year of 2018 (Table 8.1). Clearly, out

Table 8.1 Plot yield performance of ZK331-derived common wheat lines in the year of 2018.

Line code	Pedigree[a]	Yield (t/ha)	Excess CK1 (%)[b]	Excess CK2 (%)[c]
H17085	ZK331/GY 11069	7.81	11.25	18.50
H17062	ZK331/GY 11069	7.49	6.80	13.76
H17094	ZK331/GY 11069	7.44	5.97	12.87
H17013	ZK331/Henong 7069	7.39	5.27	12.13
H17003	ZK331/Zhengfeng 00530	7.38	5.14	11.98
H17063	ZK331/GY 11069	7.37	5.07	11.91
H17016	ZK331/Henong 7069	7.37	5.00	11.84
H17034	ZK331/Liangxing 99	7.33	4.51	11.32
H17082	ZK331/GY 11069	7.31	4.23	11.02
H17068	ZK331/GY 11069	7.26	3.54	10.28
H17015	ZK331/Henong 7069	7.26	3.47	10.21
H17086	ZK331/GY 11069	7.26	3.40	10.13
H17087	ZK331/GY 11069	7.25	3.26	9.99
H17067	ZK331/GY 11069	7.23	3.05	9.76
H17007	ZK331/Zhengfeng 00530	7.23	2.98	9.69
H17075	ZK331/GY 11069	7.23	2.98	9.69
H17110	ZK331/Zhongyu 04zhong36	7.22	2.91	9.62
H17066	ZK331/GY 11069	7.21	2.70	9.39
H17084	ZK331/GY 11069	7.21	2.70	9.39
H17061	ZK331/GY 11069	7.20	2.63	9.32
H17088	ZK331/GY 11069	7.19	2.50	9.17
H17109	ZK331/Zhongyu 04zhong36	7.19	2.50	9.17
H17024	ZK331/Liangxing 99	7.19	2.43	9.10
H17079	ZK331/GY 11069	7.17	2.22	8.87
H17081	ZK331/GY 11069	7.17	2.22	8.87
H17091	ZK331/GY 11069	7.15	1.94	8.58
H17008	ZK331/Zhengfeng 00530	7.14	1.80	8.43

[a]ZK331, a high-yielding and disease-resistant wheat line derived from a complex cross involved in *T. dicoccoides* germplasm and γ-ray radiation treatment.
[b]CK1, Liangxing 99, a bread wheat cultivar popularly used in the north part of Huang-Huai winter wheat region in China.
[c]CK2, Jimai 22, a bread wheat cultivar suitable for cultivation in the north part of Huang-Huai winter wheat region and the north Anhui province in China.

of the newly ZK331-derived lines, 27 were superior for yield performance in comparison with the locally adapted leading varieties, Liangxing 99 and Jimai 22 (Table 8.1). Thus elite line ZK331 has high GCA and can be extensively used in common wheat breeding programs.

As pointed out by Millet et al. [189], wild emmer wheat harbors rich genetic variation for both quantitative and qualitative traits, and the "wild" genes interact with each other in a nonadditive way in genetic background of the common wheat. The gene effects on grain yield and protein percentage is presumably enhanced when combination of genes from several "wild" genome regions are integrated into a single "domesticated" genotype. Therefore the interaction between the genes from *T. dicoccoides* and those in the recipient common wheat must be accounted for when higher yield or protein content is desired.

Concluding remarks and future perspectives

During the long-term agricultural development, early domesticates were gradually replaced first by landraces and traditional varieties, and later by genetically less-diverse modern cultivars. This has resulted in genetic bottlenecks and loss of diversity in breeding germplasm [11,12,158,162,225,288−293]. Though experiencing the diversity bottlenecks, wheat has strong adaptability to diverse environments and end uses. Wheat compensates for these bottlenecks by capturing part of the genetic diversity of its progenitors and by generating new diversity at a relatively fast pace [69]. Germplasm collections are thus essential to conserve biodiversity and thus pay big dividends to agriculture when used efficiently [10,11,21,48,158,162,225,289−294].

T. dicoccoides, is the progenitor of cultivated wheats, has the same genome formula as durum wheat and has contributed two genomes to bread wheat that contains three genomes, and is central to wheat domestication evolution. It has been proved that this wheat progenitor has high levels of genomic DNA polymorphism [22,23,295,296], methylation-based epigenetic variation [297] and chloroplast DNA polymorphism [44]. Thus *T. dicoccoides* should be subjected to in-depth studies to evaluate its structural, functional, and regulatory polymorphisms adapting it to environmental stresses [11,158,298]. The available wheat genome sequences [99,299−304] can transform today's biology, dramatically advancing both theory and application of wheat domestication study. The relationship between genomic and epigenomic diversities [305−307] could be highlighted by deciphering the regulatory function of noncoding genomes on genic components. Regulation in particular might be the key in future domestication studies. It might decipher both speciation and adaptation processes to stressful, heterogeneous, and changing environments. The *nonrandom* adaptive processes and complexes in *T. dicoccoides* and other wheat relatives could provide the basis for wheat improvement as single genes, QTL, and interacting biochemical networks.

It is essential to follow domestication processes and unravel many functional and regulatory genes that were eliminated from the cultivars during domestication, primarily by modern breeding. Identifying the polycentric sites of wild emmer domestication in the southern Levant versus monocentric ideas is feasible by tracking nonbrittle rachis remains during initial phases of the "agricultural revolution," which may have been a gradual rather than a revolutionary process. This future research could identify lost adaptive genes during domestication and their active introgression from *T. dicoccoides* back to cultivated wheat for genetic reinforcement [11].

Whole genomes of wheat and its wild progenitors have been sequenced [99,300–304,308], and the sequences should be useful in domestication genomics studies. The sequenced wheat genomes provide theoretical evolutionary perspectives and excellent tools for wheat domestication studies and for optimizing breeding practices [309]. The sequence data can be used to study the origin of genes and gene families, track rates of sequence divergence over time, and provide hints about how genes evolve and generate products with novel biological properties [310]. Now, the DSFs and other relevant genes and QTL could be isolated, and effects of wheat domestication would be accurately estimated. The improvement of bread wheat is a future challenge of mankind, based on the evidence and ideas presented above and much earlier presented by Aaronshon and Schwinfurth [71] and Aaronshon [311], based on the distinct adaptive complexes of *T. dicoccoides* to environmental stress and their direct relevance to wheat domestication.

As discussed above, *T. dicoccoides* possesses important beneficial traits, stripe rust resistance, stem rust resistance, powdery mildew resistance, soil-born wheat mosaic virus, amino acid composition, GPC, and storage protein genes (HMW glutenins), high photosynthetic yield, salt and drought tolerance, herbicide resistance, amylases and alpha-amylase inhibitors, micronutrients such as Zn and Fe, and genotypic variation for diverse traits as germination, biomass, earliness, nitrogen content, yield, short stature, and high tillering capacity [6,11,17,312]. However, *T. dicoccoides* also shows agriculturally deleterious features such as brittle rachis; no-free-threshing feature;

few, small, and light spikes; and small grains. Nevertheless, among the 75 QTL effects for 11 traits, wild QTL alleles of *T. dicoccoides* for 18 (24%) effects were agriculturally beneficial, for example, contributing to short plant, early HD, more spike number/plant, higher spike weight/plant, more kernel number per spikelet, higher GWH, and higher yield [47]. Thus this large portion of cryptic beneficial alleles together with genes for resistance or tolerance to biotic and abiotic stresses and high protein content [11,17] could substantially advance the utilization of *T. dicoccoides* for wheat improvement [3,21,47–49,312]. As of today, much of the vast adaptive potential genetic resources existing in *T. dicoccoides* remain untapped and need further investigation for wheat improvement although great progresses have been made.

Several decades of investigations have proved that *T. dicoccoides* is really a rich genetic resource with great value for improvement of the cultivated wheats. As described above, a large number of genes have been identified and mapped or cloned. The gene loci discovered in *T. dicoccoides* are mainly involved in disease resistance, tolerance to abiotic stresses, agronomically important traits, protein content, and micronutrient mineral content [17,45–47,161,204,228,235,240,241,243,268–270,277,278]. Molecular markers closely linked to the target genes in *T. dicoccoides* are helpful for transferring them into cultivated wheats. Cloning of *Gpc-B1, Yr36, Yr15, Pm41*, and other genes also opens a window for precise integration of target genes into the genome of cultivated wheats through MAS or transgenic approaches. Several *T. dicoccoides*-derived genes, namely, *Yr15, Yr35, Yr36, Lr53, Pm16, Pm26, Pm30, Pm36, Pm41, pm42, Pm64, Ml3D232, Qfhs.ndsu-3AS*, and *Gpc-B1* have been transferred to cultivated wheat lines and will play significant roles in improvement of wheat production and end-product nutrition in the near future.

To date, still only a small portion of wild emmer germplasm has been tapped although its potential for wheat improvement has been proved and wheat scientists in China has set a very good example in developing superior common wheat cultivars through exploiting *T. dicoccoides* germplasm. There are three reasons for this small scale of utilization: time shortage for application of new ideas and resources, many years needed for gene introgression from wild relatives to adapted wheat cultivars, and breeders' preference of shorter term and "easier" solutions [48]. Nevertheless, the fully annotated genome sequences, the large number of molecular markers, new genomic tools, and efficient gene cloning techniques available in wheat will greatly accelerate the application of *T. dicoccides* germplasm to wheat breeding programs.

Whole genomes of several agricultural crops including rice, maize, wheats and sorghum have been sequenced, and the fully annotated sequences have proved to be useful in genomic and genetic studies, and gene discovery. The IWGSC et al. [303] published a fully annotated reference genome of the hexaploid common wheat. This sequence resource of wheat can shift the limits in wheat research and breeding, and thus drive disruptive innovation in wheat improvement through improved understanding of wheat biology and genomics-assisted breeding. With the annotated and ordered reference genome sequence in place, researchers and breeders can now easily access sequence-level information to precisely define the necessary changes in the genomes for breeding programs. This will be realized through the implementation of new DNA marker platforms and targeted breeding technologies, including genome editing [303]. Meanwhile, alongside this annotated genome, Ramírez-González et al. [304] leveraged 850 wheat RNA sequencing samples, to determine the similarities and differences between homoeolog expression across a range of tissues, developmental stages, and cultivars. The transcriptional landscape of bread wheat is thus established and shows that homoeolog expression patterns in bread wheat have been shaped by polyploidy and are associated with both epigenetic modifications and variation in transposable elements within promoters of homoeologous genes [304]. The sequence data can be used to study the origin of genes and gene families, track rates of sequence divergence over time, and provide hints about how genes evolve and generate products with novel biological properties [307]. This would certainly help in the discovery of new functional genes. The molecular genetic maps of *T. dicoccoides* [313,314] and the nice whole genome sequence [99,308] will facilitate identification and mapping of more genes controlling agronomic traits derived from *T. dicoccoides*. These genes will be introgressed into cultivated wheats and will become good candidates for wheat improvement. Closely linked molecular markers, especially user-friendly PCR-based markers, such as SSR and STS, and the newly developed high-throughput SNP markers, provide useful tools to accelerate both the transfer and the further application of *T. dicoccoides* genes in wheat breeding programs.

The economic importance of wheats have triggered intense cytogenetic and genetic studies in the past decades that resulted in a wealth of information and tools that have been used to develop wheat varieties with increased yield, improved quality, and enhanced biotic and abiotic stress tolerance. Although genomics in wheat has lagged behind other plant species for a while due to the huge genome size (~17 Gb for the hexaploid wheat; ~12 Gb for the tetraploid wheat) and the genome complexity (high repeats and polyploidy), the situation has now changed dramatically and the convergence of several technology developments led to the whole genome sequenced for all the wheat species, *T. uratu* [301,302], *A. tauchii* [300], *T. dicoccoides* [99,305], and *T. aestivum* [7,15,303,304]. These new capabilities and genomic resources will provide a better understanding of the wheat plant biology and support the improvement of agronomically important traits in this crop species with cultural and economic importance.

Conflict of interest

The authors declare no conflicts of interest.

Acknowledgments

This work was supported in part by the National Natural Science Foundation of China (NSFC) (Grant Nos.: 31030055 and 30870233 to J.P., 31030056 to Z.L., and 31171526 to X.L.), Science and Technology Service Network Initiative of Chinese Academy of Sciences (KFJ-STS-ZDTP-024 to Z.L.), the Hu-Xiang High Level Talents Program (J.P.), and the Ancell Teicher Research Foundation for Genetics and Molecular Evolution (E.N.).

References

[1] D. Zohary, M. Hopf, Domestication of Plants in the Old World, Oxford University Press, Oxford, UK, 2000.
[2] M. Feldman, F.G.H. Lupton, T.E. Miller, Wheats. *Triticum* spp. (Gramineae-Triticinae), in: J. Smartt, N.W. Simmonds (Eds.), Evolution of Crop Plants, Longman Scientific & Technical Press, London, UK, 1995, pp. 184–192.
[3] P. Gustafson, O. Raskina, X. Ma, E. Nevo, Wheat evolution, domestication, and improvement, in: B.F. Carver (Ed.), *Wheat: Science and Trade*, Wiley-Blackwell, Ames, IA, 2009, pp. 5–30.
[4] J.H. Peng, D.F. Sun, E. Nevo, Domestication evolution, genetics and genomics in wheat, Mol. Breed 28 (2011) 281–301.
[5] J.H. Peng, D.F. Sun, E. Nevo, Wild emmer wheat, *Triticum dicoccoides*, occupies a pivotal position in wheat domestication process, Aust. J. Crop Sci 5 (2011) 1127–1143.
[6] J.H. Peng, D.F. Sun, Y.L. Peng, E. Nevo, Gene discovery in *Triticum dicoccoides*, the direct progenitor of cultivated wheats, Cereal Res. Commun 41 (1) (2013) 1–22.
[7] C. Pont, T. Leroy, M. Seidel, A. Tondelli, W. Duchemin, et al., Tracing the ancestry of modern bread wheats, Nat. Genet 51 (2019) 905–911.
[8] J.R. Harlan, D. Zohary, Distribution of wild emmer wheat and barley, Science 153 (1966) 1074–1080.
[9] B. Kilian, H. Özkan, C. Pozzi, F. Salamini, Domestication of the Triticeae in the fertile crescent, in: C. Feuillet, G.J. Muehlbauer (Eds.), Genetics and Genomics of the Triticeae, Plant Genetics and Genomics: Crops and Models 7, Springer SciencetBusiness Media, LLC, New York, 2009, pp. 81–119.

[10] E. Nevo, Ecological genomics of natural plant populations: the Israeli perspective, in: D.J. Somers, P. Langridge, J.P. Gustafson (Eds.), Methods in Molecular Biology, Plant Genomics: 513, Human Press, a part of Springer Science + Business Media, New Jersey, 2009, pp. 321–344.
[11] E. Nevo, Evolution of wild emmer wheat and crop improvement, J. Syst. Evol 52 (6) (2014) 673–696.
[12] Y.B. Fu, D.J. Somers, Genome-wide reduction of genetic diversity in wheat breeding, Crop. Sci 49 (2009) 161–168.
[13] M. Feldman, E.R. Sears, The wild gene resources of wheat, Sci. Am 244 (1981) 102–112.
[14] E. Nevo, A. Beiles, Genetic diversity of wild emmer wheat in Israel and Turkey: structure, evolution and application in breeding, Theor. Appl. Genet 77 (1989) 421–455.
[15] L. Huang, L. Feng, Y. He, Z. Tang, J. He, H. Sela, et al., Variation in stripe rust resistance and morphological traits in wild emmer wheat populations, Agronomy 9 (2019) 44. Available from: https://doi.org/10.3390/agronomy9020044.
[16] D. Zohary, The progenitors of wheat and barley in relation to domestication and agricultural dispersal in the Old World, in: P.J. Ucko, G.W. Dimbleby (Eds.), The Domestication and Exploitation of Plants and Animals, Duckworth, London, UK, 1969, pp. 47–66.
[17] E. Nevo, A.B. Korol, A. Beiles, T. Fahima, Evolution of wild emmer and wheat improvement: population genetics, genetic resources, and genome organization of wheat's progenitor, Triticum dicoccoides, Springer-Verlag Berlin, Heidelberg, Germany, 2002, p. 364.
[18] M.C. Luo, Z.L. Yang, J. Dvorak, The Q locus of Iranian and European spelt wheat, Theor. Appl. Genet. 100 (2000) 602–606.
[19] J. Dvorak, H.B. Zhang, Variation in repeated nucleotide sequences sheds light on the phylogeny of the wheat B and G genomes, Proc. Natl Acad. Sci. U S A 87 (1990) 9640–9644.
[20] J. Dvorak, P. Diterlizzi, H.B. Zhang, P. Resta, The evolution of polyploidy wheats: identification of the A genome donor species, Genome 36 (1993) 21–31.
[21] E. Nevo, G. Chen, Drought and salt tolerances in wild relatives for wheat and barley improvement, Plant Cell Environ. 33 (2010) 670–685.
[22] Y.C. Li, T. Fahima, A. Beiles, A.B. Korol, E. Nevo, Microclimatic stress and adaptive DNA differentiation in wild emmer wheat, *Triticum dicoccoides*, Theor. Appl. Genet. 98 (1999) 873–883.
[23] Y.C. Li, T. Fahima, A.B. Korol, J.H. Peng, M.S. Röder, V.M. Kirzhner, et al., Microsatellite diversity correlated with ecological-edaphic and genetic factors in three microsites of wild emmer wheat in North Israel, Mol. Biol. Evol. 17 (2000) 851–862.
[24] T. Fahima, G.L. Sun, A. Beharav, T. Krugman, A. Beiles, E. Nevo, RAPD polymorphism of wild emmer wheat populations, *Triticum dicoccoides*, in Israel, Theor. Appl. Genet. 98 (1999) 434–447.
[25] T. Fahima, M.S. Röder, V.M. Wendehake, E. Nevo, Microsatellite polymorphism in natural populations of wild emmer wheat, *Triticum dicoccoides*, in Israel, Theor. Appl. Genet. 104 (2002) 17–29.
[26] E. Nevo, "Evolution Canyon", a potential microscale monitor of global warming across life, Proc. Natl. Acad. Sci. U S A 109 (2012) 2960–2965.
[27] L. Avivi, High grain protein content in wild tetraploid wheat Triticum diccooides Korn, in: Proc. 5th International Wheat Genetics Symposium, New Delhi, India, 1978, pp. 372–380.
[28] L. Avivi, Utilization of *Triticum diccoccoides* for the improvement of grain protein quantity and quality in cultivated wheats, Monogr. Genet. Agrar 4 (1979) 27–38.

[29] L. Avivi, A.A. Levy, M. Feldman, Studies on high protein durum wheat derived from crosses with the wild tetraploid wheat *Triticum turgidum* var. dicoccoides, in: Proceedings of 6th International Wheat Genetics Symposium, Kyoto, Japan, 1983, 199–204.
[30] A. Grama, Z.K. Gerechter-Amitai, A. Blum, Wild emmer as donor of genes for resistance to stripe rust and for high protein content, in: Proceedings of 6th International Wheat Genetics Symposium, Kyoto University, Kyoto, Japan, 1983, pp. 187–192.
[31] E. Nevo, Z.K. Gerechter-Amitai, A. Beiles, E.M. Golenberg, Resistance of wild wheat to stripe rust: predictive method by ecology and allozyme genotypes, Plant Syst. Evol. 153 (1986) 13–30.
[32] A.A. Levy, Genetic studies on grain proteins in Triticum durum var. dicoccoides, Ph.D. Thesis, Weizmann Institute of Science, Rehovot, Israel, 1987.
[33] A.A. Levy, Increase in grain protein percentage in high yielding common wheat breeding lines by genes from wild tetraploid wheat, Euphytica 36 (1987) 353–359.
[34] E. Nevo, P.I. Payne, Wheat storage proteins: diversity of HMW glutenin subunits in wild emmer from Israel. I. Geographical patterns and ecological predictability, Theor. Appl. Genet. 74 (1987) 827–836.
[35] H. Kihara, Discovery of the DD-analyser, one of the ancestors of *Triticum vulgare*, Agric. Hortic. (Tokyo) 19 (1944) 13–14.
[36] E.S. McFadden, E.R. Sears, The origin of *Triticum spelta* and its free-theshing hexaploid relatives, J. Heredity 37 (81–89) (1946) 107–116.
[37] E.R. Kerber, Wheat: reconstitution of the tetraploid component (AABB) of hexaploids, Science, 143, 1964, pp. 253–255.
[38] M.E. Kislev, *Triticum parvicoccum* sp. *nov.*, the oldest naked wheat, Isr. J. Bot 28 (1980) 95–107.
[39] J. Dvorak, M.C. Luo, Z.L. Yang, Genetic evidence on the origin of *Triticum aestivum* L, in: A.B. Damania, J. Valkoun, G. Willcox, C.O. Qualset (Eds.), The Origins of Agriculture and Crop Domestication. Proceedings of the Harlan Symposium, ICARDA, Aleppo, 1998, pp. 235–251.
[40] Y. Matsuoka, S. Nasuda, Durum wheat as a candidate for the unknown female progenitor of bread wheat: an empirical study with a highly fertile F_1 hybrid with *Aegilops tauschii* Coss, Theor. Appl. Genet. 109 (2004) 1710–1717.
[41] M.E. Kislev, D. Nadel, I. Carmi, Epipalaeolithic (19,000 BP) cereal and fruit diet at Ohalo II, Sea of Galilee, Israel, Rev. Palaeobot Palynol. 73 (1992) 161–166.
[42] M. Feldman, M.E. Kislev, Domestication of emmer wheat and evolution of free-threshing tetraploid wheat, Isr. J. Plant Sci. 55 (2007) 207–221.
[43] B.S. Gill, W. Li, S. Sood, V. Kuraparthy, B.R. Friebe, K.J. Simons, et al., Genetics and genomics of wheat domestication-driven evolution, Isr. J. Plant Sci. 55 (2007) 223–229.
[44] T. Shizuka, N. Mori, H. Özkan, S. Ohta, Chloroplast DNA haplotype variation within two natural populations of wild emmer wheat (*Triticum turgidum ssp. dicoccoides*) in southern Turkey, Biotechnol. Biotechnol. Equip. 29 (3) (2015) 423–430. Available from: https://doi.org/10.1080/13102818.2015.1012648.
[45] I. Cakmak, A. Torun, E. Millet, M. Feldman, T. Fahima, A.B. Korol, et al., *Triticum dicoccoides*: an important genetic resource for increasing zinc and iron concentration in modern cultivated wheat, Soil. Sci. Plant Nutr. 50 (2004) 1047–1054.
[46] C. Uauy, A. Distelfield, T. Fahima, A. Blechl, J. Dubcovsky, A NAC gene regulating senescence improves grain protein, zinc and iron content in wheat, Science 413 (2006) 1298–1301.
[47] J.H. Peng, Y. Ronin, T. Fahima, M.S. Röder, Y.C. Li, E. Nevo, et al., Domestication quantitative trait loci in Triticum dicoccoides, the progenitor of wheat, Proc. Natl. Acad Sci. U S A 100 (2003) 2489–2494.

[48] W. Xie, E. Nevo, Wild emmer: genetic resources, gene mapping and potential for wheat improvement, Euphytica 164 (2008) 603–614.
[49] T. Krugman, V. Chagué, Z. Peleg, S. Balzergue, J. Just, A.B. Korol, et al., Multilevel regulation and signaling processes associated with adaptation to terminal drought in wild emmer wheat, Funct. Integr. Genom. 10 (2010) 167–186.
[50] E.A. Kellogg, Evolutionary history of the grasses, Plant Physiol. 125 (2001) 1198–1205.
[51] S. Huang, A. Sirikhachornkit, X. Su, J. Faris, B. Gill, R. Haselkorn, et al., Genes encoding plastid acetyl-CoA carboxylase and 3-phosphoglycerate kinase of the Triticum/Aegilops complex and the evolutionary history of polypoid wheat, Proc. Natl. Acad. Sci. U S A 99 (2002) 8133–8138.
[52] L.A. Inda, J.G. Segarra-Moragues, J. Müller, P.B. Peterson, P. Catalán, Dated historical biogeography of the temperate LoHinae (Poaceae, Pooideae) grasses in the northern and southern hemispheres, Mol. Phylogen. Evol. 46 (2008) 932–957.
[53] J. Dvorak, E. Akhunov, Tempos of gene locus deletions and duplications and their relationship to recombination rate during diploid and polyploid evolution in the *Aegilops-Triticum* alliance, Genetics 17 (2005) 323–332.
[54] M. Nesbitt, D. Samuel, From stable crop to extinction? The archaeology and history of the hulled wheats, in: S. Padulosi, K. Hammer, J. Heller (Eds.), Hulled Wheats, International Plant Genetic Resources Institute, Rome, 1996, pp. 41–100.
[55] F. Salamini, H. Özkan, A. Brandolini, R. Schäfer-Pregl, W. Martin, Genetics and geography of wild cereal domestication in the Near East, Nat. Rev. Genet. 3 (2002) 429–441.
[56] R.J. Giles, T.A. Brown, *GluDy* allele variations in *Aegilops tauschii* and *Triticum aestivum*: implications for the origins of hexaploid wheats, Theor. Appl. Genet. 112 (2006) 1563–1572.
[57] T. Sakamura, Kurze Mitteilung, über die Chromosomenzahlen und die Verwandtschaftsverhältnisse der Triticum Arten, Bot. Mag. Tokyo 32 (1918) 151–154.
[58] K. Sax, M.J. Sax, Chromosome behaviour in a genus cross, Genetics 9 (1924) 454–464.
[59] H. Kihara, Cytologische und genetische Studien bei wichtigen Getreidearten mit besonderer Rücksicht auf das Verhalten der Chromosomen und die Sterilität in den Bastarden, Mem. Coll. Sci. Univ. Kyoto Ser. B 1 (1924) 1–200.
[60] P.A. Gandilian, On wild growing *Triticum* species of Armenian SSR, Bot. Zhur 57 (1972) 173–181.
[61] B.L. Johnson, Identification of the apparent B-genome donor of wheat, Can. J. Genet. Cytol. 17 (1975) 21–39.
[62] B.L. Johnson, H.S. Dhaliwal, Reproductive isolation of *Triticum boeticum* and *Triticum urartu* and the origin of the tetraploid wheats, Am. J. Bot 63 (1976) 1088–1094.
[63] V.F. Dorofeev, A.A. Filatenko, E.F. Migushova, R.A. Udaczin, M.M. Jakubziner, in: V.F. Dorofeev, O.N. Korovina (Eds.), Wheat, 1, Flora of Cultivated Plants, Leningrad, Russia, 1979.
[64] P. Perrino, G. Laghetti, L.F. D'Antuono, M. Al Ajlouni, M. Kanbertay, A.T. Szabo, et al., Ecogeographical distribution of hulled wheat species, in: S. Padulosi, K. Hammer, J. Heller (Eds.), Hulled Wheats, International Plant Genetic Resources Institute, Rome, 1996, pp. 102–118.
[65] M. Heun, R. Schäfer-Pregl, D. Klawan, R. Castagna, M. Accerbi, B. Borghi, et al., Site of einkorn wheat domestication identified by DNA fingerprinting, Science 278 (1997) 1312–1314.
[66] B. Kilian, H. Özkan, A. Walther, J. Kohl, T. Dagan, F. Salamini, et al., Molecular diversity at 18 loci in 321 wild and 92 domesticate lines reveal no reduction of nucleotide diversity during *Triticum monococcum* (einkorn) domestication: implications for the origin of agriculture, Mol. Biol. Evol. 24 (2007). 2657–266.

[67] H. Özkan, A. Brandolini, A. Torun, S. Altintas, S. Eker, B. Kilian, et al., Natural variation and identification of microelements content in seeds of einkorn wheat (*Triticum monococcum*), in: H.T. Buck, J.E. Nisi, N. Salomon (Eds.), Wheat Production in Stressed Environments, Springer, New York, 2007, pp. 455–462.
[68] Q. Shao, C. Li, C. Basang, Semi-wild wheat from Xizang (Tibet), in: Proceedings of the 6th International Wheat Genetics Symposium, Kyoto, Japan, 1983, pp. 111–114.
[69] J. Dubcovsky, J. Dvorak, Genome plasticity a key factor in the success of polyploid wheat under domestication, Science 316 (2007) 1862–1866.
[70] A. Haudry, A. Cenci, C. Ravel, T. Bataillon, D. Brunel, C. Poncet, et al., Grinding up wheat: a massive loss of nucleotide diversity since domestication, Mol. Biol. Evol. 24 (2007) 1506–1517.
[71] A. Aaronsohn, G. Schweinfurth, Die Auffindung des wilden Emmers (*Triticum dicoccum*) in Nordpalä stina. Altneuland Monatsschrift für die irtschaft, Erschliess. Palästinas 7–8 (1906) 213–220.
[72] F.A. Körnicke, Wilde Stammformen unserer Kulturweizen. Niederrheiner Gesellsch. f. Natur- und Heilkunde in Bonn, Sitzungsber 46, 1889.
[73] H. Özkan, A. Brandolini, R. Schaefer-Pregl, F. Salamini, AFLP analysis of a collection of tetraploid wheats indicates the origin of emmer and hard wheat domestication in southeast Turkey, Mol. Biol. Evol. 19 (2002) 1797–1801.
[74] H. Özkan, A. Brandolini, C. Pozzi, S. Effgen, J. Wunder, F. Salamini, A reconsideration of the domestication geography of tetraploid wheats, Theor. Appl. Genet. 110 (2005) 1052–1060.
[75] N. Mori, T. Ishii, T. Ishido, S. Hirosawa, H. Watatani, T. Kawahara, et al., Origin of domesticated emmer and common wheat inferred from chloroplast DNA fingerprinting, in: Proceedings of the 10th International Wheat Genetics Symposium, Paestum, 2003, pp. 25–28.
[76] M.C. Luo, Z.L. Yang, F.M. You, T. Kawahara, J.G. Waines, J. Dvorak, The structure of wild and domesticated emmer wheat populations, gene flow between them, and the site of emmer domestication, Theor. Appl. Genet. 114 (2007) 947–959.
[77] A. Weide, S. Riehl, M. Zeidi, N.J. Conard, Using new morphological criteria to identify domesticated emmer wheat at the aceramic Neolithic site of Chogha Golan (Iran), J. Archaeol. Sci. 57 (2015) 109–118.
[78] P. Civáň, Z. Ivaničová, T.A. Brown, Reticulated origin of domesticated emmer wheat supports a dynamic model for the emergence of agriculture in the Fertile Crescent, PLoS One 8 (11) (2013) e81955. Available from: https://doi.org/10.1371/journal.pone.0081955.
[79] A.B. Damania, Diversity of major cultivated plants domesticated in the Near East, in: A.B. Damania, J. Valkoun, G. Willcox, C.O. Qualset (Eds.), The Origins of Agriculture and Crop Domestication. Proceedings of the Harlan Symposium, ICARDA, Aleppo, 1998, pp. 51–64.
[80] M. Tanaka, H. Ishii, Cytogenetical evidence on the speciation of wild tetraploid wheats collected in Iraq, Turkey and Iran, Proc. 4th Intern. Wheat Genetics Symp., 1973, pp. 115–121.
[81] M. Feldman, Identification of unpaired chromosomes in F_1 hybrids involving *Triticum aestivum* and *T. timopheevii*, Can. J. Genet. Cytol. 8 (1966) 144–151.
[82] N. Mori, Y. Kondo, T. Ishii, T. Kawahara, J. Valkoun, C. Nakamura, Genetic diversity and origin of timopheevi wheat inferred by chloroplast DNA fingerprinting, Breed. Sci. 59 (2009) 571–578.
[83] K. Tanno, G. Willcox, How fast was wild wheat domesticated? Science 311 (2006) 1886.
[84] G. Hillman, S. Davies, Measured domestication rates in wild wheats and barley under primitive cultivation, and their archaeological implications, J. World Prehist 4 (1990) 157–222.

[85] B.I. Honne, M. Heun, On the domestication genetics of self-fertilizing plants, Veg. Hist Archaeobot 18 (2009) 269–272.
[86] M. Balter, Seeking agriculture's ancient roots, Science 316 (2007) 1830–1835.
[87] G. Willcox, S. Fornite, L. Herveaux, Early Holocene cultivation before domestication in Northern Syria, Veg. Hist Archaeobot 17 (2008) 313–325.
[88] G. Willcox, Measuring grain size and identifying Near Eastern cereal domestication: evidence from the Euphrates valley, J. Archaeol. Sci. 31 (1) (2004) 45–150.
[89] S. Konishi, T. Izawa, S.Y. Lin, K. Ebana, Y. Fukuta, T. Sasaki, et al., An SNP caused loss of seed shattering during rice domestication, Science 312 (2006) 1392–1396.
[90] W. Cao, G.J. Scoles, P. Hucl, The genetics of rachis fragility and glume tenacity in semi-wild wheat, Euphytica 94 (1997) 119–124.
[91] N. Watanabe, K. Sogiyama, Y. Yamagashi, Y. Sakata, Comparative telosomic mapping of homoeologous genes for brittle rachis in tetraploid and hexaploid wheats, Hereditas 137 (2002) 180–185.
[92] N. Watanabe, The occurrence and inheritance of a brittle rachis phenotype in Italian durum wheat cultivars, Euphytica 142 (2005) 247–251.
[93] W. Li, B.S. Gill, Multiple genetic pathways for seed shattering in the grasses, Funct. Integr. Gen. 6 (2006) 300–309.
[94] H. Sharma, J. Waynes, Inheritance of tough rachis in crosses of *Triticum monococcum* and *Triticum boeoticum*, J. Heredity 7 (1980) 214–216.
[95] Q.F. Chen, C. Yen, J.L. Yang, Chromosome location of the gene for brittle rachis in the Tibetan weedrace of common wheat, Genet. Resour. Crop. Evol. 45 (1998) 21–25.
[96] J. Rong, E. Millet, J. Manisterski, M. Feldman, A new powdery mildew resistance gene: introgression from wild emmer into common wheat and RFLP-based mapping, Euphytica 115 (2000) 121–126.
[97] N. Watanabe, N. Ikebata, The effects of homoeologous group 3 chromosomes on grain colour dependent seed dormancy and brittle rachis in tetraploid wheat, Euphytica 115 (2000) 215–220.
[98] V. Nalam, M.I. Vales, C.J.W. Watson, S.F. Kianian, O. Riera-Lizarazu, Map based analysis of genes affecting the brittle rachis character in tetraploid wheat (*Triticum turgidum*), Theor. Appl. Genet. 112 (2006) 373–381.
[99] R. Avni, M. Nave, O. Barad, K. Baruch, S.O. Twardziok, et al., Wild emmer genome architecture and diversity elucidate wheat evolution and domestication, Science 357 (2017) 93–97.
[100] E.R. Kerber, G.G. Rowland, Origin of the threshing character in hexaploid wheat, Can. J. Genet. Cytol. 16 (1) (1974) 145–154.
[101] R. Villareal, A. Mujeeb-Kazi, S. Rajaram, Inherintance of threshability in synthetic hexaploid by *T. aestivum* crosses, Plant Breed. 115 (1996) 407–409.
[102] C. Jantasuriyarat, M.I. Vales, C.J.W. Watson, O. Riera-Lizarazu, Identifcation and mapping of genetic loci affecting the free-threshing habit and spike compactness in wheat (*Triticum aestivum* L.), Theor. Appl. Genet. 108 (2004) 261–273.
[103] V. Nalam, M.I. Vales, C.J.W. Watson, E.B. Johnson, O. Riera-Lizarazu, Map based analysis of genetic loci on chromosome 2D that affect glume tenacity and thresability, components of the free-threshing habit in common wheat (*Triticum aestivum* L.), Theor. Appl. Genet. 116 (2007) 135–145.
[104] S. Sood, V. Kuraparthy, G. Bai, B.S. Gill, The major threshability genes soft glume (*sog*) and tenacious glume (*Tg*), of diploid and polyploid wheat, trace their origin to independent mutations at non-orthologous loci, Theor. Appl. Genet. 119 (2009) 341–351.
[105] J.D. Faris, Z. Zhang, S. Chao, Map-based analysis of the tenacious glume gene *Tg-B1* of wild emmer and its role in wheat domestication, Gene 542 (2014) 198–208.

[106] R. Tzarfati, Y. Saranga, V. Barak, A. Gopher, A.B. Korol, S. Abbo, Threshing efficiency as an incentive for rapid domestication of emmer wheat, Ann. Bot 112 (5) (2013) 829–837. Available from: https://doi.org/10.1093/aob/mct148.
[107] J. McKey, Species relationships in *Triticum*, Hereditas 2 (1966) 237–276.
[108] E.R. Sears, The aneuploids of common wheat, Res. Bull. Mo. Agric. Exp. Stn 572 (1954) 1–57.
[109] J.W. Snape, W. Law, C.N. Parker, B.B. Worland, A.J. Worland, Genetical analysis of chromosome 5A of wheat and its influence on important agronomic characters, Theor. Appl. Genet. 71 (1985) 518–526.
[110] K. Kato, H. Miura, M. Akiyama, M. Kuroshima, S. Sawada, RFLP mapping of the three major genes, *Vrn1*, *Q* and *B1*, on the long arm of chromosome 5A of wheat, Euphytica 101 (1998) 91–95.
[111] K. Kato, R. Sonokawa, H. Miura, S. Sawada, Dwarfing effect associated with the threshability gene Q on wheat chromosome 5A, Plant Breed. 122 (2003) 489–492.
[112] J.D. Faris, B.S. Gill, Genomic targeting and high resolution mapping of the domestication gene Q in wheat, Genome 45 (2002) 706–718.
[113] J.D. Faris, J.P. Fellers, S.A. Brooks, B.S. Gill, A bacterial artificial chromosome contig spanning the major domestication locus Q in wheat and identification of a candidate gene, Genetics 164 (2003) 311–321.
[114] J.D. Faris, K.J. Simons, Z. Zhang, B.S. Gill, The wheat super domestication gene, Q. Front. Wheat Biosci. 100 (2005) 129–148.
[115] K. Simons, J.P. Fellers, H.N. Trik, Z. Zhang, Y.S. Tai, B.S. Gill, et al., Molecular characterization of the major wheat domestication gene, Q. Genet. 172 (2006) 547–555.
[116] M. Muramatsu, Dosage effect of the spelta gene *q* of hexaploid wheat, Genetics 48 (1963) 469–482.
[117] J.S. Sharma, K.L.D. Running, S.S. Xu, Q. Zhang, A.R. Peters Haugrud, S. Sharma, et al., Genetic analysis of threshability and other spike traits in the evolution of cultivated emmer to fully domesticated durum wheat, Mol. Genet. Gen. 294 (2019) 757–771.
[118] M. Simonetti, M.P. Bellomo, G. Laghetti, P. Perrino, R. Simeone, A. Blanco, Quantitative trait loci influencing free-threshing habit in tetraploid wheats, Genet. Resour. Crop. Evol. 46 (1999) 267–271.
[119] D.Q. Fuller, Contrasting patterns in crop domestication and domestication rates: recent archaeobotanical insights from the Old World, Ann. Bot 100 (2007) 903–924.
[120] E.M. Elias, K.D. Steiger, R.G. Cantrell, Evaluation of lines derived from wild emmer chromosome substitutions II. Agronomic traits, Crop Sci. 36 (1996) 228–233.
[121] E.S. Buckler, J.M. Thornsberry, S. Kresovich, Molecular diversity, structure and domestication of grasses, Genet. Res. 77 (2001) 213–218.
[122] A. Turner, J. Beales, S. Faure, R.P. Dunford, D.A. Laurie, The pseudo-response regulator Ppd-H1 provides adaptation to photoperiod in barley, Science 310 (2005) 1031–1034.
[123] J. Cockram, C. Norris, D.M. O'Sullivan, PCR-based markers diagnostic for spring and winter seasonal growth habit in barley, Crop Sci. 49 (2009) 403–410.
[124] L. Yan, A. Loukoianov, G. Tranquilli, M. Helguera, T. Fahima, J. Dubcovsky, Positional cloning of the wheat vernalization gene VRN1, Proc. Natl. Acad. Sci. U S A 100 (2003) 6263–6268.
[125] A. Worland, The influence of flowering time genes on environmental adaptability in European wheats, Euphytica 89 (1996) 49–57.
[126] S. Griffiths, R.P. Dunford, G. Coupland, D.A. Laurie, The evolution of CONSTANS-like gene families in barley, rice and *Arabidopsis*, Plant Physiol. 131 (2003) 1855–1867.
[127] R.P. Dunford, S. Griffiths, V. Christodoulou, D.A. Laurie, Characterisation of a barley (*Hordeum vulgare* L.) homologue of the *Arabidopsis* flowering time regulator GIGANTEA, Theor. Appl. Genet. 110 (2005) 925–931.

[128] J. Cockram, H. Jones, F.J. Leigh, D. O'Sullivan, W. Powell, D.A. Laurie, et al., Control of flowering time in temperate cereals: genes, domestication, and sustainable productivity, J. Exp. Bot 58 (2007) 1231–1244.

[129] S. Faure, J. Higgins, A. Turner, D.A. Laurie, The FLOWERING LOCUS T-like gene family in barley (*Hordeum vulgare*), Genetics 176 (2007) 599–609.

[130] A. Börner, V. Korzun, A.J. Worland, Comparative genetic mapping of loci affecting plant height and development in cereals, Euphytica 100 (1998) 245–248.

[131] M. Shah, K.S. Gill, P.S. Bezinger, Y. Yen, S.M. Kaeppler, H.M. Ariyarathne, Molecular mapping of loci for agronomic traits on chromosome 3A of bread wheat, Crop Sci. 39 (1999) 1728–1732.

[132] B. Campbell, P.S. Baenziger, K.S. Gill, K.M. Eskridge, H. Budak, M. Erayman, et al., Identification of QTL and environmental interactions associated with agronomic traits on chromosome 3A of wheat, Crop Sci. 43 (2003) 1493–1505.

[133] E. Araki, H. Miura, S. Sawada, Identification of genetic loci affecting amylose content and agronomic traits on chromosome 4A of wheat, Theor. Appl. Genet. 98 (1999) 977–984.

[134] K. Kato, H. Miura, S. Sawada, et al., QTL controlling grain yield and its components on chromosome 5A of wheat, Theor. Appl. Genet. 101 (2000) 933–943.

[135] F. Breseghello, M.E. Sorrells, Association mapping of kernel size and milling quality in wheat (*Triticum aestivum* L.) cultivars, Genetics 172 (2006) 1165–1177.

[136] P. Hedden, The genes of the green revolution, Trends Genet. 19 (2003) 5–9.

[137] J. Peng, D.E. Richards, N.M. Hartley, G.P. Murphy, K.M. Devos, J.E. Flintham, et al., Green revolution' genes encode mutant gibberellin response modulators, Nature 400 (1999) 256–261.

[138] P.M. Chandler, A. Marion-Poll, M. Ellis, F. Gubler, Mutants at the *Slender1* locus of barley *cv* Himalaya: molecular and physiological characterization, Plant Physiol. 129 (2002) 181–190.

[139] P.J. Eastmond, R.L. Jones, Hormonal regulation of gluconeogenesis in cereal aleurone is strongly cultivar-dependent and gibberellin action involves SLENDER1 but not GAMYB, Plant J 44 (2005) 483–493.

[140] W. Cai, H. Morishima, QTL clusters reflect character associations in wild and cultivated rice, Theor. Appl. Genet. 104 (2002) 1217–1228.

[141] M. Ellis, G.J. Rebetzke, F. Azanza, R.A. Richards, W. Spielmeyer, Molecular mapping of gibberellin-responsive dwarfing genes in bread wheat, Theor. Appl. Genet. 111 (2005) 423–430.

[142] D. Atsmon, E. Jacobs, A newly bred 'Gigas' form of bread wheat (*Triticum aestivum* L.): morphological features and thermo-photoperiodic responses, Crop Sci. 17 (1977) 31–35.

[143] W. Spielmeyer, R.A. Richards, Comparative mapping of wheat chromosome 1AS which contains the tiller inhibition gene (*tin*) with rice chromosome 5S, Theor. Appl. Genet. 109 (2004) 1303–1310.

[144] L. Rossini, A. Vecchietti, L. Nicoloso, N. Stein, S. Franzago, F. Salamini, et al., Candidate genes for barley mutants involved in plant architecture: an in silico approach, Theor. Appl. Genet. 112 (2006) 1073–1085.

[145] V. Kuraparthy, S. Sood, P. Chunneja, H.S. Dhaliwal, B.S. Gill, Identification and mapping of tiller inhibition gene (*tin3*) in wheat, Theor. Appl. Genet. 114 (2007) 285–294.

[146] V. Kuraparthy, S. Sood, B.S. Gill, Genomic targeting and mapping of tiller inhibition gene (*tin3*) of wheat using ESTs and synteny with rice, Funct. Integr. Genom. 8 (2008) 33–42.

[147] K.H. Hebelstrup, Differences in nutritional quality between wild and domesticated forms of barley and emmer wheat, Plant Sci. 256 (2017) 1–4.

[148] A. Frary, S. Doğanlar, Comparative genetics of crop plant domestication and evolution, Turkish J. Agric. Forestry 27 (2003) 59–69.

[149] J.R. Harlan, Crop and man, American Society of Agronomy, Crop Science Society of America. Madison, WI, 1992.
[150] A.H. Paterson, Y.R. Lin, Z. Li, K.F. Schertz, J.F. Doebley, S.R.M. Pinson, et al., Convergent domestication of cereal crops by independent mutations at corresponding genetic loci, Science 269 (1995) 1714–1718.
[151] K.S. Gill, B.S. Gill, T.R. Endo, E.V. Boyko, Identification and high-density mapping of gene-rich regions in chromo some group 5 of wheat, Genetics 143 (1996) 1001–1012.
[152] G. Kunzel, L. Korzum, A. Meister, Cytologically integrated physical restriction fragment length polymorphism maps for the barley genome based on translocation breakpoints, Genetics 154 (2000) 397–412.
[153] J.H. Peng, H. Zadeh, G.R. Lazo, J.P. Gustafson, S. Chao, et al., Chromosome bin map of expressed sequence tags in homoeologous group 1 of hexaploid wheat and homoeology with rice and *Arabidopsis*, Genetics 168 (2004) 609–623.
[154] K.S. Gill, B.S. Gill, T.R. Endo, T. Taylor, Identification and high-density mapping of gene-rich regions in chromosome 1 of wheat, Genetics 144 (1996) 1883–1891.
[155] Y.G. Liu, K. Tsunewaki, Restriction fragment length polymorphism analysis of wheat. II. Linkage maps of the RFLP sites in common wheat, Jpn. J. Genet. 66 (1991) 617–633.
[156] S.P. Otto, N.H. Barton, The evolution of recombination: removing the limits to natural selection, Genetics 147 (1997) 879–906.
[157] W.R. Rice, Experimental tests of the adaptive significance of sexual recombination, Nat. Rev. Genet. 3 (2002) 241–251.
[158] E. Nevo, Genomic diversity in nature and domestication, in: R. Henry (Ed.), Diversity and Evolution of Plants. Genotypic and Phenotypic Variation in Higher Plants, CABI Publishing CAB International, Walingford, UK, 2004, pp. 287–315.
[159] Z. Peleg, T. Fahima, S. Abbo, T. Krugman, E. Nevo, D. Yakir, et al., Genetic diversity for drought resistance in wild wheat and its ecogeographical association, Plant Cell Environ. 28 (2005) 176–191.
[160] E. Nevo, A. Beiles, Y. Gutterman, N. Storch, D. Kaplan, Genetic resources of wild cereals in Israel and vicinity: I. Phenotypic variation within and between populations of wild wheat, *Triticum dicoccoides*, Euphytica 33 (1984) 717–735.
[161] C. Uauy, J.C. Brevis, J. Dubcovsky, The high grain protein content gene *Gpc-B1* accelerates senescence and has pleiotropic effects on protein content in wheat, J. Exp. Bot 57 (2006) 2785–2794.
[162] E. Nevo, Genetic resources of wild emmer, *Triticum dicoccoides*, for wheat improvement in the third millennium, Isr. J. Plant Sci. 49 (2001) S77–S93.
[163] E. Nevo, A. Beiles, Amino-acid resources in the wild progenitor of wheats, *Triticum dicoccoides* in Israel: polymorphisms and predictability by ecology and isozymes, Plant Breed. 108 (1992) 190–201.
[164] E. Nevo, A. Grama, A. Beiles, E.M. Golenberg, Resources of high-protein genotypes in wild wheat, *Triticum dicoccoides* in Israel: predictive method by ecology and allozyme markers, Genetics 68 (1986) 215–227.
[165] M.A. Pagnotta, E. Nevo, A. Beiles, E. Porceddu, Wheat storage proteins: glutenin diversity in wild emmer, *Triticum dicoccoides*, in Israel and Turkey. 2. DNA diversity detected by PCR, Theor. Appl. Genet. 91 (1995) 409–414.
[166] S.S. Xu, K. Khan, D.L. Klindworth, J.D. Faris, G. Nygard, Chromosomal location of genes for novel glutenins and gliadins in emmer wheat (*Triticum turgidum* L. var. *dicoccoides*), Theor. Appl. Genet. 108 (2004) 1221–1228.
[167] P.F. Qi, Y.W. Yue, H. Long, Y.M. Wei, Z.H. Yan, Y.L. Zheng, Molecular characterization of α-gliadin genes from wild emmer wheat (*Triticum dicoccoides*), DNA Seq. 17 (2006) 415–421.

[168] X.H. Li, A.L. Wang, Y.H. Xiao, Y.M. Yan, Z.H. He, R. Appels, et al., Cloning and characterization of a novel low molecular weight glutenin subunit gene at the *Glu-A3* locus from wild emmer wheat (*Triticum turgidum* L. var. *dicoccoides*), Euphytica 159 (2007) 181–190.
[169] I. Cakmak, H. Özkan, H.J. Braun, R.M. Welch, V. Romheld, Zinc and iron concentrations in seeds of wild, primitive and modern wheats, Food Nutr. Bull. 21 (2000) 401–403.
[170] E. Nevo, J. Gorham, A. Beiles, Variation for ^{22}Na uptake in wild emmer wheat, *Triticum dicoccoides* in Israel: salt tolerance resources for wheat improvement, J. Exp. Bot 43 (1992) 511–518.
[171] E. Nevo, T. Krugman, A. Beiles, Genetic resources for salt tolerance in the wild progenitors of wheat (*Triticum dicoccoides*) and barley (*Hordeum spontaneum*) in Israel, Plant Breed. 110 (1993) 338–341.
[172] R. Khanna-Chopra, C. Viswanathan, Evaluation of heat stress tolerance in irrigated environment of *T. aestivum* and related species. I. Stability in yield and yield components, Euphytica 106 (1999) 169–180.
[173] Z. Peleg, T. Fahima, A.B. Korol, S. Abbo, T. Krugman, A. Avneri, et al., Genomic dissection of drought resistance in wild emmer wheat x durum wheat population (Abstr), in: Plant & Animal Genomes XIV Conference, San Diego, CA, 2006, 293.
[174] T. Krugman, B. Rubin, O. Levy, J.W. Snape, E. Nevo, RFLP mapping of chlortoluron resistance gene, *Su1*, in bread wheat *Triticum aestivum* and the wild wheat *Triticum dicoccoides* (Abstr), Annu. Wheat Newsl. 41 (1995) 125.
[175] T. Krugman, O. Levy, J.W. Snape, B. Rubin, A.B. Korol, E. Nevo, Comparative RFLP mapping of the chlorotoluron resistance gene (*Su1*) in cultivated wheat (*Triticum aestivum*) and wild wheat (*Triticum dicoccoides*), Theor. Appl. Genet. 94 (1997) 46–51.
[176] J.W. Snape, D. Leckie, B.B. Parker, E. Nevo, The genetical analysis and exploitation of differential responses to herbicides in crop species, in: J.C. Casley, G.W. Cussans, R.K. Atkin (Eds.), Herbicide Resistance in Weeds and Crops, Butterworth-Heinemann, Oxford, UK, 1991, pp. 305–317.
[177] J.G. Moseman, E. Nevo, M.A. El-Morshidy, D. Zohary, Resistance of *Triticum dicoccoides* collected in Israel to infection with *Erysiphe graminis tritici*, Euphytica 33 (1984) 41–47.
[178] E. Nevo, J.G. Moseman, A. Beiles, D. Zohary, Patterns of resistance of Israeli wild emmer wheat to pathogens I. Predictive method by ecology and allozyme genotypes for powdery mildew and leaf rust, Genetica 67 (1985) 209–222.
[179] H. Buerstmayr, M. Stierschneider, B. Steiner, M. Lemmons, M. Griessen, E. Nevo, et al., Variation for resistance to head blight caused by *Fusarium graminearum* in wild emmer (*Triticum dicoccoides*) originating from Israel, Euphytica 130 (2003) 17–23.
[180] R.W. Stack, E.M. Elias, J. Fetch, J.D. Miller, L.R. Joppa, *Fusarium* head blight reaction of Langdon durum—*Triticum dicoccoides* chromosome substitution lines, Crop. Sci. 42 (2002) 637–642.
[181] R.E. Oliver, R.W. Stack, J.D. Miller, X. Cai, Reaction of wild emmer wheat accessions to *Fusarium* head blight, Crop Sci. 47 (2007) 893–899.
[182] J.G. Moseman, E. Nevo, Z.K. Gerechter-Amitai, M.A. El-Morshidy, D. Zohary, Resistance of *Triticum dicoccoides* collected in Israel to infection with *Puccinia recondita tritici*, Crop Sci. 25 (1985) 262–265.
[183] Y. Anikster, J. Manisterski, D.L. Long, K.J. Leonard, Leaf rust and stem rust resistance in *Triticum dicoccoides* populations in Israel, Plant Dis. 89 (2005) 55–62.
[184] R.M. Hunger, J.L. Sherwood, R.E. Pennington, B.F. Carver, E. Nevo, Reaction of native populations of *Triticum dicoccoides* to wheat soilborne mosaic, Biological & Cultural Tests for Control of Plant Disease, 7, APS Press, St. Paul, MI, 1992.

[185] C.G. Chu, S.S. Xu, J.D. Faris, E. Nevo, T.L. Friesen, Seedling resistance to tan pot and *Stagonospora nodorum* leaf blotch in wild emmer wheat (*Triticum turgidum* ssp. *dicoccoides*), Plant Dis. 92 (2008) 1229–1236.

[186] P.K. Singh, J.L. Gonzalez-Hernandez, M. Mergoum, S.T.B. Ali, S.F. Adhikari, E.M. Kianian, et al., Identification and molecular mapping of a gene conferring resistance to *Pyrenophora tritici-repentis* race 3 in tetraploid wheat, Phytopathology 96 (2006) 885–889.

[187] P.K. Singh, M. Mergoum, T.B. Adhikari, S.F. Kianian, E.M. Elias, Chromosomal location of genes for seedling resistance to tan spot and *Stagonospora nodorum* blotch in tetraploid wheat, Euphytica 155 (2007) 27–34.

[188] J.R. Wang, Y.M. Wei, X.Y. Long, Z.H. Yan, E. Nevo, B.R. Baum, et al., Molecular evolution of dimeric α-amylase inhibitor genes in wild emmer wheat and its ecological association, BMC Evol. Biol. 8 (2008) 91. Available from: https://doi.org/10.1186/1471-2148-8-91.

[189] E. Millet, J.K. Rong, C.O. Qualset, P.E. Mcguire, M. Bernard, P. Sourdille, et al., Grain yield and grain protein percentage of common wheat lines with wild emmer chromosome-arm substitutions, Euphytica 195 (2014) 69–81.

[190] L. Merchuk-Ovnat, T. Fahima, T. Krugman, Y. Saranga, Ancestral QTL alleles from wild emmer wheat improve grain yield, biomass and photosynthesis across enviroinments in modern wheat, Plant Sci. 251 (2016) 23–34.

[191] P. Chhuneja, J.K. Arora, P. Kaur, S. Kaur, K. Singh, Characterization of wild emmer wheat *Triticum dicoccoides* germplasm for vernalization alleles, J. Plant Biochem. Biotechnol. 24 (2014) 249–253.

[192] W. Zhou, S. Wu, M. Ding, J. Li, Z. Shi, W. Wei, et al., Mapping of *Ppd-B1*, a major candidate gene for late heading on wild emmer chromosome arm 2BS and assessment of its interactions with early heading QTL on 3AL, PLoS One 11 (2) (2016) e0147377. Available from: https://doi.org/10.1371/journal.pone.0147377.

[193] J.D. Faris, Z. Zhang, D.F. Garvin, S. Xu, Molecular and comparative mapping of genes governing spike compactness from wild emmer wheat, Mol. Genet. Gen. 289 (2014) 641–651.

[194] R.A. McIntosh, Breeding wheat for resistance to biotic stress, Euphytica 100 (1998) 19–34.

[195] A.P. Roelfes, R.P. Singh, E.E. Saari, Rust Diseases of Wheat: Concepts and Methods of Disease Management, 81, CIMMYT, Mexico, D.F., Mexico, 1992.

[196] Z.K. Gerechter-Amitai, R.W. Stubbs, A valuable source of yellow rust resistance in Israeli population of wild emmer *Triticum dicoccoides* Koern, Euphytica 19 (1970) 12–21.

[197] Z.K. Gerechter-Amitai, C.H. van Silfhout, A. Grama, F. Kleitman, *Yr15*—a new gene for resistance to *Puccinia striiformis* in *Triticum dicoccoides* sel. G-25, Euphytica 43 (1989) 187–190.

[198] Z.K. Gerechter-Amitai, A. Grama, C.H. van Silfhout, F. Kleitman, Resistance to yellow rust in *Triticum dicoccoides*. II. Crosses with resistant *dicoccoides* sel. G25, Neth. J. Plant Pathol. 95 (1989) 79–83.

[199] C.H. van Silfhout, A. Grama, Z.K. Gerechter-Amitai, F. Kleitman, Resistance to yellow rust in *Triticum dicoccoides*. I. Crosses with susceptible *Triticum durum*, Neth. J. Plant Pathol. 95 (1989) 73–78.

[200] C.H. van Silfhout, G.H.J. Kema, Z.K. Gerechter-Amitai, Major genes for resistance to yellow rust in wild emmer wheat. Ph.D. Thesis in: C.H. Van Silfhout (Ed.), Identification and Characterization of Resistance to Yellow Rust and Powdery Mildew in Wild Emmer Wheat and Their Transfer to Bread Wheat, Research Institute for Plant Protection, Wageningen, The Netherlands, 1989, pp. 5–15.

[201] V. Chagué, T. Fahima, T. Dahan, G.L. Sun, A.B. Korol, Y.I. Ronin, et al., Isolation of microsatellite and RAPD markers flanking the *Yr15* gene of wheat using NILs and bulked segregant analysis, Genome 42 (1999) 1050–1056.

[202] R.A. McIntosh, J. Silk, T.T. The, Cytogenetic studies in wheat XVII. Monosomic analysis and linkage relationships of gene *Yr15* for resistance to stripe rust, Euphytica 89 (1996) 395–399.
[203] G.L. Sun, T. Fahima, A.B. Korol, T. Turpeinen, A. Grama, Y.I. Ronin, et al., Identification of molecular markers linked to the *Yr15* stripe rust resistance gene of wheat originated in wild emmer wheat *Triticum dicoccoides*, Theor. Appl. Genet. 95 (1997) 622–628.
[204] J.H. Peng, T. Fahima, M.S. Röder, Y.C. Li, A. Dahan, A. Grama, et al., Microsatellite tagging of stripe rust resistance gene *YrH52* derived from wild emmer wheat, *Triticum dicoccoides*, and suggestive negative crossover interference on chromosome 1B, Theor. Appl. Genet. 98 (1999) 862–872.
[205] J.H. Peng, T. Fahima, M.S. Röder, Q.Y. Huang, A. Dahan, C. Li, et al., High-density molecular map of chromosome region harboring stripe-rust resistance genes *YrH52* and *Yr15* derived from wild emmer wheat, *Triticum dicoccoides*, Genetica 109 (2000) 199–210.
[206] J.H. Peng, T. Fahima, M.S. Röder, Y.C. Li, A. Grama, E. Nevo, Microsatellite high-density mapping of stripe-rust resistance gene *YrH52* region on chromosome 1B and evaluation of its marker-assisted selection in F2 generation in wild emmer wheat, N. Phytol. 146 (2000) 141–154.
[207] C. Uauy, J.C. Brevis, X.M. Chen, I. Khan, L. Jackson, O. Chicaiza, et al., High temperature adult-plant (HTAP) stripe rust resistance gene *Yr36* from *Triticum turgidum* ssp. *dicoccoides* is closely linked to the grain protein content locus *Gpc-B1*, Theor. Appl. Genet. 122 (2005) 97–105.
[208] D. Fu, C. Uauy, A. Distelfeld, A. Blechl, L. Epstein, X. Chen, et al., A kinase-START gene confers temperature-dependent resistance to wheat stripe rust, Science 323 (2009) 1357–1360.
[209] D. Raats, E. Yaniv, A. Distelfeld, R. Ben-David, J. Shanir, V. Bocharova, et al., Application of CAPS markers for genomic studies in wild emmer wheat, in: Y. Shavrukov (Ed.), Cleaved Amplified Polymorphic Sequences, Nova Science Publishers, Inc, Hauppauge, NY, USA, 2014.
[210] E. Yaniv, D. Raats, Y. Ronin, A.B. Korol, A. Grama, H. Bariana, et al., Evaluation of marker-assisted selection for the stripe rust resistance gene *Yr15*, introgressed from wild emmer wheat, Mol. Breed. 35 (2015) 43. Available from: https://doi.org/10.1007/s11032-015-0238-0.
[211] Z. Wang, J. Xie, L. Guo, D. Zhang, G. Li, et al., Molecular mapping of *YrTZ2*, a stripe rust resistance gene in wild emmer accession TZ-2 and its comparative analyses with *Aegilops tauschii*, J. Integr. Agr. 17 (6) (2018) 1267–1275.
[212] V. Klymiuk, A. Fatiukha, D. Raats, V. Bocharova, L. Huang, et al., Three previously characterized resistances to yellow rust are encoded by a single locus *Wtk1*, J. Exp. Bot. 71 (9) (2020) 2561–2572. Available from: https://doi.org/10.1093/jxb/eraa020.
[213] A. Zakari, R.A. McIntosh, M.S. Hovmoller, C.R. Wellings, M.R. Shariflou, M. Hayden, et al., Recombination of Yr15 and Yr24 in chromosome 1BS, in: N.E. Pogna, M. Romano, E.A. Pogna, G. Galterio (Eds.), Proceedings of the 10th International Wheat Genetics Symposium, vol. 1, Instituto Sperimentale per la Cerealicoltura, Rome, Italy, 2003.
[214] B. Abdollahi Mandoulakani, E. Yaniv, R. Kalendar, D. Raats, H.S. Bariana, M.R. Bihamta, et al., Development of IRAP- and REMAP-derived SCAR markers for marker-assisted selection of the stripe rust resistance gene *Yr15* derived from wild emmer wheat, Theor. Appl. Genet. 128 (2015) 211–219.
[215] T.T. The, E. Nevo, R.A. McIntosh, Response of Israeli wild emmer to selected Australian pathotypes of *Puccinia* species, Euphytica 71 (1993) 75–81.

[216] M.S. Hovmoller, Sources of seedling and adult plant resistance to *Puccinia striiformis* f.sp. *tritici* in European wheats, Plant Breed. 126 (2007) 225–233.
[217] H. Sela, S. Ezrati, P. Ben-Yehuda, J. Manisterski, E. Akhunov, J. Dvorak, et al., Linkage disequilibrium and association analysis of stripe rust resistance in wild emmer wheat (*Triticum turgidum ssp. dicoccoides*) population in Israel, Theor. Appl. Genet. 127 (2014) 2453–2463.
[218] V. Klymiuk, E. Yaniv, L. Huang, D. Raats, A. Fatiukha, et al., Cloning of the wheat *Yr15* resistance gene sheds light on the plant tandem kinase-pseudokinase family. *Nature*, Commun 9 (2018) 3735. Available from: https://doi.org/10.1038/s41467-018-06138-9.
[219] G.F. Marais, Z.A. Pretorius, C.R. Wellings, B. McCallum, A.S. Marais, Leaf rust and stripe rust resistance genes transferred to common wheat from *Triticum dicoccoides*, Euphytica 143 (2005) 115–123.
[220] N.A. Dadkhodaie, H. Karaoglou, C.R. Wellings, R.F. Park, Mapping genes *Lr53* and *Yr35* on the short arm of chromosome 6B of common wheat with microsatellite markers and studies of their association with *Lr36*, Theor. Appl. Genet. 122 (2011) 479–487.
[221] L. Huang, H. Sela, L. Feng, Q. Chen, T. Krugman, J. Yan, et al., Distribution and haplotype diversity of *WKS* resistance genes in wild emmer wheat natural populations, Theor. Appl. Genet. 129 (2016) 921–934.
[222] D. Bai, D.R. Knott, Genetic studies of leaf and stem rust resistance in six accessions of *Triticum turgidum var. dicoccoides*, Genome 37 (1994) 405–409.
[223] E. Nevo, Z.K. Gerechter-Amitai, A. Beiles, Resistance of wild emmer wheat to stem rust: ecological, pathological and allozyme associations, Euphytica 53 (1991) 121–130.
[224] F.A.G. Bennet, Resistance to powdery mildew in wheat: a review of its use in agriculture and breeding programs, Plant Pathol. 33 (1984) 279–300.
[225] E. Nevo, Genetic resources of wild emmer *Triticum dicoccoides* for wheat improvement: news and views, in: Z.S. Li, Z.Y. Xin (Eds.), Proceedings of the 8th International Wheat Genetics Symposium, China Agricultural Scientech Press, Beijing, China, 20–25 July 1993, 1995, pp 79–87.
[226] S.M. Reader, T.E. Miller, The introduction into bread wheat of a major gene for resistance to powdery mildew from wild emmer wheat, Euphytica 53 (1991) 57–60.
[227] H. Yin, Y. Ben-Abu, H. Wang, A. Li, E. Nevo, L. Kong, Natural selection causes adaptive genetic resistance in wild emmer wheat against powdery mildew at "Evolution Canyon" microsite, Mt. Carmel, Israel, PLoS One 10 (4) (2015) e0122344. Available from: https://doi.org/10.1371/journal.pone.0122344.
[228] Z.Y. Liu, Q.X. Sun, Z.F. Ni, E. Nevo, T.M. Yang, Molecular characterization of a novel powdery mildew resistance gene *Pm30* in wheat originating from wild emmer, Euphytica 123 (2002) 21–29.
[229] V. Mohler, F.J. Zeller, G. Wenzel, S.L.K. Hsam, Chromosomal location of genes for resistance to powdery mildew in common wheat (*Triticum aestivum* L. em Thell.). 9. Gene *MlZec1* from the *Triticum dicoccoides*-derived wheat line Zecoi-1, Euphytica 142 (2005) 161–167.
[230] X.M. Chen, Y.H. Luo, X.C. Xia, L.Q. Xia, X. Chen, Z.L. Ren, et al., Chromosomal location of powdery mildew resistance gene *Pm16* in wheat using SSR marker analysis, Plant Breed. 124 (2005) 225–228.
[231] R.A. McIntosh, K.M. Devos, J. Dubcovsky, W.J. Rogers, C.F. Morris, R. Appels, et al., Catalogue of gene symbols for wheat: 2007 supplement. <http://wheat.pw.usda.gov/ggpages/awn/53/Textfile/WGC.html>, 2007. Accessed on 28 March 2021.
[232] A. Blanco, A. Gadaleta, A. Cenci, A.V. Carluccio, A.M.M. Abdelbacki, R. Simeone, Molecular mapping of the novel powdery mildew resistance gene *Pm36*

introgressed from *Triticum turgidum* var. *dicoccoides* in durum wheat, Theor. Appl. Genet. 117 (2008) 135–142.

[233] M.D. Ashraful Alam, F. Xue, M. Ali, C. Wang, W. Ji, Identification and molecular mapping of powdery mildew resistance gene *PMG25* in common wheat originated from wild emmer (*Triticum Turgidum var. Dicoccoides*), Pak. J. Bot 45 (1) (2013) 203–208.

[234] M. Geng, J. Zhang, F. Peng, X. Liu, X. Lv, Y. Mi, et al., Identification and mapping of *MLIW30*, a novel powdery mildew resistance gene derived from wild emmer wheat, Mol. Breed. 36 (2016) 130. Available from: https://doi.org/10.1007/s11032-016-0553-0.

[235] D. Zhang, K. Zhu, L. Dong, Y. Liang, G. Li, T. Fang, et al., Wheat powdery mildew resistance gene *Pm64* derived from wild emmer (*Triticum turgdium var. dicoccoides*) is tightly linked in repulsion with stripe rust resistance gene *Yr5*, Crop. J. 7 (2019) 761–770.

[236] C.J. Xie, Q.X. Sun, Z.F. Ni, T. Yang, E. Nevo, T. Fahima, Chromosomal location of a *Triticum dicoccoides*-derived powdery mildew resistance gene in common wheat by using microsatellite markers, Theor. Appl. Genet. 106 (2003) 341–345.

[237] W.L. Xie, Identification and molecular mapping of powdery mildew resistance genes derived from wild relatives of wheat, Ph.D. Thesis, University of Haifa, Israel, 2006.

[238] R. Ben-David, W. Xie, Z. Peleg, Y. Saranga, A. Dinoor, T. Fahima, Identification and mapping of *PmG16*, a powdery mildew resistance gene derived from wild emmer wheat, Theor. Appl. Genet. 121 (2010) 499–510.

[239] X. Ji, C.J. Xie, Z.F. Ni, T. Yang, E. Nevo, T. Fahima, et al., Identification and genetic mapping of a powdery mildew resistance gene in wild emmer (*Triticum dicoccoides*) accession IW72 from Israel, Euphytica 159 (2008) 385–390.

[240] S. Ouyang, D. Zhang, J. Han, X. Zhao, Y. Cui, et al., Fine physical and genetic mapping of powdery mildew resistance gene *MlIW172* originating from wild emmer (*Triticum dicoccoides*), PLoS One 9 (6) (2014) e100160. Available from: https://doi.org/10.1371/journal.pone.0100160.

[241] M. Li, L. Dong, B. Li, Z. Wang, J. Xie, et al., A CNL protein in wild emmer wheat confers powdery mildew resistance, New Phytol. 228 (3) (2020) 1027–1037. Available from: https://doi.org/10.1111/nph.16761.

[242] W. Hua, Z. Liu, J. Zhu, C. Xie, T. Yang, Y. Zhou, et al., Identification and genetic mapping of *pm42*, a new recessive wheat powdery mildew resistance gene derived from wild emmer (*Triticum turgidum* var. *dicoccoides*), Theor. Appl. Genet. 119 (2009) 223–230.

[243] H. Zhang, H. Guan, J. Li, J. Zhu, C. Xie, Y. Zhou, et al., Genetic and comparative genomics mapping reveals that a powdery mildew resistance gene *Ml3D232* originating from wild emmer co-segregates with an NBS-LRR analog in common wheat (*Triticum aestivum L.*), Theor. Appl. Genet. 121 (2010) 1613–1621.

[244] D. Zhang, S. Ouyang, L. Wang, Y. Cui, Q. Wu, Y. Lang, et al., Comparative genetic mapping revealed powdery mildew resistance gene *MlWE4* derived from wild emmer is located in same genomic region of *Pm36* and *Ml3D232* on chromosome 5BL, J. Integr. Agric. 14 (4) (2015) 603–609.

[245] M. Saidou, C. Wang, M.A. Alam, C. Chen, W. Ji, Genetic analysis of powdery mildew resistance gene using SSR markers in common wheat originated from wild emmer (*Triticum dicoccoides* Thell), Turkish J. Field Crop. 21 (1) (2016) 10–15.

[246] T. Meidaner, Breeding wheat and rye for resistance to *Fusarium* diseases, Plant Breed. 116 (1997) 201–220.

[247] A. Mesterhazy, Breeding for resistance to FHB in wheat. Mexico in: H.J. Duben, et al. (Eds.), Fusarium Head Scab: Global Status and Future Prospects, CIMMYT, Mexico, DF, 1997, pp. 79–85.

[248] J.D. Miller, R.W. Stack, L.R. Joppa, Evaluation of *Triticum turgidum* L. var. dicoccoides for resistance to Fusarium head blight and stem rust, in: A.E. Slinkard (Ed.), Proceedings of the 9th International Wheat Genetics Symposium, vol. 3, University Extension Press, University of Saskatchewan, Saskatoon, Canada, 1998.
[249] C.D. Otto, S.F. Kianian, E.M. Elias, R.W. Stack, L.R. Joppa, Genetic dissection of a major Fusarium head blight QTL in tetraploid wheat, Plant Mol. Biol. 48 (2002) 625–632.
[250] S. Kumar, R.W. Stack, T.L. Friesen, J.D. Faris, Identification of a novel Fusarium head blight resistance quantitative trait locus on chromosome 7A in tetraploid wheat, Phytopathology 97 (2007) 592–597.
[251] D.F. Garvin, R.W. Stack, J.M. Hansen, Quantitative trait locus mapping of increased *Fusarium* head blight susceptibility associated with a wild emmer wheat chromosome, Phytopathology 99 (2009) 447–452.
[252] K.F. Finney, W.T. Yamazaki, V.L. Youngs, G.L. Rubenthaler, Quality of hard, soft, and durum wheats, in: E.G. Heyne (Eds.), Wheat and Wheat Improvement, Agron. Monogr. 13, 2nd ed. ASA, CSSA, and SSSA, Madison, WI, 1987, pp. 677–748.
[253] J.W. Dick, V.L. Youngs, Evaluation of durum wheat, semolina, and pasta in the United States, in: G. Fabriani, C. Lintas (Eds.), Durum Wheat: Chemistry and Technology, American Association of Cereal Chemists, St. Paul, MN, 1988, pp. 237–248.
[254] Z.K. Gerechter-Amitai, A. Grama, Use of alien genes in wheat breeding, Annu. Wheat Newsl. 23 (1977) 57–58.
[255] A.A. Levy, M. Feldman, Location of genes for high grain protein percentage and other quantitative traits in wild wheat *Triticum turgidum* var. *dicoccoides*, Euphytica 41 (1989) 113–122.
[256] L.R. Joppa, R.G. Cantrell, Chromosomal location of genes for grain protein content of wild tetraploid wheat, Crop. Sci. 30 (1990) 1059–1064.
[257] L.R. Joppa, G.E. Hart, G.A. Hareland, Mapping a QTL for grain protein in tetraploid wheat (*Triticum turgidum* L.) using a population of recombinant inbred chromosome lines, Crop. Sci. 37 (1997) 1586–1589.
[258] A. Mesfin, R.C. Frohberg, J.A. Anderson, RFLP markers associated with high grain protein from *Triticum turgidum* L. var. *dicoccoides* introgressed into hard red spring wheat, Crop. Sci. 39 (1999) 508–513.
[259] P.W. Chee, E.M. Elias, J.A. Anderson, S.F. Kianian, Evaluation of a high grain protein QTL from *Triticum turgidum* L. var. *dicoccoides* in an adapted durum wheat background, Crop. Sci. 41 (2001) 295–301.
[260] S. Olmos, A. Distelfeld, O. Chicaiza, A.R. Schlatter, T. Fahima, V. Echenique, et al., Precise mapping of a locus affecting grain protein content in durum wheat, Theor. Appl. Genet. 107 (2003) 1243–1251.
[261] A. Distelfeld, C. Uauy, S. Olmos, A.R. Schlatter, J. Dubcovsky, T. Fahima, Microcolinearity between a 2-cM region encompassing the grain protein content locus *Gpc-6B1* on wheat chromosome 6B and a 350-kb region on rice chromosome 2, Funct. Integr. Genom. 4 (2004) 59–66.
[262] A. Distelfeld, C. Uauy, T. Fahima, J. Dubcovsky, Physical map of the wheat high-grain protein content gene *Gpc-B1*and development of a high-throughput molecular marker, N. Phytol. 169 (2006) 753–763.
[263] J.L. Gonzalez-Hernandez, E.M. Elias, S.F. Kianian, Mapping genes for grain protein concentration and grain yield on chromosome 5B of *Triticum turgidum* (L.) var. *dicoccoides*, Euphytica 139 (2004) 217–225.
[264] A. Blanco, R. Simeone, A. Gadaleta, Detection of QTL for grain protein content in durum wheat, Theor. Appl. Genet. 112 (2006) 1195–1204.
[265] A. Distelfeld, T. Fahima, Wild emmer wheat as a source for high-grain protein genes: map-based cloning of *Gpc-B1*, Isr. J. Plant Sci. 55 (2007) 297–306.

[266] D.L. Klindworth, G.A. Hareland, E.M. Elias, J.D. Faris, S. Chao, S. Xu, Agronomic and quality characteristics of two new sets of Langdon durum–wild emmer wheat chromosome substitution lines, J. Cereal Sci. 50 (2009) 29–35.
[267] B.P. Salmanowicz, M. Langner, B. Mrugalska, D. Ratajczaka, A.G. Gorny, Grain quality characteristics and dough rheological properties in Langdon durum-wild emmer wheat chromosome substitution lines under nitrogen and water deficits, J. Sci. Food Agric. 97 (2017) 2030–2041.
[268] L. Qin, Y. Liang, D. Yang, L. Sun, G. Xia, S. Liu, LMW glutenin subunit genes from wild emmer wheat (*Triticum turgidum ssp. dicoccoides*) in relation to *Glu-3* evolution, Dev. Genes Evol. 225 (2015) 31–37.
[269] D. Zhang, Y. Yuan, Y. Su, S. Li, Analysis of dough rheological property and gluten quality characteristics in wild emmer wheat (*Triticum dicoccoides* (Körn. ex Asch. et Graebn.) Schweinf, Genet. Resour. Crop. Evol. 63 (2016) 675–683.
[270] D. Zhang, T. He, H. Liang, L. Huang, Y. Su, Y. Li, et al., The flour quality and related molecular characterization of HMW-GS genes from wild emmer wheat accession TD-256, J. Agric. Food Chem. 64 (24) (2016) 5128–5136.
[271] A. Distelfeld, I. Cakmak, Z. Peleg, L. Ozturk, A.M. Yazici, H. Budak, et al., Multiple QTL-effects of wheat *Gpc-B1* locus on grain protein and micronutrient concentrations, Physiol. Plant 129 (2007) 635–643.
[272] Z. Peleg, I. Cakmak, L. Ozturk, A. Yazici, J. Yan, H. Budak, et al., Quantitative trait loci conferring grain mineral nutrient concentrations in durum wheat × wild emmer wheat RIL population, Theor. Appl. Genet. 119 (2009) 353–369.
[273] M. Chatzav, Z. Peleg, L. Ozturk, A. Yazici, T. Fahima, I. Cakmak, et al., Genetic diversity for grain nutrients in wild emmer wheat: potential for wheat improvement, Ann. Bot 105 (2010) 1211–1220.
[274] A. Fatiukha, V. Klymiuk, Z. Peleg, Y. Saranga, I. Cakmak, T. Krugman, et al., Variation in phosphorus and sulfur content shapes the genetic architecture and phenotypic associations within the wheat grain ionome, Plant J. (2019). Available from: https://doi.org/10.1111/tpj.14554.
[275] S. Husain, R. Munns, A.G. Condon, Effect of sodium exclusion trait on chlorophyll retention and growth of durum wheat in saline soil, Aust. J. Agric. Res. 54 (2003) 589–597.
[276] K. Feng, L. Cui, S. Lv, J. Bian, M. Wang, W. Song, et al., Comprehensive evaluating of wild and cultivated emmer wheat (*Triticum turgidum* L.) genotypes response to salt stress, Plant Growth Regul. 84 (2018) 261–273.
[277] L. Chen, J. Ren, H. Shi, X. Chen, M. Zhang, Y. Pan, et al., Physiological and molecular responses to salt stress in wild emmer and cultivated wheat, Plant Mol. Biol. Rep 31 (2013) 1212–1219.
[278] L. Chen, J. Ren, H. Shi, Y. Zhang, Y. You, J. Fan, et al., *TdCBL6*, a calcineurin B-like gene from wild emmer wheat (*Triticum dicoccoides*), is involved in response to salt and low-K^+ stresses, Mol. Breed. 35 (2015) 50. Available from: https://doi.org/10.1007/s11032-015-0229-1.
[279] N.Z. Ergen, J. Thimmapuram, H.J. Bohnert, H. Budak, Transcriptome pathways unique to dehydration tolerant relatives of modern wheat, Funct. Integr. Genom. 9 (2009) 377–396.
[280] Z. Peleg, T. Fahima, T. Krugman, S. Abbo, D. Yakir, A.B. Korol, et al., Genomic dissection of drought resistance in durum wheat x wild emmer wheat recombinant inbred line population, Plant Cell Environ. 32 (2009) 758–779.
[281] L. Merchuk-Ovnat, T. Fahima, J.E. Ephrath, T. Krugman, Y. Saranga, QTL alleles from wild emmer wheat enhance root development under drought in modern wheat, Front. Plant Sci. 8 (2017) 703. Available from: https://doi.org/10.3389/fpls.2017.00703.

[282] S. Lucas, E. Durmaz, B.A. Akpınar, H. Budak, The drought response displayed by a DRE-binding protein from *Triticum dicoccoides*, Plant Physiol. Biochem. 49 (2011) 346–351.
[283] S. Lucas, E. Dogan, H. Budak, TMPIT1 from wild emmer wheat: first characterisation of a stress-inducible integral membrane protein, Gene 483 (2011) 22–28.
[284] B.A. Akpinar, M. Kantar, H. Budak, Root precursors of microRNAs in wild emmer and modern wheats show major differences in response to drought stress, Funct. Integr. Genom. 15 (2015) 587–598.
[285] F. He, R. Pasam, F. Shi, S. Kant, G. Keeble-Gagnere, et al., Exome sequencing highlights the role of wild-relative introgression in shaping the adaptive landscape of the wheat genome, Nat. Genet. 51 (2019) 896–904.
[286] Z. Jiang, H. Wu, Z. Wang, J. Hu, J. Yuan, H. Chen, et al., Enriching novel *Glu-Ax* alleles and significantly strengthening gluten properties of common wheat through wide hybridization with wild emmer, J. Cereal Sci. 76 (2017) 271–279.
[287] Z. Wang, L. Huang, B. Wu, J. Hu, Z. Jiang, P. Qi, et al., Characterization of an integrated active *Glu-1Ay* allele in common wheat from wild emmer and its potential role in flour improvement, Int. J. Mol. Sci. 19 (2018) 923. Available from: https://doi.org/10.3390/ijms19040923.
[288] S.D. Tanksley, S.R. McCouch, Seed banks and molecular maps: unlocking genetic potential from the wild, Science 277 (1997) 1063–1066.
[289] E. Nevo, Genetic resources of wild emmer wheat: structure, evolution, and application in breeding, in: S. Sakamoto (Eds.), Proceedings of the 6th International Wheat Genetics Symposium. Kyoto University, Kyoto, Japan, 1983, pp. 421–431.
[290] E. Nevo, Genetic resources of wild cereals and crop improvement: Israel, a natural laboratory, Israeli J. Bot 235 (1986) 255–278.
[291] E. Nevo, Genetic resources of wild emmer wheat revisited: genetic evolution, conservation and utilization, in: T.E. Miller, R.M.D. Koebner (Eds.), Proceedings of the 7th International Wheat Genetics Symposium, Inst of Plant Sci Res, Cambridge, UK, 1989, pp. 121–126.
[292] E. Nevo, Genetic diversity in wild cereals: regional and local studies and their bearing on conservation *ex-situ* and *in-situ*, Genet. Resour. Crop. Evol. 45 (1998) 355–370.
[293] E. Nevo, Evolution of wild wheat and barley and crop improvement: studies at the Institute of Evolution, Plant Sci. 55 (2007) 251–263.
[294] R.C. Johnson, Gene banks pay big dividends to agriculture, the environment, and human welfare, PLoS Biol. 6 (6) (2008) e148. Available from: https://doi.org/10.1371/journal.pbio.0060148.
[295] J. Ren, L. Chen, D. Sun, F.M. You, J. Wang, Y. Peng, et al., SNP-revealed genetic diversity in wild emmer wheat correlates with ecological factors, BMC Evol. Biol. 13 (2013) 169.
[296] J. Ren, L. Chen, X. Jin, M. Zhang, F.M. You, J. Wang, et al., Solar radiation-associated adaptive SNP genetic differentiation in wild emmer wheat, Triticum dicoccoides, Front. Plant Sci. 8 (2017) 258.
[297] A. Venetsky, A. Levy-Zamir, V. Khasdan, K. Domb, K. Kashkush, Structure and extent of DNA methylation-based epigenetic variation in wild emmer wheat (*T. turgidum ssp. dicoccoides*) populations, BMC Plant Biol. 15 (2015) 200. Available from: https://doi.org/10.1186/s12870-015-0544-z.
[298] P. Parsons, Environments and evolution: interactions between stress, resource inadequacy and energetic efficiency, Biol. Rev. 80 (2005) 589–610.
[299] S.C. Schuster, Next-generation sequencing transforms today's biology, Nat. Methods 5 (2008) 16–18.
[300] J. Jia, S. Zhao, X. Kong, Y. Li, G. Zhao, et al., *Aegilops tauschii* draft genome sequence reveals a gene repertoire for wheat adaptation, Nature 496 (2013) 91–95.

[301] H. Ling, S. Zhao, D. Liu, J. Wang, H. Sun, et al., Draft genome of the wheat A-genome progenitor *Triticum urartu*, Nature 496 (2013) 87–90.
[302] H. Ling, B. Ma, X. Shi, H. Liu, L. Dong, et al., Genome sequence of the progenitor of wheat A subgenome *Triticum urartu*, Nature 557 (2018) 424–428.
[303] The International Wheat Genome Sequencing Consortium (IWGSC), R. Appels, K. Eversole, N. Stein, C. Feuillet, et al., Shifting the limits in wheat research and breeding using a fully annotated reference genome, Science 361 (2018) eaar7191. Available from: https://doi.org/10.1126/science.aar7191.
[304] R.H. Ramírez-González, P. Borrill, D. Lang, S.A. Harrington, J. Brinton, et al., The transcriptional landscape of polyploid wheat, Science 361 (2018) eaar6089. Available from: https://doi.org/10.1126/science.aar6089.
[305] K. Kashkush, M. Feldman, A.A. Levy, Gene loss, silencing and activation in a newly synthesized wheat allotetraploid, Genetics 160 (2002) 1651–1656.
[306] A.A. Levy, M. Feldman, Genetic and epigenetic reprogramming of the wheat genome upon allopolyploidization, Biol. J. Linn. Soc. 82 (2004) 607–613.
[307] K. Kashkush, Genome-wide impact of transcriptional activation of long terminal repeat retrotransposons on the expression of adjacent host genes, Isr. J. Plant Sci. 55 (2007) 241–250.
[308] T. Zhu, L. Wang, J.C. Rodriguez, K.R. Deal, R. Avni, A. Distelfeld, et al., Improved genome sequence of wild emmer wheat Zavitan with the aid of optical maps, G3 (Bethesda) 9 (3) (2019) 619–624. Available from: https://doi.org/10.1534/g3.118.200902.
[309] C. Feuillet, K. Eversole, Physical mapping of the wheat genome: a coordinated effort to lay the foundation for genome sequencing and develop tools for breeders, Isr. J. Plant Sci 55 (2007) 307–313.
[310] J.F. Hancock, Contributions of domesticated plant studies to our understanding of plant evolution, Ann. Botany 96 (2005) 953–963.
[311] A. Aaronshon, Agricultural and Botanical Explorations in Palestine, Bureau of Plant Industry Bull 80, USDA, No 180, 1910, 66 pp.
[312] L. Huang, D. Raats, H. Sela, V. Klymiuk, G. Lidzbarsky, L. Feng, et al., Evolution and adaptation of wild emmer wheat populations to biotic and abiotic stresses, Annu. Rev. Phytopathol. 54 (2016) 279–301.
[313] J.H. Peng, A.B. Korol, T. Fahima, M.S. Röder, Y.I. Ronin, Y.C. Li, et al., Molecular genetic maps in wild emmer wheat, *Triticum dicoccoides*: genome-wide coverage, massive negative interference, and putative quasi-linkage, Genome Res. 10 (2000) 1509–1531.
[314] C. Jorgensen, M.C. Luo, R. Ramasamy, M. Dawson, B.S. Gill, A.B. Korol, et al., A high-density genetic map of wild Emmer wheat from the Karaca Dağ region provides new evidence on the structure and evolution of wheat chromosomes, Front. Plant Sci. 8 (2017) 1798. Available from: https://doi.org/10.3389/fpls.2017.01798.

CHAPTER 9

Evolutionary Modeling of Protein Families by Chromosomal Translocation Events

Gon Carmi*, Alessandro Gorohovski* and Milana Frenkel-Morgenstern
Cancer Genomics and BioComputing of Complex Diseases Lab, The Azrieli Faculty of Medicine, Bar-Ilan University, Safed, Israel

Introduction

Proteins are assembled from a collection of domains that coincide with conserved regions with distinct structural and functional characteristics [1]. According to the domain-oriented view of proteins, domains groups together to form domain architectures (DAs), that is, ordered sequences of domains. Some domains participate in specific DAs, while others are found in diverse DAs. This latter property is termed as "domain promiscuity" or "domain mobility." Domain promiscuity analysis can uncover the mechanisms for gain and loss of domains [2]. Mechanisms for protein domain gain, described by Marsh and Teichmann [1], include: (1) gene fusion, namely, the merge of adjacent genes by the removal of noncoding intergenic regions via alternative splicing; (2) exon extension, elogation of exon into introns to code new domain; (3) exon recombination, merge of two exons originating from two genes; (4) intron recombination or exon shuffling, insertion of a exon from one gene into an intron of a second gene; and (5) retroposition, insertion of genetic sequence from one gene into second gene. The specific mechanism responsible for the addition of a given domain to a protein can be identified by defining the properties of the gained domain, for example, the position of the domain in a protein sequence and the number of exons in the encoding message. For example, multiexon domain gain reflected by extension at the C-terminus reflects gene fusion. Additionally, formation of new protein-protein interactions (PPIs), e.g., during metazoan evolution, can result as by product of

* These authors contributed equally to this work.

New Horizons in Evolution
DOI: https://doi.org/10.1016/B978-0-323-90752-1.00003-1

exchange of exons which encode protein domains mediating PPIs [3]. Bornberg-Bauer and Mar Albà [4], added to the models intrinsically disordered regions, and suggested links between formation of de novo domains and emergence of novel genes [4].

We now present a novel model, "evolution of protein domains" (EvoProDom), to describe the evolution of proteins by means of chromosomal translocation events. The model is based on an exchange of protein domains in a "mix and merge" manner. We obtained and combined genomic and proteomic data from 109 organisms. Part of these data was orthologous protein content and evolutionary events, such as chromosomal translocations that were defined as reciprocal exchanges of protein domains between orthologs in EvoProDom. We found that protein domains, appearing frequently in translocation events, were overrepresented in proteins obtained from the translation of a "slippage" of two different genes [5]. A general method was developed in EvoProDom to collect protein domain content and orthologous protein annotation. This method is based on protein sequences using the Pfam search tool [6,7] for protein domain content and KoFamKOALA [8] for protein annotation. The EvoProtDom method can be implemented in other research fields, such as proteomics [9], design of proteins [10], and exploring host–virus interactions [11].

Materials and methods

The EvoProDom model is build on complete genomic and annotated proteome data. Additionally, the model uses orthologous protein annotation and protein domain content. Proteins from different organisms were combined based on orthologous protein groups, collecting changes in protein domains between orthologous proteins within matching groups of organisms. Orthologous proteins were collected from the Kyoto Encyclopedia of Genes and Genomes (KEGG) orthologs namely, KEGG ortholog (KO) [12,13]. Protein domain content was implemented as Pfam domains. Proteins were thus regarded as a collection of Pfam domains, with orthologous proteins that were proteins of the same KO number. The Pfam search tool [6,7] and KoFamKOALA [8] were used to identify protein domains and KO assignments respectively. Statistical analysis was performed using R (R: A language and environment for statistical computing, 3.3.2,2016).

Data resources

The model was tested on an assembly of 109 organisms with complete genomes and annotated proteomes (Entrez/NCBI [14]). Organisms were arranged as follows: (i) 15 fish; (ii) 4 subterranean, 8 fossorial, and 21 above-ground animals [15,16]; (iii) 65 organisms with known PPIs (BioGrid version 3.5.173, [17,18]); (iv) 17 organisms with HiC datasets; (v) 4 cats; and (vi) 15 pathogenic organisms [19]. The NCBI GEO database was searched for a "HiC" keyword to retrieve organisms with the available HiC datasets (Table 9.1).

Orthologous protein annotation

A Hidden Markov Model (HMM) profile-based search tool, KoFamKOALA [8], was used to assign proteins to KO groups. Since KoFamKOALA operates on protein sequences, an in-house script was written to automate the assignment procedure. In this study, proteins without KO annotation were excluded from the analysis. In addition, a 3-4 letter code for organism was generated from organism's name. A lowercase code represents the KEGG organisms (Table 9.1).

Protein domain detection

Using a dedicated HMM-based search tool, the Pfam domains (realease 32.0) [6,7] were annotated from protein sequences using an in-house written script. In addition, Pfam domains belong to super-familes or clans in Pfam nomenclature, and these clans were used in our classification method. Moreover, the clan classification had a supplement protein domain content.

EvoProDomDB

The core data, genomic and proteomic including orthologous proteins and protein domains content were related by shared data (e.g., protein id). This reflects a data structure and thus the data was organized in a database using MySQL (EvoProDomDB). The MySQL MariaDB (10.0.26) has provided an efficient search engine. EvoProDomDB contained orthologous protein and protein content for the 1,835,600 protein products divided in 23,147 KO groups, comprised 17,929 unique Pfam domains. The Pfam domains were divided among 629 super-families, and EvoProDomDB combined

Table 9.1 EvoProDom model was tested on an assembly of 109 organisms from divergent taxa. These organisms were grouped according to six categories: (i) 15 fish; (ii) 4 subterranean (S), 8 fossorial (F), and 21 aboveground (A) animals (SFA) [15,16]; (iii) 65 organisms with known PPIs (BioGrid version 3.5.173, [17,18]); (iv) 17 organisms with HiC datasets (GEO_hic); (v) 4 cats; and (vi) 15 pathogenic organisms [19]. The NCBI GEO database was searched with "HiC" to retrieve organisms with HiC datasets. Data on taxonomy ID, organism code, organism name, ecology, and common name were provided. In addition, assembly id and group category are shown. A lowercase (uppercase) code represents (non-) KEGG organisms.

Taxonomy ID	Organism code	Organism name	Common name	Ecology	Category	Assembly ID
32536	aju	*Acinonyx jubatus*	Cheetah	na	cats	GCF_001443585.1_aciJub1
180454	aga	*Anopheles gambiae PEST*	African malaria mosquito	na	biogrid_3.5.173	GCF_000005575.2_AgamP3
7460	ame	*Apis mellifera*	Honey bee	na	biogrid_3.5.173	GCF_000002195.4_Amel_4.5
3702	ath	*Arabidopsis thaliana*	Thale cress	na	GEO_hic, biogrid_3.5.173	GCF_000001735.4_TAIR10.1
7994	ASM	*Astyanax mexicanus*	Mexican tetra	na	fish	GCF_000372685.2_Astyanax_mexicanus-2.0
224308	bsu	*Bacillus subtilis 168*	na	na	biogrid_3.5.173	GCF_002009135.1_ASM200913v1
1415167	bsp	*B. subtilis PY79*	na	na	GEO_hic	GCF_000497485.1_ASM49748v1
9913	bta	*Bos taurus*	Cattle	A	biogrid_3.5.173, SFA	GCF_002263795.1_ARS-UCD1.2
6239	cel	*Caenorhabditis elegans*	Nematode	na	GEO_hic, biogrid_3.5.173	GCF_000002985.6_WBcel235
237561	cal	*Candida albicans SC5314*	na	na	biogrid_3.5.173, Jones, et al. 2008	GCF_000182965.3_ASM18296v3
9615	cfa	*Canis familiaris*	Dog	na	biogrid_3.5.173	GCF_000002285.3_CanFam3.1
7957	CAA	*Carassius auratus*	Goldfish	na	fish	GCF_003368295.1 ASM336829v1
190650	ccr	*Caulobacter vibrioides*	na	na	GEO_hic	GCF_000006905.1_ASM690v1
10141	CAP	*Cavia porcellus*	Domestic guinea pig	A	biogrid_3.5.173, SFA	GCF_000151735.1 Cavpor3.0
34839	CHL	*Chinchilla lanigera*	Long-tailed chinchilla	A	SFA	GCF_000276665.1_ChiLan1.0

3055	cre	*Chlamydomonas reinhardtii*	Green algae	na	biogrid_3.5.173	GCF_000002595.1_v3.0
60711	csab	*Chlorocebus sabaeus*	Green monkey	na	biogrid_3.5.173	GCF_000409795.2_Chlorocebus_sabeus_1.1
185453	CHA	*Chrysochloris asiatica*	Cape golden mole	S	SFA	GCF_000296735.1_ChrAsi1.0
143302	COC	*Condylura cristata*	Star-nosed mole	F	SFA	GCF_000260355.1_ConCri1.0
56716	COG	*Cottoperca gobio*	Channel bull blenny	na	fish	GCF_900634415.1 fCotGob3.1
10029	cge	*Cricetulus griseus*	Chinese hamster	A	biogrid_3.5.173, SFA	GCF_000419365.1_C_griseus_v1.0
7962	ccar	*Cyprinus carpio*	Common carp	na	fish	GCF_000951615.1_common_carp_genome
7955	dre	*Danio rerio*	Zebrafish	na	fish, biogrid_3.5.173, GEO_hic	GCF_000002035.6_GRCz11
9361	DAN	*Dasypus novemcinctus*	Nine-banded armadillo	F	SFA	GCF_000208655.1_Dasnov3.0
352472	ddi	*Dictyostelium discoideum AX4*	na	na	biogrid_3.5.173	GCF_000004695.1_dicty_2.7
10020	DIO	*Dipodomys ordii*	Ord's kangaroo rat	F	SFA	GCF_000151885.1_Dord_2.0
7227	dme	*Drosophila melanogaster*	Fruit fly	A	SFA,GEO_hic, biogrid_3.5.173	GCF_000001215.4_Release_6_plus_ISO1_MT
9371	ECTE	*Echinops telfairi*	Small Madagascar hedgehog	A	SFA	GCF_000313985.1 EchTel2.0
28737	ELE	*Elephantulus edwardii*	Cape elephant shrew	A	SFA	GCF_000299155.1 EleEdw1.0
227321	ani	*Emericella nidulans FGSC A4*	*Aspergillus nidulans*	na	biogrid_3.5.173	GCF_000149205.2_ASM14920v2
9796	ecb	*Equus caballus*	Horse	na	biogrid_3.5.173	GCF_002863925.1_EquCab3.0

(Continued)

Table 9.1 (Continued)

Taxonomy ID	Organism code	Organism name	Common name	Ecology	Category	Assembly ID
9365	ERE	*Erinaceus europaeus*	Western European hedgehog	A	SFA	GCF_000296755.1_EriEur2.0
511145	eco	*Escherichia coli str. K-12 substr. MG1655*	na	na	biogrid_3.5.173	GCF_001566335.1_ASM156633v1
9685	fca	*Felis catus*	Domestic cat	A	SFA,cats	GCF_000181335.3_Felis_catus_9.0
885580	FUD	*Fukomys damarensis*	Damara mole-rat	S	SFA	GCF_000743615.1_DMR_v1.0
9031	gga	*Gallus gallus*	Chicken	A	biogrid_3.5.173, SFA,GEO_hic	GCF_000002315.5_GRCg6a
3847	gmx	*Glycine max*	Soybean	na	biogrid_3.5.173	GCF_000004515.4_Glycine_max_v2.0
11103	HCV	*Hepatitis C virus*	HCV	na	biogrid_3.5.173, Jones, et al. 2008	GCF_000861845.1_ViralProj15432
10181	hgl	*Heterocephalus glaber*	Naked mole-rat	S	SFA	GCF_000247695.1_HetGla_female_1.0
9606	hsa	*Homo sapiens*	Human	A	biogrid_3.5.173, SFA,GEO_hic	GCF_000001405.37_GRCh38.p11
10376	HHV4	*Human gammaherpesvirus 4*	EBV	na	GEO_hic, biogrid_3.5.173	GCF_002402265.1_Decoy
37296	HHV8	*Human gammaherpesvirus 8*	KSHV	na	GEO_hic, biogrid_3.5.173, Jones, et al. 2008	GCF_000838265.1_ViralProj14158
10298	HHV1	*Human Herpesvirus 1*	Herpes simplex virus type 1	na	biogrid_3.5.173, Jones, et al. 2008	GCF_000859985.2_ViralProj15217
10310	HHV2	*Human Herpesvirus 2*	HHV2	na	biogrid_3.5.173	GCF_000858385.2_ViralProj15218

10335	HHV3	*Human Herpesvirus 3*	Varicella-zoster virus	na	biogrid_3.5.173, Jones, et al. 2008	GCF_000858285.1_ViralProj15198
10359	HHV5	*Human Herpesvirus 5*	Human		cytomegalovirus	na
		biogrid_3.5.173, Jones, et al. 2008				GCF_000845245.1_ViralProj14559
32603	HHV6A	*Human Herpesvirus 6A*	HHV6A	na	biogrid_3.5.173	GCF_000845685.1_ViralProj14462
32604	HHV6B	*Human Herpesvirus 6B*	HHV6B	na	biogrid_3.5.173	GCF_000846365.1_ViralProj14422
10372	HHV7	*Human Herpesvirus 7*	HHV7	na	biogrid_3.5.173, Jones, et al. 2008	GCF_000848125.1_ViralProj14625
11676	HIV1	*Human Immunodeficiency Virus 1*	HIV1	na	biogrid_3.5.173, Jones, et al. 2008	GCF_000864765.1_ViralProj15476
11709	HIV2	*Human Immunodeficiency Virus 2*	HIV2	na	biogrid_3.5.173, Jones, et al. 2008	GCF_000856385.1_ViralProj14991
333760	HPV16	*Human papillomavirus 16*	HPV16	na	biogrid_3.5.173, GEO_hic,Jones, et al. 2008	GCF_000863945.3_ViralProj15505
333759	HPV10	*Human papillomavirus type 10*	HPV10	na	biogrid_3.5.173, Jones, et al. 2008	GCF_000864905.1_ViralProj15504
10600	HPV6b	*Human papillomavirus type 6b*	HPV6b	na	biogrid_3.5.173, Jones, et al. 2008	GCF_000861945.1_ViralProj15454

(Continued)

Table 9.1 (Continued)

Taxonomy ID	Organism code	Organism name	Common name	Ecology	Category	Assembly ID
43179	ICT	*Ictidomys tridecemlineatus*	Thirteen-lined ground squirrel	F	SFA	GCF_000236235.1 SpeTri2.0
8187	lcf	*Lates calcarifer*	Barramundi perch	na	fish	GCF_001640805.1_ASM164080v1
7897	lcm	*Latimeria chalumnae*	Coelacanth	na	fish	GCF_000225785.1_LatCha1
7918	LEO	*Lepisosteus oculatus*	Spotted gar	na	fish	GCF_000242695.1 LepOcu1
9785	lav	*Loxodonta Africana*	African savanna elephant	A	SFA	GCF_000001905.1_Loxafr3.0
9544	MAM	*Macaca mulatta*	Rhesus monkey	na	biogrid_3.5.173, GEO_hic	GCF_003339765.3 Mmul_10
9993	MARM	*Marmota marmota*	European marmot	F	SFA	GCF_001458135.1 marMar2.1
9103	mgp	*Meleagris gallopavo*	Turkey	na	biogrid_3.5.173	GCF_000146605.2_Turkey_5.0
79684	MIO	*Microtus ochrogaster*	Prairie vole	F	SFA	GCF_000317375.1_MicOch1.0
10090	mmu	*Mus musculus*	House mouse	A	biogrid_3.5.173, SFA,GEO_hic	GCF_000001635.26_GRCm38.p6
83332	mtv	*Mycobacterium tuberculosis H37Rv*	na	na	biogrid_3.5.173, Jones, et al. 2008	GCF_000195955.2_ASM19595v2
243273	mge	*Mycoplasma genitalium*	na	na	Jones, et al. 2008	GCF_000027325.1_ASM2732v1
1026970	ngi	*Nannospalax galili*	Upper Galilee mountains blind mole rat	S	SFA	GCF_000622305.1_S.galili_v1.0
56216	NEL	*Neotoma lepida*	Desert woodrat	A	SFA	GCF_001675575.1 ASM167557v1
367110	ncr	*Neurospora crassa OR74A*	na	na	biogrid_3.5.173	GCF_000182925.2_NC12

4098	nto	*Nicotiana tomentosiformis*	Tobacco	na	biogrid_3.5.173	GCF_000390325.2_Ntom_v01
61853	nle	*Nomascus leucogenys*	Northern white-cheeked gibbon	na	GEO_hic	GCF_000146795.2_Nleu_3.0
105023	nfu	*Nothobranchius furzeri*	Turquoise killifish	na	fish	GCF_001465895.1_Nfu_20140520
8208	ncc	*Notothenia coriiceps*	Black rockcod	na	fish	GCF_000735185.1_NC01
10160	OCD	*Octodon degus*	Degu	F	SFA	GCF_000260255.1_OctDeg1.0
9258	oaa	*Ornithorhynchus anatinus*	Platypus	A	SFA	GCF_000002275.2_Ornithorhynchus_anatinus_5.0.1
1230840	ORA	*Orycteropus afer afer*	Aardvark	*A*	SFA	GCF_000298275.1_OryAfe1.0
9986	ocu	*Oryctolagus cuniculus*	Rabbit	na	biogrid_3.5.173, GEO_hic	GCF_000003625.3_OryCun2.0
39947	osa	*Oryza sativa Japonica*	Rice	na	biogrid_3.5.173	GCF_001433935.1_IRGSP-1.0
8090	ola	*Oryzias latipes*	Japanese medaka	na	fish	GCF_002234675.1_ASM223467v1
30732	ORM	*Oryzias melastigma*	Indian medaka	na	fish	GCF_002922805.1 Om_v0.7.RACA
9940	oas	*Ovis aries*	Sheep	na	biogrid_3.5.173	GCF_000298735.2_Oar_v4.0
9598	ptr	*Pan troglodytes*	Chimpanzee	A	biogrid_3.5.173, SFA	GCF_002880755.1_Clint_PTRv2
9691	PAP	*Panthera pardus*	Leopard	na	cats	GCF_001857705.1_PanPar1.0
74533	ptg	*Panthera tigris altaica*	Tiger	na	cats	GCF_000464555.1_PanTig1.0
121224	phu	*Pediculus humanus corporis*	Human body louse	na	biogrid_3.5.173	GCF_000006295.1_JCVI_LOUSE_1.0
8167	PEF	*Perca flavescens*	Yellow perch	na	fish	GCF_004354835.1 PFLA_1.0
230844	PEM	*Peromyscus maniculatus bairdii*	Prairie deer mouse	A	SFA	GCF_000500345.1_Pman_1.0

(*Continued*)

Table 9.1 (Continued)

Taxonomy ID	Organism code	Organism name	Common name	Ecology	Category	Assembly ID
36329	pfa	*Plasmodium falciparum 3D7*	Malaria parasite P. falciparum	na	biogrid_3.5.173, Jones, et al. 2008	GCF_000002765.4_ASM276v2
8081	pret	*Poecilia reticulata*	Guppy	na	fish	GCF_000633615.1_Guppy_female_1.0_MT
10116	rno	*Rattus norvegicus*	Norway rat	A	biogrid_3.5.173, SFA	GCF_000001895.5_Rnor_6.0
3988	rcu	*Ricinus communis*	Castor bean	na	biogrid_3.5.173	GCF_000151685.1_JCVI_RCG_1.1
4932	sce	*Saccharomyces cerevisiae S288c*	Baker's yeast	na	biogrid_3.5.173, GEO_hic	GCF_000146045.2_R64
8030	sasa	*Salmo salar*	Atlantic salmon	na	fish	GCF_000233375.1_ICSASG_v2
4896	spo	*Schizosaccharomyces pombe*	Fission yeast	na	GEO_hic, biogrid_3.5.173	GCF_000002945.1_ASM294v2
88036	smo	*Selaginella moellendorffii*	na	na	biogrid_3.5.173	GCF_000143415.4_v1.0
11723	SIV	*Simian Immunodeficiency Virus*	SIV	na	biogrid_3.5.173	GCF_000863925.1_ViralProj15501
1891767	SV40	*Simian Virus 40*	Macaca mulatta polyomavirus 1	na	biogrid_3.5.173	GCF_000837645.1_ViralProj14024
4081	sly	*Solanum lycopersicum*	Tomato	na	biogrid_3.5.173	GCF_000188115.3_SL2.50
4113	sot	*Solanum tuberosum*	Potato	na	biogrid_3.5.173	GCF_000226075.1_SolTub_3.0
42254	SOA	*Sorex araneus*	European shrew	A	SFA	GCF_000181275.2 SorAra2.0
7668	spu	*Strongylocentrotus purpuratus*	Purple sea urchin	na	biogrid_3.5.173	GCF_000002235.4_Spur_4.2
9823	ssc	*Sus scrofa*	Pig	A	biogrid_3.5.173, SFA	GCF_000003025.6_Sscrofa11.1

12242	TMV	*Tobacco Mosaic Virus*	TMV	na	biogrid_3.5.173	GCF_000854365.1_ViralProj15071
9999	URP	*Urocitellus parryii*	Arctic ground squirrel	F	SFA	GCF_003426925.1 ASM342692v1
237631	USM	*Ustilago maydis 521*	na	na	biogrid_3.5.173	GCF_000328475.2 Umaydis521_2.0
10245	VAV	*Vaccinia Virus*	na	na	biogrid_3.5.173	GCF_000860085.1_ViralProj15241
29760	vvi	*Vitis vinifera*	Wine grape	na	biogrid_3.5.173	GCF_000003745.3_12X
8355	xla	*Xenopus laevis*	African clawed frog	na	biogrid_3.5.173	GCF_001663975.1_Xenopus_laevis_v2
4577	zma	*Zea mays*	Maize	na	biogrid_3.5.173	GCF_000005005.2_B73_RefGen_v4

data for 109 organisms from divergent taxa. EvoProDomDB was constructed from six relational tables sharing common features, e.g., organism identity (Fig. 9.1). Relational tables, pfam_domain, clan_domain, ko_annotation and taxonomy provided the annotation data for domain and superfamily description, KO assignments and taxonomy rankings, e.g., genus and species, respectively.

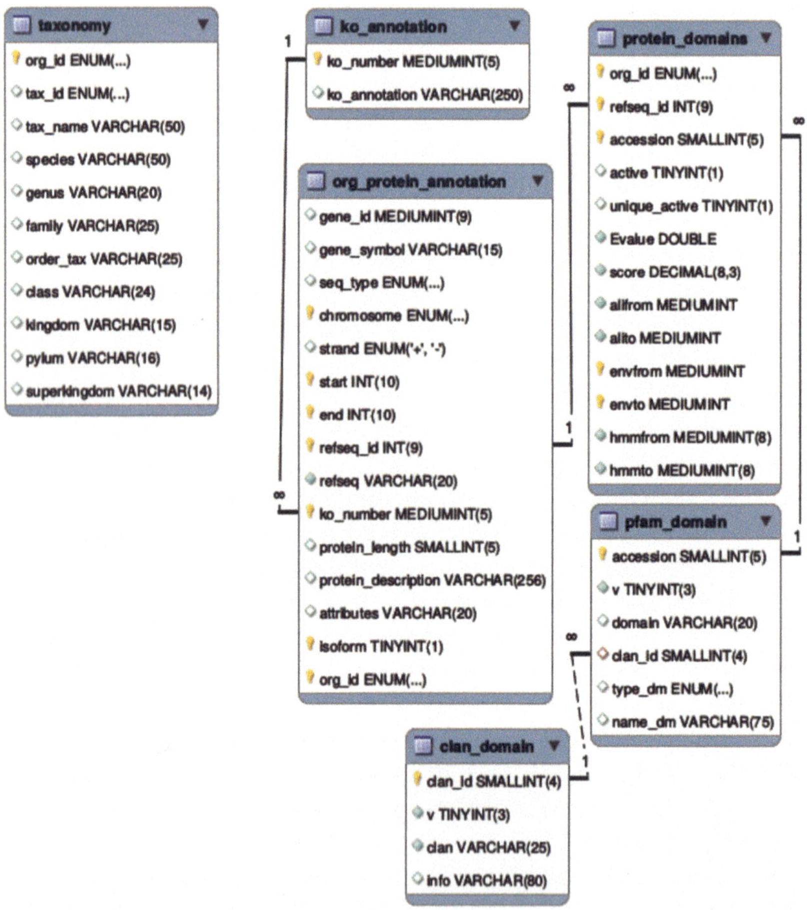

Figure 9.1 EvoProDomDB relational tables (MySQL scheme). Six relation tables were included. Of these, four contained data regarding taxonomy (taxonomy) e.g., genus and species, KO (ko_annotation), super-families (clan_domain), pfam domains (pfam_domain). The remaining tables contain protein, genomic and proteomic data (org_protein_annotation), as well as protein domain content (pfam data) (see the main text for details).

Relational tables included protein genomic and proteomic data (org_protein_annotation), as well as protein domain content (Pfam data). Common genomic and proteomic data were also included, e.g., gene_symbol, chromosome, strand, refseq_id, protein length, and protein description. To theses data KO numbers (ko_number) were added. The longest isoform provided a link between genomic and proteomic data. Protein domain content contained standard Pfam domains as obtained from the output of Pfam search tool [6,7]. To these data, calculated data was added. This data identified nonoverlapping Pfam domains with maximal scores (active), delineated by "envfrom" and "envto" coordinates. Such data corresponds to the largest region within protein sequnce were a pfam domain was predicted to be located. The highest scoring domain, among multiple copies of active domain, was identified (unique active). These data and construction of EvoProDomDB were collected using both custom and standard scripts written perl and bash to form a in-house pipeline. In particular, EvoProDom model was implemented in Perl with MySQL queries to obtain data from the database. Data sources and databases were outlined in the study workflow (Fig. 9.2).

Results

The EvoProDom model

We hypothesized that evolution of proteins proceed by "mix and merge" or "shuffling" of protein domains which correspond to well-defined functional units [1,20]. This evolutionary model was build to reflect the exchangeable fuction of protein domains and their dynamics in the evolution of proteins. This model formulates standard evolutionary mechanisms, for example, translocations, duplications, and indel (insertion and deletion) events, which acted upon protein domains (Pfam domains [6,7]). Proteins, according to the EvoProDom model, changed their fuction as a result of presence or absence of exchangeable domains. Therefore proteins were modeled as protein domain sets or DAs and evolutionary events, e.g., translocations, were defined as an exchange of protein domains among DAs. The KEGG database catalogs divergent taxa and generated groups of orthologous proteins (KOs) based on common function [8,12,13]. In EvoProDom, proteins were designated to KO groups (see Materials and methods). Translocation events, which acted upon protein domain and reflected changes in DA, were mapped to groups of organisms. Consequently translocation events defined on

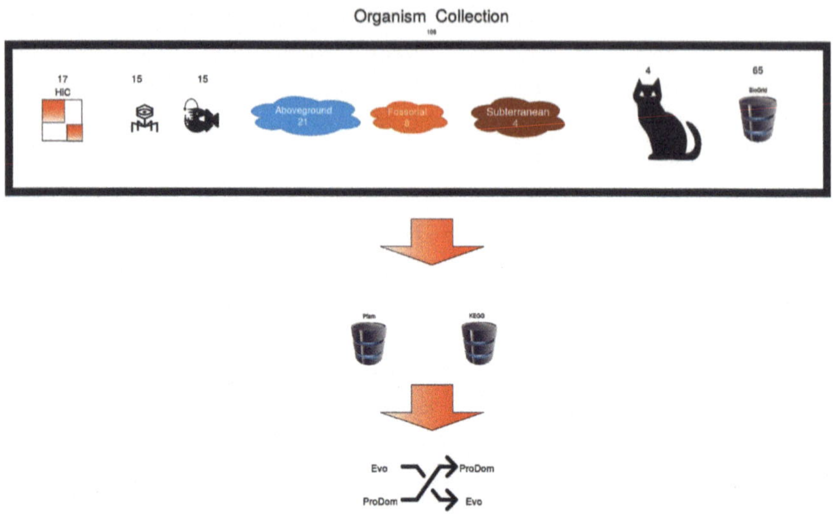

Figure 9.2 Study overview. An assembly of 109 organisms was used to implement and test the EvoProDom model. The assembly included: (i) 15 fish; (ii) 4 subterranean, 8 fossorial, and 21 aboveground animals [15,16]; (iii) 65 organisms with known PPIs (BioGrid version 3.5.173, [17,18]); (iv) 17 organisms with HiC datasets; (v) 4 cats; and (vi) 15 pathogenic organisms [19]. Protein domains were predicted using the Pfam (release 32.0) database, in combination with the search tool [6,7]. Orthologous proteins were defined as members of a Kyoto Encyclopedia of Genes and Genomes (KEGG) [12,13] ortholog (KO) group, and assigned to KO group using KofamKOALA [8].

protein domains were manifested at the organism level. EvoProDom was implemented with and tested on the EvoProDomDB (see Materials and methods). Overall, 5,548 translocation events, involving 94 protein super-families, excluding one "unknown" super-family, were found. This result points to the existence of numerous evolutionary translocation events, as defined by the model.

Mapping of genes to proteins and alternative splicing

EvoProDom integrated genomic data of proteins, and in turn, proteins with Pfam domain content. Additionally, assignment to KO groups was also included [8,12,13]. Genes were mapped to more than a single mRNA transcript and, in turn, to more than a single protein product, identified by their Refseq id. These transcripts coded isoforms of a gene product and originate from alternative splicing, i.e., the inclusion of gene exons. As protein domains mainly correspond to exons [1,3,5,20], alterations in protein domain composition can account for alterations in DA as a consequence of translocation events. However, alterations in protein domain composition could arise as a result of alternative splicing. Hence, to avoid this confounding effect due to alternative splicing, only the longest isoform was used in the model (see Materials and methods). As a result, each gene was mapped to a single protein product.

Protein domain content

Predicted Pfam domains, for a particular protein, can appear in multiple copies and or overlap with other Pfam domains. In our models, overlapping or multiple copies Pfam domains were removed as they do not adhere to the linear order of protein domains within a protein. To resolve this issue, a single domain (active domain) with a maximal score, was determined for each set of overlapping protein domains. The resulting set of nonoverlapping Pfam domains confirms to the linear order of protein domains in protein sequence. This procedure ensured a set of nonoverlapping domains, yet multiple copies are present. For translocation events, a unique set was required. Therefore, a unique set of nonoverlapping active domains was detremined by a similar procedure, namely, selection of domain with maximal score for each set of multiple copies of same domains. These domains were referred to as unique active domains.

DA as a basic unit in EvoProDom

In EvoProDom model evolutionary events, such as translocations require protein domains, DAs, orthologous groups (KOs) and organisms. EvoProDomDB enables organizing these data elements around DAs. In short, each orthologous group (KO) was divided into distinct sets (items), i.e., a list of domains (DA), and corresponding lists of organisms and proteins. Of note, within these corresponding lists, duplicated organisms reflect paralogous proteins. In addition, for each DA, present and lacking domains were resolved from all DAs associated with a particular KO. Mobile and translocation domains, i.e., domains participating in translocation events were detremined from these data. Overall, we found a total of 5,548 translocation events involving 94 protein super-families, excluding an "unknown" super-family (Table 9.2). We identified 2,041 mobile domains associated with translocation events, 259 which had undergone translocation and 1,782 that participated in indel events (Table 9.3).

Table 9.2 Frequency of translocation events per super family (Frequency). Super family (clan) description is shown.

Clan ID	Clan Name	Description	Frequency
0010.21	SH3	Src homology-3 domain	630
0011.26	Ig	Immunoglobulin superfamily	616
0465.3	Ank	Ankyrin repeat superfamily	529
0001.27	EGF	EGF superfamily	414
0361.4	C2H2-zf	Classical C2H2 and C2HC zinc fingers	390
0022.32	LRR	Leucine Rich Repeat	282
0020.25	TPR	Tetratrico peptide repeat superfamily	246
0229.11	RING	Ring-finger/U-box superfamily	242
0186.14	Beta_propeller	Beta propeller clan	222
0221.11	RRM	RRM-like clan	210
9999.0	Unknown	null	210
0159.16	E-set	Ig-like fold superfamily (E-set)	187
0466.3	PDZ-like	PDZ domain-like peptide-binding superfamily	165
0016.22	PKinase	Protein kinase superfamily	164
0266.9	PH	PH domain-like superfamily	141
0023.34	P-loop_NTPase	P-loop containing nucleoside triphosphate hydrolase superfamily	121
0220.12	EF_hand	EF-hand like superfamily	115
0511.3	Retroviral_zf	Retrovirus zinc finger-like domains	95
0271.7	F-box	F-box-like domain	79
0003.21	SAM	Sterile Alpha Motif (SAM) domain	74
0390.4	zf-FYVE-PHD	FYVE/PHD zinc finger superfamily	47
0063.25	NADP_Rossmann	FAD/NAD(P)-binding Rossmann fold Superfamily	37
0357.4	SMAD-FHA	SMAD/FHA domain superfamily	37
0123.18	HTH	Helix-turn-helix clan	34
0680.1	WW	WW domain	34
0167.15	Zn_Beta_Ribbon	Zinc beta-ribbon	33
0006.20	C1	Protein kinase C, C1 domain	25
0214.13	UBA	UBA superfamily	24
0306.4	HeH	LEM/SAP HeH motif	24
0188.10	CH	Calponin homology domain	23
0459.3	BRCT-like	BRCT like	23
0537.2	CCCH_zf	CCCH-zinc finger	22
0004.20	Concanavalin	Concanavalin-like lectin/glucanase superfamily	20
0072.20	Ubiquitin	Ubiquitin superfamily	19
0033.14	POZ	POZ domain superfamily	17
0154.11	C2	C2 superfamily	11

(*Continued*)

Table 9.2 (Continued)

Clan ID	Clan Name	Description	Frequency
0007.18	KH	K-Homology (KH) domain Superfamily	9
0021.18	OB	OB fold	8
0029.20	Cupin	Cupin fold	8
0049.15	Tudor	Tudor domain 'Royal family'	8
0164.13	CUB	CUB clan	8
0172.17	Thioredoxin	Thioredoxin-like	8
0212.9	SNARE	SNARE-like superfamily	8
0392.4	Chaperone-J	Chaperone J-domain superfamily	8
0124.15	Peptidase_PA	Peptidase clan PA	7
0137.15	HAD	HAD superfamily	7
0364.4	Leu-IlvD	LeuD/IlvD-like	7
0575.2	EFTPs	Translation proteins of elongation factors superfamily	7
0541.2	SH2-like	SH2, phosphotyrosine-recognition domain superfamily	6
0173.11	STIR	STIR superfamily	5
0192.13	GPCR_A	Family A G protein-coupled receptor-like superfamily	5
0244.9	PGBD	PGBD superfamily	5
0602.2	Kringle	Kringle/FnII superfamily	5
0671.1	AAA_lid	AAA + ATPase lid domain superfamily	5
0041.13	Death	Death domain superfamily	4
0178.16	PUA	PUA/ASCH superfamily	4
0183.14	PAS_Fold	PAS domain clan	4
0642.1	SOCS_box	SOCS-box like superfamily	4
0015.20	MFS	Major facilitator superfamily	3
0030.16	Ion_channel	Ion channel (VIC) superfamily	3
0084.13	ADP-ribosyl	ADP-ribosylation superfamily	3
0198.16	HHH	Helix-hairpin-helix superfamily	3
0497.3	GST_C	Glutathione S-transferase, C-terminal domain	3
0661.1	Gain	GPCR autoproteolysis inducing	3
0027.15	RdRP	RNA dependent RNA polymerase	2
0028.22	AB_hydrolase	Alpha/Beta hydrolase fold	2
0055.13	Nucleoplasmin	Nucleoplasmin-like/VP (viral coat and capsid proteins) superfamily	2
0107.12	KOW	KOW domain	2
0202.11	GBD	Galactose-binding domain-like superfamily	2
0492.3	S4	S4 domain superfamily	2

(*Continued*)

Table 9.2 (Continued)

Clan ID	Clan Name	Description	Frequency
0531.2	AMP-binding_C	AMP-binding enzyme C-terminal domain superfamily	2
0005.27	Kazal	Kazal like domain	1
0025.14	His_Kinase_A	His Kinase A (phospho-acceptor) domain	1
0026.20	CU_oxidase	Multicopper oxidase-like domain	1
0070.13	ACT	ACT-like domain	1
0088.16	Alk_phosphatase	Alkaline phosphatase-like	1
0109.12	CDA	Cytidine deaminase-like (CDA) superfamily	1
0110.12	GT-A	Glycosyl transferase clan GT-A	1
0113.13	GT-B	Glycosyl transferase clan GT-B	1
0114.12	HMG-box	HMG-box like superfamily	1
0117.11	uPAR_Ly6_toxin	uPAR/Ly6/CD59/snake toxin-receptor superfamily	1
0125.15	Peptidase_CA	Peptidase clan CA	1
0144.13	Periplas_BP	Periplasmic binding protein like	1
0196.12	DSRM	DSRM-like clan	1
0236.17	PDDEXK	PD-(D/E)XK nuclease superfamily	1
0381.4	Metallo-HOrase	Metallo-hydrolase/oxidoreductase superfamily	1
0426.4	HRDC-like	HRDC-like superfamily	1
0449.3	G-PATCH	DExH-box splicing factor binding site	1
0505.3	Pentapeptide	Pentapeptide repeat	1
0547.2	GF_recep_C-rich	Growth factor receptor Cys-rich	1
0552.2	Hect	Hect, E3 ligase catalytic domain	1
0607.2	TNF_receptor	TNF receptor-like superfamily	1
0630.1	PSI	Plexin fold superfamily	1
0672.1	p35	Baculovirus p35 protein superfamily	1
0677.1	GHMP_C	GHMP C-terminal domain superfamily	1

Evolutionary mechanism in EvoProDom

Implementation of domain architecture

Information for a given DA, uniquely characterized by a (ko, item) pair, were retrieved from EvoProDomDB, while screening for active and unique active domains (see Materials and Methods). Each DA contained: (i) a ko:item; (ii) a Pfam domain list; (ii) a list of organisms (org_id); (iii) a list of refseq_ids; (iv) a list of missing domains; and (v) a list of gained

Table 9.3 Frequency of Indel events per super family (Frequency). Super family (clan) description is shown.

Clan ID	Clan Name	Description	Frequency
9999.0	Unknown	Null	8536
0023.34	P-loop_NTPase	P-loop containing nucleoside triphosphate hydrolase superfamily	688
0020.25	TPR	Tetratrico peptide repeat superfamily	520
0123.18	HTH	Helix-turn-helix clan	453
0219.14	RNase_H	Ribonuclease H-like superfamily	347
0361.4	C2H2-zf	Classical C2H2 and C2HC zinc fingers	228
0665.1	BET	BET superfamily	216
0021.18	OB	OB fold	196
0072.20	Ubiquitin	Ubiquitin superfamily	187
0063.25	NADP_Rossmann	FAD/NAD(P)-binding Rossmann fold superfamily	173
0148.11	Viral_Gag	Viral Gag protein	157
0186.14	Beta_propeller	Beta propeller clan	152
0016.22	PKinase	Protein kinase superfamily	148
0266.9	PH	PH domain-like superfamily	143
0031.13	Phosphatase	Phosphatase superfamily	142
0523.3	GAG-polyprotein	LTR-copia-type polyprotein segment	140
0671.1	AAA_lid	AAA + ATPase lid domain superfamily	140
0550.2	SRCR	SRCR-like	123
0114.12	HMG-box	HMG-box like superfamily	109
0010.21	SH3	Src homology-3 domain	106
0137.15	HAD	HAD superfamily	106
0214.13	UBA	UBA superfamily	100
0172.17	Thioredoxin	Thioredoxin-like	98
0128.12	vWA-like	von Willebrand factor type A	95
0208.11	UBC	Ubiquitin conjugating enzyme like superfamily	90
0154.11	C2	C2 superfamily	88
0220.12	EF_hand	EF-hand like superfamily	87
0001.27	EGF	EGF superfamily	84
0081.13	MBD-like	MBD-like DNA-binding domain	83
0390.4	zf-FYVE-PHD	FYVE/PHD zinc finger superfamily	82
0168.15	PAN	PAN-like	77
0049.15	Tudor	Tudor domain 'Royal family'	75
0183.14	PAS_Fold	PAS domain clan	70
0459.3	BRCT-like	BRCT like	67
0159.16	E-set	Ig-like fold superfamily (E-set)	66
0575.2	EFTPs	Translation proteins of elongation factors superfamily	65
0126.18	Peptidase_MA	Peptidase clan MA	58
0029.20	Cupin	Cupin fold	53
0129.14	Peptidase_AA	Peptidase clan AA	52
0221.11	RRM	RRM-like clan	51
0451.3	FnI-like	von Willebrand Factor like superfamily	51

(Continued)

Table 9.3 (Continued)

Clan ID	Clan Name	Description	Frequency
0041.13	Death	Death domain superfamily	48
0280.7	PIN	PIN domain superfamily	42
0202.11	GBD	Galactose-binding domain-like superfamily	37
0188.10	CH	Calponin homology domain	36
0464.3	5_3_exonuc_C	5'-3'-exonuclease C-terminal subregion	36
0537.2	CCCH_zf	CCCH-zinc finger	35
0004.20	Concanavalin	Concanavalin-like lectin/glucanase superfamily	34
0556.2	PapD-like	PapD-like superfamily,immunoglobulin-like beta sandwich	32
0678.1	CLIP	CLIP domain superfamily	31
0167.15	Zn_Beta_Ribbon	Zinc beta-ribbon	30
0074.13	Matrix	Retroviral matrix superfamily	28
0240.8	PFK	PFK-like superfamily	27
0589.2	KIX_like	Kix domain of CBP (creb binding protein) and MED13/15	27
0113.13	GT-B	Glycosyl transferase clan GT-B	26
0257.9	Acetyltrans	N-acetyltransferase like	25
0087.13	Acyl-CoA_dh	Acyl-CoA dehydrogenase, C-terminal domain-like	23
0497.3	GST_C	Glutathione S-transferase, C-terminal domain	21
0011.26	Ig	Immunoglobulin superfamily	20
0179.14	ATP-grasp	ATP-grasp superfamily	20
0145.15	Golgi-transport	Golgi-transport	19
0229.11	RING	Ring-finger/U-box superfamily	19
0511.3	Retroviral_zf	Retrovirus zinc finger-like domains	19
0544.2	AcylCoA_ox_dh_N	Acyl-coenzyme A oxidase/dehydrogenase N-terminal	17
0573.2	KA1-like	Kinase associated domain 1-like	17
0036.24	TIM_barrel	Common phosphate binding-site TIM barrel superfamily	16
0103.12	Gal_mutarotase	Galactose Mutarotase-like superfamily	16
0630.1	PSI	Plexin fold superfamily	16
0039.12	HUP	HUP - HIGH-signature proteins, UspA, and PP-ATPase.	15
0125.15	Peptidase_CA	Peptidase clan CA	15
0306.4	HeH	LEM/SAP HeH motif	15
0028.22	AB_hydrolase	Alpha/Beta hydrolase fold	14
0109.12	CDA	Cytidine deaminase-like (CDA) superfamily	14
0175.11	TRASH	TRASH superfamily	13
0181.10	ABC-2	ABC-2-transporter-like clan	13
0600.2	S15_NS1	S15/NS1 RNA-binding domain superfamily	13
0616.1	SPOC	SPOC domain-like superfamily	13
0012.18	Histone	Histone superfamily	12
0192.13	GPCR_A	Family A G protein-coupled receptor-like superfamily	12

(*Continued*)

Table 9.3 (Continued)

Clan ID	Clan Name	Description	Frequency
0265.7	HIT	HIT superfamily	12
0357.4	SMAD-FHA	SMAD/FHA domain superfamily	11
0369.4	GHD	Glycosyl hydrolase domain superfamily	11
0381.4	Metallo-HOrase	Metallo-hydrolase/oxidoreductase superfamily	11
0614.1	L27	L27 Domain superfamily	11
0632.1	FERM_M	FERM middle domain superfamily	11
0005.27	Kazal	Kazal-like domain	10
0056.12	C_Lectin	C-type lectin-like superfamily	10
0058.16	Glyco_hydro_tim	Tim barrel glycosyl hydrolase superfamily	10
0178.16	PUA	PUA/ASCH superfamily	10
0223.8	MACRO	MACRO domain superfamily	10
0389.4	TRAF	TRAF domain-like superfamily	10
0014.22	Glutaminase_I	Class-I Glutamine amidotransferase superfamily	9
0092.12	ADF	Actin depolymerizing Factor	9
0298.5	tRNA_bind_arm	tRNA-binding arm superfamily	9
0512.3	CRAL_TRIO	CRAL-TRIO domain superfamily	9
0561.2	LisH	LisH-like	9
0644.1	Fz	Frizzled cysteine-rich domain-related superfamily	9
0015.20	MFS	Major facilitator superfamily	8
0062.13	APC	APC superfamily	8
0064.12	CPA_AT	CPA/AT transporter superfamily	8
0070.13	ACT	ACT-like domain	8
0105.13	Hybrid	Barrel sandwich hybrid superfamily	8
0258.7	DALR	DALR superfamily	8
0314.4	PP-binding	ACP-like superfamily	8
0329.4	S5	Ribosomal protein S5 domain 2-like superfamily	8
0332.4	AcetylDC-like	Acetyl-decarboxylase like superfamily	8
0479.3	PLD	Phospholipase D superfamily	8
0483.3	PreATP-grasp	Probable substrate-binding preceding ATP-grasp domain	8
0673.1	GYF	GYF domain	8
0263.8	His-Me_finger	His-Me finger endonuclease superfamily	7
0343.4	MHC	MHC antigen-recognition domain	7
0344.4	4Fe-4S	4Fe-4S ferredoxins	7
0375.4	Transporter	Transporter superfamily, four TM region	7
0396.4	Marvel-like	MARVEL domain containing superfamily	7
0533.2	PRTase-like	PRPP synthetase-associated protein 1	7
0647.1	FG_rpt	Nucleoporin FG repeat	7
0661.1	Gain	GPCR autoproteolysis inducing	7
0034.15	Amidohydrolase	Amidohydrolase superfamily	6
0050.12	HotDog	HotDog superfamily	6
0110.12	GT-A	Glycosyl transferase clan GT-A	6
0190.12	HSP20	HSP20-like chaperone superfamily	6

(Continued)

Table 9.3 (Continued)

Clan ID	Clan Name	Description	Frequency
0198.16	HHH	Helix-hairpin-helix superfamily	6
0236.17	PDDEXK	PD-(D/E)XK nuclease superfamily	6
0268.7	Pec_lyase-like	Pectate lyase-like beta helix	6
0296.5	GroES	GroES-like superfamily	6
0364.4	Leu-IlvD	LeuD/IlvD-like	6
0409.4	GAP	GTPase activation domain superfamily	6
0445.4	SNARE-fusion	SNARE-fusion membrane complex superfamily	6
0458.3	IIaaRS-ABD	Class II aaRS Anticodon-binding domain-like	6
0492.3	S4	S4 domain superfamily	6
0502.3	STAS	STAS domain superfamily	6
0506.3	Succ_CoA_synth	Succinyl-CoA synthetase flavodoxin domain superfamily	6
0545.2	APCOP-app_sub	subdomain	6
0547.2	GF_recep_C-rich	Growth factor receptor Cys-rich	6
0003.21	SAM	Sterile Alpha Motif (SAM) domain	5
0026.20	CU_oxidase	Multicopper oxidase-like domain	5
0052.18	NTN	NTN hydrolase superfamily	5
0065.15	Cyclin	Cyclin-like superfamily	5
0116.13	Calycin	Calycin superfamily	5
0212.9	SNARE	SNARE-like superfamily	5
0226.9	M6PR	Mannose 6-phosphate receptor	5
0260.8	NTP_transf	Nucleotidyltransferase superfamily	5
0274.8	WRKY-GCM1	WRKY-GCM1 superfamily	5
0340.4	PTase-anion_tr	Phosphotransferase/anion transport protein superfamily	5
0466.3	PDZ-like	PDZ domain-like peptide-binding superfamily	5
0542.2	RAS_GEF_N	Ras guanyl-nucleotide exchange factor activity N-term	5
0660.1	SHOCT	SHOCT superfamily	5
0022.32	LRR	Leucine Rich Repeat	4
0035.16	Peptidase_MH	Peptidase clan MH/MC/MF	4
0094.12	Peptidase_ME	LuxS/MPP-like metallohydrolase	4
0118.12	Ribokinase	Ribokinase-like superfamily	4
0187.11	LysM	LysM-like domain	4
0196.12	DSRM	DSRM-like clan	4
0261.7	NUDIX	NUDIX superfamily	4
0291.7	KNTase_C	Nucleotidyltransferase substrate binding domain	4
0350.4	PRC-barrel	PRC-barrel like superfamily	4
0399.4	Asp-glut_race	Aspartate/glutamate racemase superfamily	4
0540.2	GCP	Gamma-tubulin complex superfamily	4
0570.2	PPP-I	Protease propeptides/inhibitors	4
0025.14	His_Kinase_A	His Kinase A (phospho-acceptor) domain	3
0033.14	POZ	POZ domain superfamily	3

(*Continued*)

Table 9.3 (Continued)

Clan ID	Clan Name	Description	Frequency
0040.17	tRNA_synt_II	Class II aminoacyl-tRNA and Biotin synthetases	3
0051.14	NTF2	NTF2-like superfamily	3
0088.16	Alk_phosphatase	Alkaline phosphatase-like	3
0093.14	Peptidase_CD	Peptidase clan CD	3
0209.11	Bet_V_1_like	Bet V 1 like	3
0272.7	RGS	RGS-like superfamily	3
0301.6	PA14	PA14 superfamily	3
0322.4	RND_permease	RND permease superfamily	3
0336.4	FMN-binding	FMN-binding split barrel superfamily	3
0402.4	Cdc48_2-like	Cdc48 domain 2-like	3
0412.4	Frag1-like	Frag1 like	3
0416.4	Anoctamin-like	Transmembrane protein families of the Anoctamin type	3
0442.4	Tubulin_C	Tubulin, FtsZ and Misato and their C-termini	3
0448.3	Cargo_bd_muHD	Second domain of Mu2 adaptin subunit (ap50) of ap2 adaptor	3
0530.2	DNase_I-like	DNase I-like	3
0007.18	KH	K-Homology (KH) domain superfamily	2
0018.15	bZIP	bZIP-like leucine zipper	2
0046.16	Thiolase	Thiolase-like superfamily	2
0053.15	4H_Cytokine	4-helical cytokine superfamily	2
0066.14	Trefoil	Beta-trefoil superfamily	2
0068.12	RIIa	RIIa-like fold	2
0069.12	GFP	GFP-like superfamily	2
0108.16	Actin_ATPase	Actin-like ATPase superfamily	2
0139.11	GADPH_aa-bio_dh	Amino acid biosynthesis and glycosomal dehydrogenase	2
0147.11	Traffic	Trafficking protein	2
0163.11	Calcineurin	Calcineurin-like phosphoesterase superfamily	2
0173.11	STIR	STIR superfamily	2
0206.11	TRB	Transcriptional repressor beta-barrel domain	2
0231.9	MazG	all-alpha NTP pyrophosphohydrolase superfamily	2
0233.8	SufE_NifU	SufE/NifU superfamily	2
0277.7	FAD-oxidase_C	FAD-linked oxidase C-terminal domain superfamily	2
0295.7	Vps51	Vps51 domain superfamily	2
0302.6	Arginase	Arginase/deacetylase superfamily	2
0311.4	SCP2	SCP-2 sterol transfer superfamily	2
0316.4	Acyl_transf_3	Membrane acyl transferase superfamily	2
0321.4	PLAT	PLAT domain like superfamily	2
0334.4	THBO-biosyn	Tetrahydrobiopterin biosynthesis-like enzyme superfamily	2
0353.4	TIMP-like	TIMP-like superfamily	2

(Continued)

Table 9.3 (Continued)

Clan ID	Clan Name	Description	Frequency
0356.4	AMP_N-like	Creatinase/prolidase N-terminal domain superfamily	2
0387.4	DHFred	Dihydrofolate reductase-like	2
0422.4	Fibrinogen_C	Fibrinogen C-terminal domain-like	2
0441.4	AlbA	RNA-DNA binding Alba-like superfamily	2
0462.3	TPA-repeat	Transcription elongation factor C-terminal nonapeptide repeat	2
0487.3	FKBP	FKBP-like superfamily	2
0516.3	ISP-domain	Rieske-like iron-sulfur domain	2
0526.3	SUKH	SUKH superfamily	2
0527.3	Sm-like	Sm (Small RNA binding protein domain)	2
0536.2	HEXAPEP	Hexapeptide repeat superfamily	2
0566.2	Tubulin	Tubulin nucleotide-binding domain-like, GTPase	2
0586.2	CTC1	CST, telomere maintenance, complex subunit CTC1	2
0608.2	Reductase_C	FAD/NAD-linked reductase C-terminal domain superfamily	2
0609.2	sPC4_like	PC4-like superfamily	2
0662.1	Triple_barrel	Triple barrel superfamily	2
0006.20	C1	Protein kinase C, C1 domain	1
0009.20	ENTH_VHS	ENTH/ANTH/VHS superfamily	1
0059.15	6_Hairpin	Six-hairpin glycosidase superfamily	1
0061.13	PLP_aminotran	PLP dependent aminotransferase superfamily	1
0067.13	SIS	SIS domain fold	1
0071.12	His_phosphatase	Histidine phosphatase superfamily	1
0076.12	FAD_Lum_binding	Riboflavin synthase/Ferredoxin reductase FAD binding domain	1
0079.13	Cystine-knot	Cystine-knot cytokine superfamily	1
0085.14	FAD_DHS	DHS-like NAD/FAD-binding domain	1
0091.12	NAD_Ferredoxin	Ferredoxin / Ferric reductase-like NAD binding	1
0098.14	SPOUT	SPOUT Methyltransferase Superfamily	1
0100.13	C1q_TNF	C1q and TNF superfamily	1
0101.12	PELOTA	Pelota - RNA ribose binding superfamily	1
0104.13	Glyoxalase	VOC superfamily	1
0106.13	6PGD_C	6-phosphogluconate dehydrogenase C-terminal-like superfamily	1
0121.12	Cystatin	Cystatin-like superfamily	1
0151.12	PK_TIM	Pyruvate kinase-like TIM barrel superfamily	1
0155.11	CBM_14_19	Carbohydrate binding domain 14/19 clan	1
0157.11	Kleisin	Kleisin superfamily	1
0161.12	GAF	GAF domain-like	1
0165.11	Cache	Cache-like domain	1

(Continued)

Table 9.3 (Continued)

Clan ID	Clan Name	Description	Frequency
0171.11	Phospoesterase	Inositol polyphosphate 1 phosphatase like superfamily	1
0177.16	PBP	Periplasmic binding protein clan	1
0182.13	IT	IT (Ion Transporter) superfamily	1
0191.11	POTRA	POTRA domain superfamily	1
0194.10	DNA_pol_B-like	DNA polymerase B like	1
0213.9	ShK-like	Sea anemone toxin k like	1
0235.9	PspA	PspA/ESCRT-III	1
0243.10	AEP	Archaeo-eukaryotic primase	1
0247.9	2H	2H phosphoesterase superfamily	1
0251.8	MORN	MORN repeat	1
0255.9	ATP_synthase	ATP synthase F0 subunit	1
0271.7	F-box	F-box-like domain	1
0276.8	Nucleot_cyclase	Nucleotide cyclase superfamily	1
0287.7	Transthyretin	Transthyretin superfamily	1
0290.6	EPT_RTPC	EPT/RTPC-like superfamily	1
0299.5	Peptidase_SF	Peptidase clan SF	1
0303.5	H2TH	Helix-two-turns-helix superfamily	1
0318.4	Cytochrome-c	Cytochrome c superfamily	1
0323.4	Patatin	Patatin/FabD/lysophospholipase-like superfamily	1
0325.4	Form_Glyc_dh	Formate/glycerate dehydrogenase catalytic domain-like superfamily	1
0341.4	LDH_C	LDH C-terminal domain-like superfamily	1
0378.4	ANL	ANL superfamily	1
0382.4	DNA-mend	DNA breaking-rejoining enzyme superfamily	1
0383.4	PheT-TilS	Phenylalanine- and lysidine-tRNA synthetase domain superfamily	1
0391.4	CAP_C-like	Adenylate cyclase associated (CAP) C terminal like	1
0392.4	Chaperone-J	Chaperone J-domain superfamily	1
0395.4	Tubby_C	Tubby C-terminal domain-like	1
0407.4	TBP-like	TATA-binding protein like	1
0413.4	Toprim-like	Toprim domain	1
0417.4	BIR-like	BIR-like domains	1
0437.4	EF-G_C	Transcription elongation factor G C-terminal	1
0475.3	Cyclophil-like	Cyclophilin-like superfamily	1
0476.3	tRNA-IECD_N	tRNA-intron endonuclease catalytic domain-like N-term	1
0484.3	Peroxisome	Peroxisome-like domain	1
0489.3	SAF	SAF-like beta clip superfamily	1
0509.3	RBP11-like	RBP11-like subunits of RNA polymerase	1
0513.3	LCCL-domain	LCCL-domain like	1
0521.3	GOLD-like	Sec14-like superfamily of golgi trafficking	1
0534.2	YjgF-like	YjgF-like superfamily	1

(*Continued*)

Table 9.3 (Continued)

Clan ID	Clan Name	Description	Frequency
0539.2	RNase_III	RNase III domain-like superfamily	1
0546.2	Hexosaminidase	beta-N-acetylhexosaminidase-like domain	1
0585.2	Nucleoporin_A	Nucleoporin superfamily	1
0599.2	GH57_38_middle	Families 57/38 glycoside transferase middle domain superfamily	1
0603.2	AA_dh_N	Aminoacid dehydrogenase-like, N-terminal domain superfamily	1
0629.1	PLA2	Phospholipase A2 superfamily	1
0642.1	SOCS_box	SOCS-box like superfamily	1
0648.1	Aha1_BPI	Aha1/BPI domain-like superfamily	1
0649.1	PseudoU_synth	Pseudouridine synthase superfamily	1
0664.1	UB2H	UB2H/UvrB interaction domain superfamily	1

domains. Notably, the lists of organisms (ii) and refseq_ids (iii) were lists, i.e., the first refseq originated from the first organism and the second refseq originated from the second organism and so forth. The remaining information, that is refseqs presented similar domain content (item) were shared by all organisms and belonged to the same KO group. Gained and lost domains [(iv) and (v), above] were calculated for each KO group across all DAs as items. The minimal number of items was two.

Next sections formally define DA, active domain and unique active.

Definition: domain architecture (DA)

Algorithm: Let $p_1, p_2, \ldots, p_n \subseteq D$, where $D = \{d_1, d_2, \ldots, d_m\}$, is a set of protein domains and p_i is DA. Organization of DAs as distinct groups is a partition of $p_1, p_2, \ldots p_n$

Definition: active domains and unique active domains

Assumptions: Protein$p = \{d_1, d_2, \ldots d_m\}$, is DA, $c(d) \in \mathbb{R}$ is a score

Algorithm: Domain $d \in p$ is an active domain if $c(d)$ is maximal among overlapping including nested domains. A unique active domain is the maximal scoring active domain among multiple copies of the same domain within p.

Translocation and indel events of a mobile domain

The main objective of the EvoProDom model was to reflect changes in domain content, namely, at the protein level, with the organism level. This highlights groups of organisms with orthologous proteins that share similar patterns of protein domain gain/loss. Protein domain content was linked with organisms by defining mobile and translocation domains with consraints on number of organisms. In other words, mobile domain required minimal number of organims sharing patterns of gain/loss. A mobile domain was defined as follows:

Assumptions: Let A, B, T be sets of organisms with proteins members of KO group, k, such that $T = A \cup B, A \cap B = \emptyset, O \in A\{p \in O|d_x \in p\}, O \in B\{p \in O|d_x \notin p\}$. Organisms members of A contain domain d_x, while organisms members of B lack domain d_x.

Algorithm: Unique active domain d_x is a mobile domain between organisms members of A and members of B if $4 \leq |A| < |T| - 4$.

Next, translocations and indel events of mobile domains were defined such that these events are mutually exclusive for a given domain. Translocation events of mobile domains expanded the patterns of domain gain and/or loss in a single orthologous group to two orthologous groups, such that these patterns generated a reciprocal event. A reciprocal event was characterized by a mobile domain that was gained and lost in the first and second orthologous groups, and vice versa (Fig. 9.3). Translocation

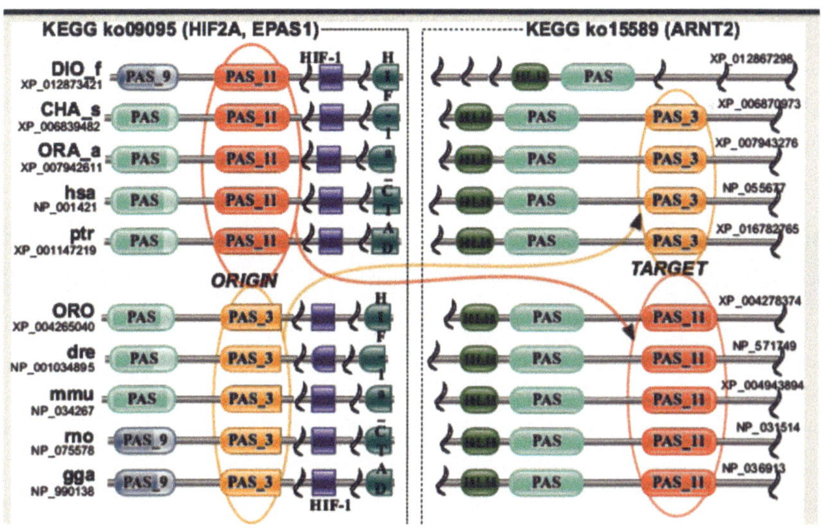

Figure 9.3 Illustration of a translocation event for PAS_11. PAS_11 (red domain) underwent a reciprocal translocation event between two orthologous protein groups K09095 (HIF2A) and K15589 (ARNT2). Accordingly, PAS_11 is present in HIF2A and absent from ARNT2 for organisms DIO, etc., while for organisms ORO, etc., PAS_11 is present in ARNT2 and missing from HIF2A.

event criterion was defined for groups of organisms with at least four members, similar to the definition of a mobile domain. For example, a translocation event of Pfam domain PAS_11 is shown in Fig. 9.3.

In this translocation event, PAS_11 was present in the KEGG orthologous group number 09095, corresponding to HIF2A. PAS_11 was absent from the orthologous protein group number 15589 (ARNT2). This gain and loss pattern of PAS_11 was observed among ten orthologous proteins from five different organisms (A^*), namely, DIO to ptr, such that each organism includes a protein from HIF2A and one from the ARNT2 orthologous group. Reciprocal gain and loss patterns of PAS_11 were observed in orthologous proteins from a second group of organisms (B^*), namely, ORO to gaa. A reciprocal gain and loss pattern means that PAS_11 is absent from the HIF2A orthologous group and gained in the ARNT2 orthologous group. Since these reciprocal patterns of domain gain/loss between two orthologous proteins involved two groups of organisms (A^* and B^*) with more than four organisms each, it can be determined that a translocation event occurred for PAS_11. Note that an additional translocation event occurred for PAS_3 among the same organisms and orthologous proteins (Fig. 9.3).

Translocations and indel events were formally defined:

Assumptions: Let d_x be a mobile domain between A_i and B_i in k_i, where $i = 1,2$, A_i, B_i are sets of organisms and k_i is a KO group. Let $A^* = A_1 \cap B_2$ and $B^* = A_2 \cap B_1$.

Algorithm: Mobile domain d_x undergoes translocation if $|A^*|, |B^*| \geq 4$. Otherwise, an indel event occurred.

Duplication of domains

Translocation and indel events were defined for mobile domains. Mobile domains, in turn, were derived from unique active Pfam domains. For duplication events, active domains were considered in order to retain non-overlapping duplicates (see Materials and methods). These active domains were calculated for each orthologous protein group (KO group). Therefore a duplicate status, "duplicated" or "non-duplicated," was determined for a particular orthologous KO group and, consequently, varied among orthologous KO groups. Accordingly, the overall duplicate status of a particular Pfam domain was determined by the majority from individual assignments of duplicate status. For example, the overall duplicate status of a Pfam domain was defined as "duplicated" if the difference between the number of KOs with "duplicated" and the number of KOs with "non-duplicated" status was significant, i.e., in the 99% percentile of the cumulative sum of the differences. Similarly, overall "non-duplicated" status was determined when considering "non-duplicated" and "duplicated" differences. The duplicate status, namely, "duplicated" or "non-duplicated," for a particular KO group was determined based on whether the copy number of particular domains varied or was constant across all DAs, respectively. For example, if two copies of a particular domain appeared in one KO while three copies of a domain appeared in a second KO, the domain was assigned as "duplicated." However, if a particular domain had the same number of copies within all DAs, for example, two copies, then the domain was considered "non-duplicated." Duplication was formally defined:

Assumptions: Let d_x be an active domain, *ko* be the KO group with $da_1, da_2, \ldots, da_m$ DAs of active domains. Then, d_x is "non-duplicated" in *ko* if the copy number of *dx* is the same in each DA, otherwise d_x is "duplicated."

Algorithm: d_x is duplicated if the difference between the number of KO groups where d_x is "duplicated" and the number of KO groups where it is "non-duplicated" is significant (above 99% of the cumulative sum of the differences). A nonduplicated domain is similarly defined.

Translocation domains are enriched in chimeric transcripts

Previously, Frenkel-Morgenstern and Valencia [5], analyzed enrichment of the domain content encoded by chimeric transcripts combined from two distinct genes. This effort found that encoded domains were enriched within chimeric transcripts that belonged to the following super-families (super-family name): ANK (Ank), EFh (EF_hand), EGF-like (EFG), GTP_EFTU (P-loop_NTPase), IG-like (E-set), LRR (LRR), PH (zf-FYVE-PHD), Pkinase (PKinase), RING (RING), RRM (RRM), SH2 (SH2-like), SH3 (SH3), WD40 (Beta_propeller) and ZnF (C2H2-zf) [5]. Of these, EFh (EF_hand), EGF-like (EFG), GTP_EFTU (P-loop_NTPase), IG-like(E-set), Pkinase (PKinase), RRM (RRM), SH2 (SH2-like), SH3 (SH3), WD40 (Beta_propeller) and ZnF (C2H2-zf), all confirmed by RNA-seq data [5]. These domains appeared in high copy numbers within proteins, for example, Ank [21–23] and WD40 [24], as repeats or highly abundant within proteins, e.g., SH3 [25,26]. Therefore, we hypothesized that highly abundant domains might experience multiple translocation events. To this end, we applied EvoProDom to the assembly of organisms (EvoProDomDB) and identified a total of 2,041 mobile domains. Of these, 259 had undergone translocation events and 1,782 participated in indel events (Tables 9.2 and 9.3).

The Pfam domains were classified into super-families [6,7], and translocation events and indel event frequencies were grouped according to super-family (Tables 9.2 and 9.3, respectively). Among the 10 most abundant domain super-families were SH3 (Src homology-3 domain), Ig (Immunoglobulin super-family) and Ank (Table 9.2). Likewise, the most abundant super-families of mobile domains that participated in indel events were "Unknown," P-loop_NTPase and TPR (Table 9.3).

The SH3 super-family contains SH3_2 (239 translocations), SH3_1 (198 translocations) and SH3_9 (193 translocations). SH3 (src Homology-3) domains are short protein domains roughly 50 amino acids in length [27,28] and are found in diverse intracellular or membrane-associated proteins [29–31], for example, fodrin and yeast actin binding protein (ABP-1). SH3 domains mediate PPIs by facilitating assembly of protein complexes [25]. The Ig super-family contains Ig (219 translocations), I-set (135 translocations), V-set (117 translocations) and Ig_2 (116 translocations). These domains are found in the vertebrate immune system (V-set), cell surface proteins and in intracellular muscle proteins (I-set) [32,33]. The Ank repeats super-family comprises Ank_2 (231 translocations), Ank_4 (184 translocations) and Ank_5 (94 translocations). These domains are found in copies, which form arrays. These repeats participate in PPIs that regulate cell cycle transition from G1 to S [21–23]. Such regulation is accomplished by INK4 (inhibitors of cyclin-dependent kinase 4) protein complex formation and inhibition of cyclin-dependent kinase 4 and 6 (CDK4/6) proteins [23]. These findings showed that protein domains overrepresented in chimeric transcripts underwent multiple translocations. This reinforces a link between chimeric transcripts and EvoProDom translocations. Additionally, P kinase and ubiquitin domains are known to form new fusions, i.e., appear in novel transcripts, are found in multiple tanslocation events [5]. Note that super-families with maximal and minimal number of translocations, SH3 (630) and SH2-like (6), were overrepresented in chimeric transcripts (Table 9.2).

Discussion

Here, we presented a novel, evolution model for proteins, EvoProDom, which was developed according to the "mix and merge" view of protein domains. The EvoProDom model combined genomic and proteome data, including orthologous protein and protein domain data from 109 organisms from divergent taxa and was implemented with and tested on EvoProDomDB. In the model, evolutionary events, e.g., translocations were defined to reflect changes in protein domain composition, which are coupled with orgnasims. By this method protein level dynamics are manifested at the organism level. Thus SH3, which binds ligands [25,26] and mediates PPIs [34], was found to be a highly frequent domain in translocations. Repetitive domains, e.g., Ank [21–23] and WD40 [24], were found in multiple copies in proteins. These domains fold into 3D structures and generally mediate PPIs [21–24] by means of novel chimeric proteins that form novel PPIs and alter PPI networks of parent proteins [35]. Accordingly, such domains, for example, SH3_2, Ig and Ank_2 and others (see Results), were overrepresented in many fusion event chimeric transcripts [5]. As hypothesized, these domains were involved in multiple evolutionary translocation events. Repetition of domains provides a plausible explanation for the high frequency of their translocation events. Fusions are produced upon slippage of two parent genes. These parent genes are truncated at the junction point, resulting in domain loss. Consequently, the proper function of a chimeric protein is impaired [35]. As an example, fusion within a catalytic domain would render the protein nonfunctional, and such fusion would be selected against. Naturally, due to selection, high copy number repetitive domains would appear in chimeras at higher frequencies than expected based on their frequency alone, albeit, with fewer repeats. Indeed, an average copy number of these domains was reduced in chimeric transcripts [5]. In EvoProDom, repetitive domains or abundant domains, for example, SH3, within KO groups resulted in a greater number of DAs. This translates into a greater number of (ko, item) pairs (see Materials and methods). As a result, repetitive and abundant domains contribute more to the pool of mobile domains from which translocation events were derived, and therefore were highly frequent in translocations. Together, these results indicate that translocation events of highly frequent and repetitive domains modulate PPI networkts to bring about adaptive evolution.

The inclusion of new organisms into EvoProDomDB required only complete genome sequences and annotated proteomes. Protein domain composition and orthologous protein data were obtained from protein sequences using the Pfam search tool [6,7] and KoFamKOALA [8], respectively. The use of these tools enabled the applicability of

EvoProDom to any new organism with a sequenced genome and annotated proteome and a general method of obtaining these data from only protein sequence. This method may be utilized in other research fields, for example, proteomics [9], protein design [10], and defining host–virus interactions [11]. To conclude, EvoProdom is a novel model for evolution of proteins which is founded on dynamics of protein domains in a "mix and merge" manner, and is markedly different from DNA-based models. This highlights the advantage of introducing chromosomal alterations into evolutionary events.

Conflict of interest

The authors declare no conflicts of interest.

Author contributions

G.C. and A.G. performed the study, verified the results, and wrote the paper. M.F.M. designed and supervised the study and wrote the paper

Acknowledgments

M.F.M. is a member of the Dangoor Center for Personalized Medicine and the Data Science Institute (DSI), Bar-Ilan University, Israel.

We thank Dr. Eivatar Nevo for his expertise and helpful comments on the manuscript.

This work was supported by a grant for Biomarkers for Treatment of Arthritis Patients (Israel Innovation Authority, 66824, 1.7.2019 to 30.6.2020) and The Roland and Dawn Arnall Foundation (Research Grant, 205227 1.9.2018 to 31.8.2019).

Abbreviations

DA Domain architecture
EvoProDom Evolution of protein domains
PPI Protein–protein interaction
KEGG Kyoto Encyclopedia of Genes and Genomes
KO KEGG ortholog

References

[1] J.A. Marsh, S.A. Teichmann, How do proteins gain new domains? Genome Biol. 11 (7) (2010) 126.

[2] M.K. Basu, E. Poliakov, I.B. Rogozin, Domain mobility in proteins: functional and evolutionary implications, Brief. Bioinforma. 10 (3) (2009) 205–216.
[3] D.V. Cancherini, G.S. França, S.J. de Souza, The role of exon shuffling in shaping protein–protein interaction networks, BMC Genomics 11 (Suppl 5)) (2010) S11. Available from: https://doi.org/10.1186/1471-2164-11-S5-S11.
[4] E.A. Bornberg-Bauer, M. Mar, Dynamics and adaptive benefits of modular protein evolution, Curr. Opin. Struct. Biol. 23 (3) (2013) 459–466.
[5] M. Frenkel-Morgenstern, A. Valencia, Novel domain combinations in proteins encoded by chimeric transcripts, Bioinformatics 28 (12) (2012) i67.
[6] R.D. Finn, et al., The Pfam protein families database, Nucleic Acids Res. 36 (2008) D281–D288. Available from: https://doi.org/10.1093/nar/gkm960. Database issue.
[7] R.D. Finn, et al., The Pfam protein families database: towards a more sustainable future, Nucleic Acids Res. 44 (D1) (2016) D279–D285. Available from: https://doi.org/10.1093/nar/gkv1344.
[8] T. Aramaki, et al., KofamKOALA: KEGG ortholog assignment based on profile HMM and adaptive score threshold, bioRxiv, (2019), 602110, doi: 10.1101/602110.
[9] J.A. Jadwin, M. Ogiue-Ikeda, K. Machida, The application of modular protein domains in proteomics, FEBS Lett. 586 (17) (2012) 2586–2596. Available from: https://doi.org/10.1016/j.febslet.2012.04.019.
[10] D. Setiawan, J. Brender, Y. Zhang, Recent advances in automated protein design and its future challenges, Expert. Opin. Drug. Discov. 13 (7) (2018) 587–604. Available from: https://doi.org/10.1080/17460441.2018.1465922.
[11] A.F. Brito, J.W. Pinney, Protein–protein interactions in virus–host systems, Front. Microbiol. 8 (2017) 1557. Available from: https://doi.org/10.3389/fmicb.2017.01557.
[12] M. Kanehisa, Y. Sato, M. Kawashima, M. Furumichi, M. Tanabe, KEGG as a reference resource for gene and protein annotation, Nucleic Acids Res. 44 (D1) (2016) D457–D462. Available from: https://doi.org/10.1093/nar/gkv1070.
[13] M. Kanehisa, M. Furumichi, M. Tanabe, Y. Sato, K. Morishima, KEGG: new perspectives on genomes, pathways, diseases and drugs, Nucleic Acids Res. 45 (D1) (2017) D353.
[14] D.A. Benson, et al., GenBank, Nucleic Acids Res. 46 (D1) (2018) D41–D47. Available from: https://doi.org/10.1093/nar/gkx1094.
[15] K.T.J. Davies, N.C. Bennett, C.G. Faulkes, S.J. Rossiter, Limited evidence for parallel molecular adaptations associated with the subterranean niche in mammals: a comparative study of three superorders, Mol. Biol. Evol. (2018). Available from: https://doi.org/10.1093/molbev/msy161.
[16] W.C. Tavares, H.N. Seuánez, Changes in selection intensity on the mitogenome of subterranean and fossorial rodents respective to aboveground species, Mamm. Genome 29 (5-6) (2018) 353–363. Available from: https://doi.org/10.1007/s00335-018-9748-5.
[17] A. Chatr-Aryamontri, et al., The BioGRID interaction database: 2015 update, Nucleic Acids Res. 43 (2015) D470–D478. Available from: https://doi.org/10.1093/nar/gku1204. Database issue.
[18] A. Chatr-Aryamontri, et al., The BioGRID interaction database: 2017 update, Nucleic Acids Res. 45 (D1) (2017) D369–D379. Available from: https://doi.org/10.1093/nar/gkw1102.
[19] K.E. Jones, et al., Global trends in emerging infectious diseases, Nature 451 (2008) 990. Available from: http://doi.org/10.1038/nature06536. Available from: https://www.nature.com/articles/nature06536#supplementary-information.
[20] J.A. Marsh, S.A. Teichmann, Structure, dynamics, assembly, and evolution of protein complexes, Annu. Rev. Biochem. 84 (1) (2015) 551–575.

[21] S.G. Sedgwick, S.J. Smerdon, The ankyrin repeat: a diversity of interactions on a common structural framework, Trends Biochem. Sci. 24 (8) (1999) 311–316. Available from: https://doi.org/10.1016/s0968-0004(99)01426-7.

[22] L.K. Mosavi, T.J. Cammett, D.C. Desrosiers, Z.Y. Peng, The ankyrin repeat as molecular architecture for protein recognition, Protein Sci. 13 (6) (2004) 1435–1448. Available from: https://doi.org/10.1110/ps.03554604.

[23] J. Li, A. Mahajan, M.D. Tsai, Ankyrin repeat: a unique motif mediating protein–-protein interactions, Biochemistry 45 (51) (2006) 15168–15178. Available from: https://doi.org/10.1021/bi062188q.

[24] C. Xu, J. Min, Structure and function of WD40 domain proteins, Protein Cell 2 (3) (2011) 202–214. Available from: https://doi.org/10.1007/s13238-011-1018-1.

[25] C.J. Morton, I.D. Campbell, SH3 domains. Molecular 'Velcro', Curr. Biol. 4 (7) (1994) 615–617. Available from: https://doi.org/10.1016/s0960-9822(00)00134-2.

[26] T. Kaneko, L. Li, S.S. Li, The SH3 domain--a family of versatile peptide- and protein-recognition module, Front. Biosci. 13 (2008) 4938–4952. Available from: https://doi.org/10.2741/3053.

[27] T. Pawson, J. Schlessingert, SH2 and SH3 domains, Curr. Biol. 3 (7) (1993) 434–442. Available from: https://doi.org/10.1016/0960-9822(93)90350-w.

[28] B.J. Mayer, SH3 domains: complexity in moderation, J. Cell Sci. 114 (Pt 7) (2001) 1253–1263.

[29] A. Musacchio, T. Gibson, V.P. Lehto, M. Saraste, SH3—an abundant protein domain in search of a function, FEBS Lett. 307 (1) (1992) 55–61. Available from: https://doi.org/10.1016/0014-5793(92)80901-r.

[30] B.J. Mayer, D. Baltimore, Signalling through SH2 and SH3 domains, Trends Cell Biol. 3 (1) (1993) 8–13. Available from: https://doi.org/10.1016/0962-8924(93)90194-6.

[31] T. Pawson, Protein modules and signalling networks, Nature 373 (6515) (1995) 573–580. Available from: https://doi.org/10.1038/373573a0.

[32] A.F. Williams, A.N. Barclay, The immunoglobulin superfamily—domains for cell surface recognition, Annu. Rev. Immunol. 6 (1988) 381–405. Available from: https://doi.org/10.1146/annurev.iy.06.040188.002121.

[33] Y. Harpaz, C. Chothia, Many of the immunoglobulin superfamily domains in cell adhesion molecules and surface receptors belong to a new structural set which is close to that containing variable domains, J. Mol. Biol. 238 (4) (1994) 528–539. Available from: https://doi.org/10.1006/jmbi.1994.1312.

[34] F. Malagrinò, F. Troilo, D. Bonetti, A. Toto, S. Gianni, Mapping the allosteric network within a SH3 domain, Sci. Rep. 9 (1) (2019) 8279. Available from: https://doi.org/10.1038/s41598-019-44656-8.

[35] M. Frenkel-Morgenstern, A. Gorohovski, S. Tagore, V. Sekar, M. Vazquez, A. Valencia, ChiPPI: a novel method for mapping chimeric protein–protein interactions uncovers selection principles of protein fusion events in cancer, Nucleic Acids Res. 45 (12) (2017) 7094–7105.

CHAPTER 10

Evolution Canyons model: biodiversity, adaptation, and incipient sympatric ecological speciation across life: a revisit

Eviatar Nevo
Institute of Evolution, University of Haifa, Haifa, Israel

The Evolution Canyon model

Environmental stress is a major driving force of evolution [1,2]. *Geological, climatic, edaphic, abiotic, and biotic* stresses are major environmental factors affecting evolution. They often lead to similar outcomes at *macro* and *micro*scales highlighted by *genomic, proteomic phenomic, epigenomic, metagenomic,* and *metabolomic* processes [3,4]. Such biological patterns due to environmental stresses may be accentuated at *ecologically sharply divergent microsites*, such as in the "Evolution Canyon" (EC) model (Fig. 10.1). The four *microclimatic* ECs, in Carmel, Galilee, Golan, and Negev mountains in Israel (Fig. 10.2), have been extended to two edaphically divergent microsites models in Israel, dubbed "Evolution Plateau" (EP), and "Evolution Slope" (ES) (Figs. 10.3 and 10.4).

The EC research project is a long-term project that started in 1990, exploring *evolution in action* (see 250 publications including four books in Nevo list of EC publications at http://evolution.haifa.ac.il, and reviews listed in the full list of Nevo 1995, 1997, 2001, 2006, 2009, 2011, 2012, 2014, and 2015). Remarkably, interslope incipient sympatric ecological speciation was found in EC I across life from bacteria to mammals. The "EC" model represents the Israeli ecological analog of the Galapagos Islands, and we dubbed it the "Israeli Galapagos" [4,5]. In these *micro*climatic or *micro*-edaphic microsites, selection overrides gene flow and drift and drives both interslope adaptive divergence and incipient sympatric ecological speciation at a microscale. The EC model could potentially highlight many mysteries of evolutionary biology including the genetic

New Horizons in Evolution
DOI: https://doi.org/10.1016/B978-0-323-90752-1.00009-2

basis of adaptation and speciation, especially now in the age of genomics [6,7] with the rapid high-throughput techniques of whole-genome analysis across life, and the universality of DNA transfer across distant taxa (Fig. 10.1).

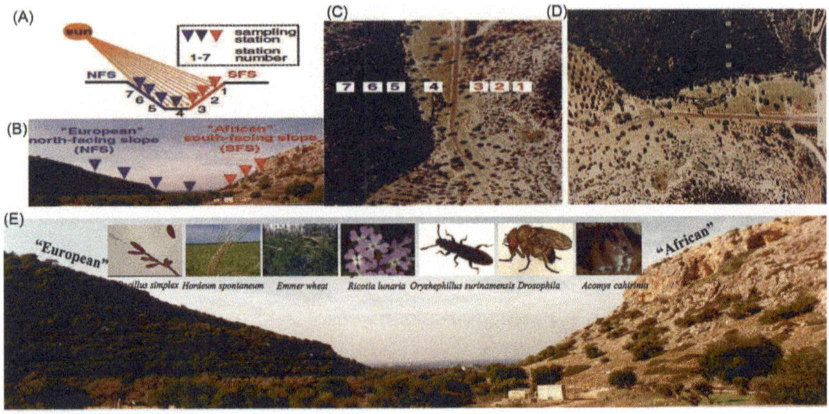

Figure 10.1 "Evolution Canyon" I (EC I) at Mount Carmel, Israel. (A) The model of Evolution Canyon I (EC I): ***Microclimatic*** interslope divergence: stations (populations) 1–3 are on the tropical, savannoid, South-Facing Slope (SFS), also dubbed the "African" slope (AS; *red triangles* in A and B, SFS = AS, and red numbers in C, SFS = AS), characterized by high solar radiation, temperature, and drought; station 4, in the crick (*gray triangle*), and stations 5–7 (*blue triangles* in B, NSF, and blue numbers in C, NSF) are on the temperate, forested, north-facing slope (NFS), also dubbed the "European" slope (ES, *blue triangles* in B, and *blue numbers* in C), characterized by low solar radiation and temperature, high humidity, and forested. AS is distant ~250 m from ES. (B) Cross section of Evolution Canyon I (EC I), covered, on the left-hand side, by a green forest with live oaks, *Alon* in Hebrew, *Quercus calliprinos*, and *Pistacia palaestina*, *elah* in Hebrew, on the "European" slope (ES = NFS), versus the savannoid "African" slope (AS = SFS). The blue and red triangles represent experimental stations (populations) colored as in A. (C) Air view of Evolution Canyon I (EC I), Mount Carmel, showing the south European forested slope (ES = NFS) versus the savannoid African slope, with open park forest (AS = SFS), with carob trees, *Ceratonia silqua* and *Pistacia lentiscus* bushes, and African grasses, *Hyparrhenia hirta, Andropogon distachion*, and *Pennisetum asperifolium*. (D) Cross section of Evolution Canyon I (EC I), Mount Carmel, with the forested slope ES = NFS, and opposite, abutting savannoid, park forest (AS = SFS), with seven distant organisms from bacteria to mammals, that represent incipient adaptive sympatric ecological speciation across life: from left to right: soil bacterium, *Bacillus simplex*; wild barley, *Hordeum spontaneum*; wild emmer wheat, *Triticum dicoccoides*, crucifer, Maltese Cross *Ricotia lunaria*, in Hebrew Carmelite Na-ah, fruit flies, *Drosophila melanogaster*, saw-toothed grain beetle, *Oryzaephilus surinamensis;* and spiny mouse, *Acomys cahirinus*.

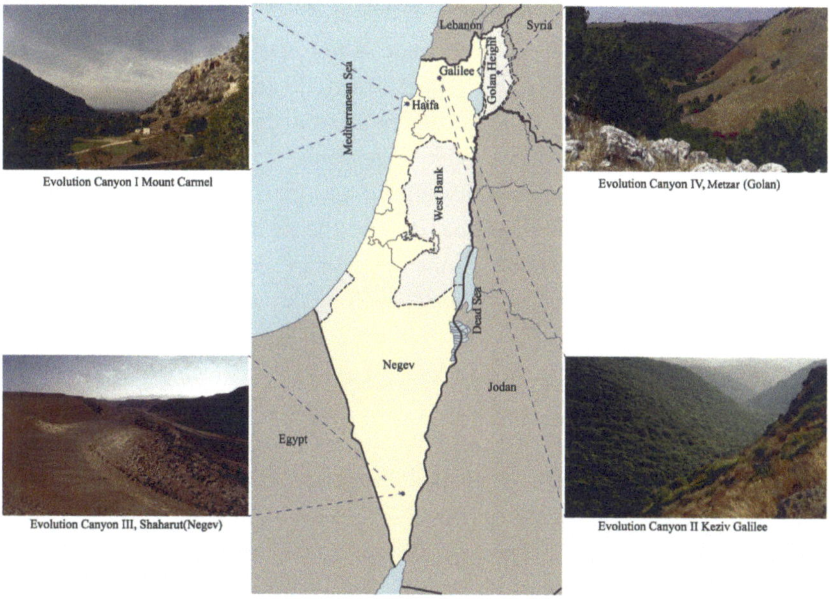

Figure 10.2 The four "Evolution Canyons" studied in Israel: lower Nahal Oren (Mount Carmel)-EC I; Nahal Keziv (Galilee)-EC II; Nahal Shaharut (Negev desert)-EC III, and Nahal Metzar (Golan)-IV. Note that in ECs I, II, IV, the north-facing slope (NFS), also dubbed the European slope (ES), is on the left-hand side, representing temperate, cool, humid, and forested biome. The opposite, abutting slope, on the right-hand side, is tropical, hot, dry, and savannoid. By contrast, in Nahal Shaharut (south Negev desert)-EC III, the slope orientation in the picture is reversed: the SFS is on the left, covered by cyanobacteria, and the NFS is on the right, darker in color, covered by lichens, with angiosperm bushes only in the creek.

Four ECs have been studied in Israel in the mountains of Carmel (EC I), Galilee (EC II), Negev desert (EC III), and Golan (EC IV) (Fig. 10.2 [5]). About 4000 species have been described to date, from bacteria to mammals, in the four ECs. Most studies have been conducted in EC I, Mount Carmel (Fig. 10.1), at an area of 7000 m^2, representing a 250-m interslopes, airway distance, between stations 2 and 6, and a transect of increasing aridity, extending from the north-facing slope (NFS), also dubbed "European" slope (ES) station 7, down to the creek (station 4) and up to the "AS" station 1, across ~300 m groundway, yielding to date 2500 species: 100 species of bacteria, 500 fungi, 350 flowering plants, 1500 insects, and 50 vertebrates. The EC represents multiple biological-evolutionary models: *biodiversity, adaptation, ecological sympatric speciation (SS), monitor of global warming (extinction, or interslope migration from warm AS to cool ES), and host–pathogen, or disease resistance* (Fig. 10.2).

Notably, the opposite, abutting, slopes at three ECs (EC I–III), share the same geology (Upper Cenomanian limestones), terra rossa soils, and Mediterranean *macro*-climate (EC III is under desert climate), but they all differ sharply ecologically due to ***microclimatic*** divergence (Fig. 10.1A), resulting locally, in continental scale divergence of hot and dry tropical, savannoid [AS = SFS (south-facing slope)], and cool and humid temperate, forested (ES = NFS) ***microclimates***, despite the short ***interslope*** distance of only hundreds of meters [8,9]. Only the Metzar, EC IV in the southern Golan Heights, is geologically ***basaltic*** on both slopes, and is the least studied to date of the four ECs. The distinctly higher solar radiation (up to 800% more) on the African slope (AS = SFS) at EC I, than on the opposite and abutting ES (=NFS), results in a continental scale climatic divergence: the AS = SFS at EC I, II, III, and IV, represent a tropical, hot and dry, savannoid biome (stations 1–3), versus temperate, cool and humid, forested biome (stations 5–7) on the ES = NFS. Fig. 10.1A–D represents the model (1), cross section (2), air view (3), and species that speciated sympatrically (4), with the seven research stations (populations), three on each slope, and one, 4, at the creek. Notably, station no 2 on the AS (=SFS) is separated at EC I by only ~250 m, air-distance, from station 6 on the ES (=NFS). Despite the short interslope distance, the opposite slopes represent continental scale divergent climates, ***tropical*** savannoid, dry African, versus ***temperate***, forested, humid south European forest (Fig. 10.1A–D), that is, representing two divergent *biomes*. The EC model provides a unique microsite sharply unfolding *Evolution in action*, of adaptation and differential degrees of SS, across life from viruses and bacteria to mammals. It reveals *adaptive convergence within* each slope, and *adaptive divergence (incipient or complete SS)* between slopes (Figs. 10.1A–D and 10.2, EC I–IV) (Fig. 10.3).

Evolution Plateau

EP (Fig. 10.3) is an edaphic microsite in eastern Upper Galilee, between the Alma and Dalton basalt Plateaus (Fig. 10.3) comprised two abutting divergent soils, Senonian chalk weathered to rendzina soil, and Pleistocene volcanic basalt weathered to basalt soil. The basalts in the eastern upper Galilee form two nearby basalt plateaus: the Alma basalt plateau and the Dalton basalt Plateau (see geological map in Fig. 10.12B). This site has been studied for mammalian *micro*-evolution [10], and then, for adaptive ecological SS in blind subterranean mole rats, ***Spalax galili*** [11–17], and in wild barley [18]. EP is paralleling EC as a microsite

Figure 10.3 "Evolution Plateau" in the eastern Upper Galilee, near Dalton and Alma basalt plateaus, a microsite of edaphic rather than microclimatic divergence, displaying hot spot of sympatric speciation due to abutting divergent rock formations and soil types, chalk versus basalt, and their respective soils, rendzina versus basalt soils. The rendzina soil is covered by bushlets of *Sarcopterium spinosum*, and the basalt is the brown soil with stones and mounds of blind mole rats *Spalax galili*, that originated on the Pleistocene basalt rock and soil as a new species called temporarily ***S. galili* basalt**, originating by sympatric speciation, with gene flow. The sharp line between chalk and basalt is on a geological fault, and separates the new sympatrically evolving species, *S. galili* basalt, since it speciated on the basalt soil (mole rat mounds are visible). Both species of blind mole rats of the *S. galili* ($2n = 52$), the ancestor on chalk and the derivative on basalt, belong to the *Spalax ehrenbergi* superspecies, comprising five species in Israel, four chromosomal and peripatric, one genic and sympatric.

highlighting adaptive evolution and sympatric ecological speciation with gene flow. The EP is, however, a ***geological-edaphic*** microsite comprised Senonian Chalk-versus Pleistocene volcanic basalt, or their respective soils, rendzina, abutting with basalt soil. It differs from the EC microsites that derive from opposing slopes sharply divergent ***microclimatically***, *tropical* versus *temperate* microclimates, separated from few meters to hundreds of meters apart [4] (Fig. 10.3).

Evolution Slope

Evolution Slope (Tabigha), is an edaphic microsite, similar to EP, but instead of Senonian chalk abutting with the Pleistocene basalt as in EP, at

Figure 10.4 **Evolution Slope,** (Tabigha) showing in front the dry terra rossa with yellow vegetation on the hill on the left side, generated on hard Middle Eocene limestone (big boulders, covered with a surface of gray cyanobacteria). In the back, and right-hand side the abutting humid green vegetation with isolated *Zizypus spina christi* trees on Pleistocene volcanic basalt dipping southward towards the lake of Galilee (Lake Kinneret). The sharp line between the yellow and green vegetation is a geological-edaphic line, separating dramatically divergent wild barley (*Hordeum spontaneum*), and wild emmer wheat (*Triticum dicoccoides*) populations, the progenitors of cultivated barley and wheat, respectively. Both wild cereals are incipiently sympatrically speciating in this microsite dipping towards the Lake of Galilee (Hence designated "Evolution Slope"). This site, named earlier Tabigha, has been studied since 1980s on adaptive evolution in wild barley, *Hordeum spontaneum* [19,20], wild emmer wheat [21], *Aegilops peregrina* [22], and, recently, wild barley genome [23].

ES, terra rossa soil weathered from hard Middle Eocene limestones, abuts with the Pleistocene volcanic basalt (Fig. 10.4). In both edaphic microsites, EP and ES, the divergent soils are abutting and the SS of wild barley and wild emmer wheat (WEW) (the latter only at Evolution Slope), on the Pleistocene basalt at both microsites derived from Senonian chalk (rendzina) population at EP, but from middle Eocene hard limestones, weathered to terra rossa, at Evolution Slope.

Microclimatic interslope divergence underlying biodiversity contrasts in EC

Microclimatic interslope differences of illuminance, temperature, and humidity were measured on the AS = SFS and ES = NFS of EC I microsite, Lower Nahal Oren, Mt. Carmel, Israel [9]. Two stations (populations) were measured on each slope: SFS 1 and SFS 2 on the AS, and NSF 5 and NSF 6 on the ES. Preliminary results show that (1) illuminance on the AS = SFS was significantly higher (up to 800% more than on the ES = NFS) during April-October 1997, (2) mean daily temperatures, as well as daily temperature ranges, were higher on the AS than on the ES, and (3) except under the high summer sun, relative humidity was 1%–7% higher on the ES than on the AS. We concluded that microclimatic stress is responsible for the drastic *interslope biodiversity divergence* across life at EC I [9].

Biodiversity evolution

Biodiversity colonization to EC microsites, as to Israel at large, arrived from northern *temperate* Europe, and southern tropical dry Africa and Asia, and from eastern Mediterranean climates. The European colonizers invaded first the **temperate ES** also dubbed NFS, characterized by low solar radiation, humid and cool microclimate, and forested. By contrast, the African and Asian colonizers invaded first the tropical AS, or SFS, characterized by high solar radiation (200%–800% higher than on ES, at EC I in Mount Carmel), temperature, and drought [8,9]. Only later some AS colonizers may have moved to the ES and originated there new sympatric species such as spiny mouse, *Acomys cahirinus* [24,25] or even the fruit fly, cosmopolitan *Drosophila melanogaster* [26].

The four thousands of identified species at the four ECs include new species to science, Asia, Israel, and to each of the EC biomes (AS = SFS, and ES = NFS). For detailed list of EC, EP, and ES publications, see Nevo list of ECs at http://evolution.haifa.ac.il. Cyanobacteria, earlier called cyanophyta [27–30], soil bacteria [31–33], Coccidia [34] algae, fungi, mosses, and lichens [35–43], Phytoplankton [44], ants, interslope divergent at both EC I and EC II, with higher species number on the AS = SFS [45,46], soil microfungi [41,47–55], earthworms [56], yeasts [57,58], vascular plants [59,60], isopods, springtails (*Collembola*) [61], beetles [62–69], mites [70], grasshoppers [71], aphids [72], land snails. The local physical microclimatic sharp divergence leads to gastropod adaptive interslope biotic divergence [73], *Drosophila* [74], butterflies [75], wasps [76,77], lizards [78], rodents [79], blind mole rats [11–13,15,16,80]. Beetles, represent a remarkable example in both EC I (Carmel) and EC II (western upper Galilee), on which we wrote 3 books on coleoptera in

EC I [65,69], and in EC II [68], they highlight sharp interslope species divergence. Out of 633 beetle species (including 350 genera and 56 families), reported in EC I, 318 species were reported in Ref. [69] and 315 species in Ref. [65]. One species was new to science, 10 species were new to Israel, and 30 species were new to Mount Carmel. In EC II, in western upper Galilee [68], out of 513 beetles reported, 136 species were reported to the Galilee Mountains, 19 species were new to Israel, and one species was new to science [68]. The recent study on parasitic hymenoptera by Sonia Bigalk from Hohenheim University, Germany, yielded remarkable results. During two weeks of collecting, 9323 Hymenoptera specimens were collected at EC I, in Mount Carmel, comprising 42 families. The study revealed an amazingly high diversity as several rarely collected families like Orussidae, Heloridae, Stephanidae, Embolemidae, and Sclerogibbidae were recorded. Furthermore, one unknown species of *Pristapenesia* sp. nov. of the extremely rarely collected Scolebythidae was recorded, representing the first West Palaearctic record of this family. All recorded hymenopteran families are discussed with reference to their biological, climatic and biogeographic preferences in the MSc of Bigalk. In addition, 53 Pteromalidae species in 27 genera were identified, representing 49 new species and 18 new generic records for Israel. Furthermore, a first checklist of all newly and previously recorded Pteromalidae species from Israel is provided (Bigalk Sonia, MSc Thesis, Hohenheim University, Germany). These results significantly contributed to evolutionary theory, to the fauna of Israel and the Near East.

Yeast pioneering discovery in *micro*- and *macroscales* in Israel

We have first recorded natural populations of yeasts *locally* at EC I [57], and regionally, across Israel [58]. At EC I, a total of 25 species, belonging to 14 genera were morphologically described,19 species of soil yeasts from 11 genera and 14 species of plant yeasts from 10 genera, including 8 species and 7 genera common to plants and soil. The total number, and total frequency of yeast species, was higher on the mesic ES than on the xeric AS in samples from both plants and soil. Our regional studies in Israel [58] involved collections from both soil and fallen leaves. A total of 69 species belonging to 27 genera were molecularly described [58]. Totally, 53 species and 13 genera were newly recorded taxa for Israel. The general part of the book described the climate, geology, soil composition, and flora of

Israel, followed by material and molecular methods of identification. The special taxonomic part, describes the ecology, distribution, and diversity of soil and fallen leaves species, and discussion of yeast communities in Israel. The book was published by Koeltz scientific books, Germany [58].

Continental biome interslope divergence at a microsite EC I, Mount Carmel

European and African biomes predominate on the ES and AS, respectively, due to the microclimatic interslope dramatic divergence, tropical hot-dry microclimate on AS versus temperate cool-humid microclimate on the ES [8]. The ***dry tropical*** AS involves significantly more terrestrial species than the ***temperate humid*** ES which involves more species associated with humid habitats [8].

In all, 2500 taxa have been identified in EC I, during near to 30 research years, on both slopes in 7000 m^2, including 100, 500, 350, 1500, and 50 species of bacteria, soil fungi, plants, insects, and vertebrates, respectively. African species and European species characterize AS and ES, respectively, adapted to each continental origin, ***dry and hot Africa and humid and cold Europe*** [8].

Interslope opposite floras appear in the caption of Fig. 10.1. African faunal key taxa are represented by the land snail *Sphincterochila cariossa*, and the lizards *Stelagama stellio* and *Ptyodactylus guttata*, by the birds. *Pycnonotus barbatus*, *Nectarinia ossea*, and by the spiny mouse *A. cahirinus*. European key taxa by the land snail *Pomatias olivieri*, lizard, *Lacerta laevis*, birds *Carduelis carduelis*, *Turdus merula*, and *Troglodytes troglodytes*, and the wood mice *Apodemus mystacinus* and *A. flavicollis* [8].

Adaptation to environmental stresses

The tropical dry and hot savannoid slope (AS = SFS) of EC I (AS = SFS) is heavily stressed by solar radiation (up to 800% more than ES = NFS), causing high temperature and drought [9]. Adaptive convergent responses to abiotic stresses occur across life on AS = SFS demonstrated by the following examples. Germination of caryopses of wild barley, *Hordeum spontaneumn*, decreased germination rate following decrease in humidity ES > AS > Negev [81]. Genetic diversity across 16 distant terrestrial taxa, based on allozymes and DNA markers, was significantly higher on the abiotically more stressful AS slope due to higher sun radiation, causing

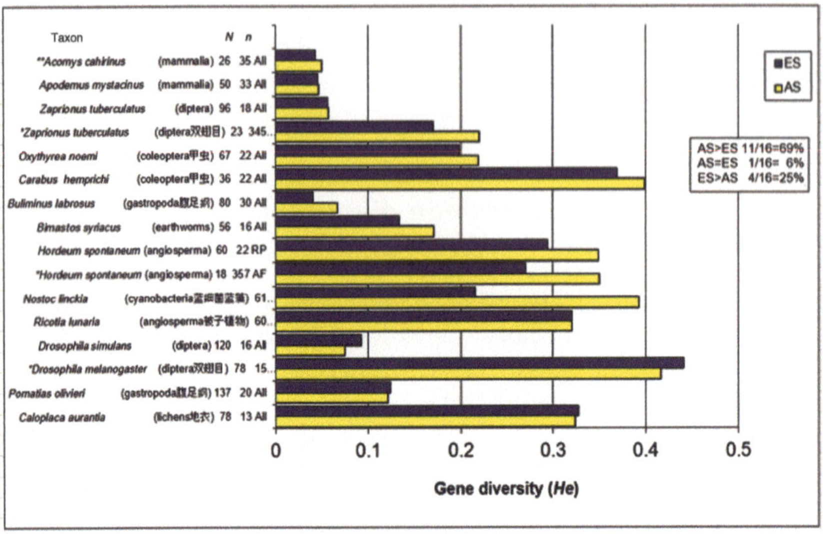

Figure 10.5 Gene diversity is higher in allozymes and DNA markers in 16 distant taxa at Evolution Canyon I (EC I) Mount Carmel, Israel. *Original figure of Nevo, done by Dr Alex Beharav.*

higher temperature, hence higher drought (Fig. 10.4), as has been shown earlier for molecular evolution and ecological stress at global, regional, and local scales [82]. Higher genetic polymorphisms, either genic or chromosomal [8,83–85] provide fertile basis for the origin and evolution of adaptations to ecological stresses as was shown at EC I, in Mount Carmel, Israel (Fig. 10.5). In some model taxa, we found largely higher levels of mutation rates, gene conversion, recombination, DNA repair, genome size (GS), small sequence repeats (SSRs), single nucleotide polymorphisms (SNPs), retrotransposons, transposons, candidate gene diversity, splice variation, gene transplant and genome-wide gene expression and regulation on the more stressful AS = SFS. Higher terrestrial species richness was found on the AS [8]. Aquatic species richness prevails on the ES.

Cyanobacteria evolution at Evolution Canyon I

The relationship between genomic diversity and environmental heterogeneity and stress was assessed in the generalist and hardy cyanobacterium species *Nostoc linckia*, across the sharp microclimatic contrasts of opposing slopes at EC I. Sharing a limestone lithology, the slopes contrast biotically, primarily due to interslope differences in solar radiation, resulting in higher rock and soil temperatures on the AS = SFS. The cyanobacterium *N. linckia*, a sessile prokaryote that grows in the canyon on limestone rock surfaces, is exposed to extreme fluctuations of solar radiation, temperature and desiccation.

We found remarkable interslope genomic diversity of *N. linckia* populations and within elevations on SFS [86]. These differences were assessed by amplification of genomic DNA with primers based on the highly iterated palindrome (*HIP1*) (5′-GCGATCGC-3′), thereby revealing HIP1 polymorphism across the genome of *N. linckia*. The interslope divergence was demonstrated by significantly higher diversity (*He* and v) and polymorphism (*P*) indices on the heterogeneous SFS, particularly in populations SFS 1 and SFS 2. *Nostoc linckia* diversity [86] increases upward with drought in parallel to retrotransposon BARE-1 in wild barley [87]. Correlations were found between *P* and *He* and variables influencing aridity stress: solar radiation, temperature and day—night temperature differences, substantiating ***microclimate*** as the main driving selective force within and between slopes. This may imply that the major cause for the ***inter***- and ***intraslope*** genetic divergence is the differences in their ***microclimatic*** conditions. Clearly, HIPI polymorphism is selected by water stress, highlighting the importance of ecological stress and selection in adaptive evolution and its remarkable effect on the genetic system across the prokaryotic genome. Thus HIPI is not only a genetic marker, but a genetic monitor of climatic stress. It promotes genome diversity in cyanobacteria.

In a parallel genomic study [88], we demonstrated remarkable *interslope* and *intraslope* genetic divergence of the genome (including both *coding* and *noncoding* regions) of *Nostoc linckia*, by using 211 AFLP (amplified fragment length polymorphisms) DNA molecular marker loci. Genetic polymorphism of *N. linckia* populations on the ecologically harsher SFS was significantly ($p < 0.05$) higher ($p = 99.53\%$) than was that of the populations on the climatically milder NFS ($p = 85.78\%$). Genetic polymorphism (*P*) and gene diversity (*He*) were significantly correlated with variables influencing aridity stress: solar radiation (Sr) (rp = 0.956; p = 0.046), temperature (Tm) (rp = 0.993; p = 0.0068), and day—night temperature difference (Tdd) (rp = 0.975; p = 0.025). As in other tested organisms from EC, but even more exceptionally because of its completely sedentary nature, we suggest that the climatically stressed SFS environment is responsible for this marked increase of genetic polymorphism, which is maintained by the combined evolutionary forces of diversifying and balancing selection [88].

Long-term microclimatic stress causes rapid adaptive radiation of *kai*ABC clock gene family in a cyanobacterium, *Nostoc linckia*, from both ECs I and II, Israel.

Cyanobacteria are the only prokaryotes known till recently to possess regulation of physiological functions with approximate daily periodicity, or circadian rhythms that are controlled by a cluster of three genes, *kaiA, kaiB, and kaiC* (but see later). We demonstrated considerably higher genetic polymorphism and extremely rapid evolution of the *kaiABC* gene family in a filamentous cyanobacterium, *Nostoc linckia*, permanently exposed to the acute natural environmental stress in the two ecological replicates of EC I in Mount Carmel, and EC II (western Upper Galilee) in Israel, separated by 38 km [89] (Fig. 10.2). The family consists of five distinct subfamilies (*kai*I−*kai*V) comprising at least 20 functional genes and pseudogenes. The obtained data suggest that the duplications of *kai* genes have adaptive significance, and some of them are evolutionarily quite recent (80,000 years ago). The observed patterns of within- and between-subfamily polymorphisms indicate that positive diversifying, balancing, and purifying selections are the principal driving forces of the ***kai*** gene family's evolution [89].

Origin and evolution of circadian clock genes in prokaryotes

Regulation of physiological functions with approximate daily periodicity, or circadian rhythms, is a characteristic feature of eukaryotes. Until recently, cyanobacteria were the only prokaryotes reported to possess circadian rhythmicity. It is controlled by a cluster of three genes: *kaiA, kaiB*, and *kaiC*. Using sequence data of approximately 70 complete prokaryotic genomes from the various public depositories, we showed [90] that the *kai* genes and their homologs have quite a different evolutionary history and occur in Archaea and Proteobacteria as well. Among the three genes, *kaiC* is evolutionarily the oldest, and *kaiA* is the youngest and likely evolved only in cyanobacteria. Our data suggest that the prokaryotic circadian pacemakers have evolved in parallel with the geological history of the earth, and that natural selection, multiple lateral transfers, and gene duplications and losses have been the major factors shaping their evolution [90].

Genetic polymorphism of cyanobacteria under permanent natural stress: a lesson from the Evolution Canyons

Cyanobacteria have high adaptive potential and occur in the most extreme habitats [91]. The available literature data indicate that the versatility of cyanobacteria is related to their *higher polymorphism under stress.* The studies of the filamentous cyanobacterium, *Nostoc linckia,* from the ecological microsite replicates known as ECs I and II, showed that, among the evolutionary forces maintaining the higher polymorphism and genome diversity under permanent natural stress, the various types of natural selection, diversifying, balanced, and positive selection regimes, play a key role.

Evolution of wild barley: adaptation, sympatric ecological speciation, and domestication at EC I

Wild barley, *Hordeum spontaneum,* the progenitor of cultivated barley, a generalist and widespread species, is a major model organism at EC I. Interslope adaptive complexes of *H. spontaneum* at EC include genetic diversity of proteins and DNA markers and sequences [87,92,93]. Dehydrins are one of the major stress-induced gene families, and the expression of dehydrins 1 and 6 (*Dhn1,6*) is strictly related to drought in barley. Drought resistance by *dehydrin 1* [94], at EC I revealed that most of the haplotypes, 25 out of 29 (86.2%), were represented by one genotype; hence, unique to one population. Only a single haplotype was common to both slopes (biomes). Nucleotide diversity was higher on the AS, whereas haplotype diversity was higher on the ES. Interslope divergence was significantly higher than intraslope divergence. The applied Tajima's *D* rejected neutrality of the SNP diversity. In *dehydrin 6* [95] both nucleotide and haplotype diversity of SNP in coding regions were higher on the AS (or dry slope) than on the ES (or humid slope), and the applied Tajima's *D* and Fu-Li tests rejected neutrality of SNP diversity [95]. Duplicated heat shock proteins, the older copy *hsp17a* and younger copy, *hsp 17 b* diverged adaptively both ***intra- and interspecifically*** [96,97].

Genomic nuclear microsatellite and chloroplast microsatellites display adaptive interslope divergence in wild barley at EC I [98]. Out of 19 microsatellites 17 (89.5%) of nuDNA, and 3 chDNA SSRs out of 4 were polymorphic in 7 populations at EC I. A total of 216 nuDNA SSRs, with a maximum of 23 alleles in a nuclear locus, and 10 chDNA SSRs with maximum 4 alleles in a locus were strikingly and significantly registered both ***inter-*** and ***intraslopes,*** with a remarkable genetic distance between midslope populations on opposite slopes (DA = 0.481) across 250 m. This genetic distance is as large as that of wild barley populations between Jerusalem and Sede-Boker, which are separated by 100 km. Slope unique alleles (103 = 45.6%) were higher on ES than on AS. Slope-specific (predominant) alleles [98] were equal on opposite slopes. SSR gene diversity was higher on the ES and the opposite was found for chDNA. NuDNA SSR interslope divergence was very high, with *Gst* −0.187, and 0.127 for chDNA SSR. The results indicate that diversifying selection is overriding migration and drift. Ecological local, regional, and global stress, or ***micro***- and ***macroscales*** [3] can generate similar adaptive patterns leading to similar results by natural selection at both *coding* and *noncoding* SSRs linking *micro*- and *macro*-evolutionary processes [98].

H. spontaneum rhizosphere bacteria, *Paenibacillus polymixa*, at the AS = SFS harbors significantly higher populations of ACCd producing biofilm forming phosphorus solubilizing osmotic stress tolerant bacteria, promoting drought resistance of the wild barley host plant [33]. Remarkably, wheat inoculation with bacteria that had lost their Sfp-type PPTase gene resulted in two times higher plant survival and about three times increased biomass under severe drought stress compared to wild type [99]. Transcriptome comparative profiling of barley *Eibi* mutant revealed pleiotropic effects of *HvABCG31* gene on cuticle biogenesis and stress responsive pathways [100]; *Isa* defense locus [101], and vitamin E components [102] were highly variable. The intergenic spacer region of the ribosomal DNA repeat unit [103] was also extremely variable *regionally* and *locally*. High divergence were found between AS and ES at EC I, and between terra rossa and basalt soils at ES (Tabigha). This sharp microsite ecogeographic variation in ribosomal DNA appears adaptive in nature, and is presumably driven by climatic and edaphic natural selection [103]. All the aforementioned genetic variables diverge significantly between the opposite ***tropical*** and ***temperate*** abutting slopes, expressing higher levels on the hot-dry tropical AS = SFS than on the mild cool-humid ES = NFS. A highly likely preagricultural collection, and domestication site of *H. spontaneum* was described on the Natufian cemetery of the Oren and Um Usba caves in EC I [104]. Finally, wild barley, *Hordeum spontaneum* [105], and its ***genomic*** sequence indicate that it evolved sympatrically at EC 1. Wild barley speciated sympatrically also at EP [18]. ***Brachyopodium stacei*** also incipiently or completely speciated sympatrically at EC I. This is true for other organisms from viruses (unpublished) and bacteria to fruit flies, and spiny mice [4,105].

Finally, we have recently shown by analyzing the whole genome that wild barley speciated *primarily* sympatrically at three microsites with contrasting ecologies. The first microsite at **EC I**, Mount Carmel, is due to ***microclimatic*** *divergence* at a distance of ~250 m [4] also analyzed gnomically. The second microsite at "**EP**," in eastern Upper Galilee, due to ***edaphic divergence*** of Senonian chalk and rendzina soil abutting with Pleistocene volcanic basalt and soil [18], and the third microsite at "**Evolution Slope**" (Tabigha), north of the lake of Galilee, due to ***edaphic divergence*** but in this case of Middle Eocene hard limestone and terra rossa soil, abutting with Pleistocene volcanic basalt, see later [23]. These microsites of SS of wild barley in north Israel are hot spots of SS. They highlight the commonality of SS in free breeding populations with gene flow due to sharply contrasting ecologies, selecting for the origin of new species. The third microsite awaits additional evidence of other taxa besides wild barley, as well as a future in depth analysis of SS.

Genomic adaptation to drought in wild barley caused by edaphic natural selection at Evolution Slope (Tabigha), AS, and microclimate at EC I

Ecological divergence at a microsite suggests adaptive evolution. We examined two abutting wild barley populations, each 100 m across, differentially adapted to drought tolerance on two contrasting and abutting soil types, terra rossa and basalt at ES (Tabigha), Israel [23], studied earlier for allozyme variation [19]. We identified a total of 69,192,653 single nucleotide variants (SNVs) and insertions/deletions in comparison with a reference barley genome. Comparative genomic analysis between these abutting wild barley populations involved 19,615,087 high quality SNVs. The results revealed dramatically different selection sweep regions relevant to drought tolerance driven by edaphic natural selection within these regions, including key drought-responsive genes associated with ABA synthesis and degradation (such as Cytochrome P450 protein) and ABA receptor complex (such as *PYL2, SNF1*-related kinase). The genetic diversity of the wild barley population inhabiting hot and dry terra rossa soil is much higher than that from the cooler and humid basalt soil. *Additionally*, we identified different sets of genes for drought adaptation in the wild barley populations from terra rossa soil and from wild barley populations from EC I at Mount Carmel. These genes are associated with abscisic acid signaling, signaling and metabolism of reactive oxygen species, detoxification and antioxidative systems, rapid osmotic adjustment, and deep root morphology. The unique mechanisms for drought *edaphic* adaptation of the wild barley from the ES at Tabigha in eastern upper Galilee, and those from EC I in Mount Carmel, derived by *microclimatic* divergence, may be useful *complementarily* for crop improvement, particularly for breeding of barley cultivars with high drought tolerance.

Evolution of tetraploid wild emmer wheat, *Triticum dicoccoides*: adaptive evolution and sympatric speciation at EC I, Mount Carmel

Allotetraploid WEW, *Triticum dicoccoides*, $2n = 28$, the progenitor of modern cultivated wheat, is an ecological specialist and excellent model organism for advancing evolutionary theory, wheat polyploid evolution, and wheat improvement [106]. It is a major investigated organism at the Institute of Evolution, University of Haifa [107]. It is also a major investigated organism at EC I, where it *primarily* sympatrically speciated dramatically into *three species of WEW*. Two new species speciated sympatrically, 30 m apart, on the hot and dry AS, SFS = AS, adapted differently against the same harsh abiotic stressful African microclimate. They, reproductively isolated both prezygotically by earlier flowering time (station 2), and postzygotically by Robertsonian chromosome rearrangements (station 1). A third new species of WEW speciated sympatrically on the humid and cold ES, NSF = ES, adapted against pathogens [108]. The center of origin and diversity of WEW is the northeastern upper Galilee and the Golan [106]. Elsewhere, in the Fertile Crescent, it occurs in semi-isolated and isolated populations. Regional and local genetic patterns are partly or largely adaptive both at the *coding* and *noncoding* genomes, correlated with, and predictable by, environmental abiotic and biotic stresses. WEW is a rich, mostly untapped, genetic resource for improving cultivated wheat, harboring drought, salt, mineral and disease resistance, grain protein, and high variation in photosynthetic yield. WEW has been recently mapped genomically [109,110]. Its large genome is ~12 GB, with 80% repeated elements, as against 17 GB in hexaploid bread wheat.

Natural selection of allozyme polymorphisms: a microgeographical differentiation by edaphic, topographical, and temporal factors in wild emmer wheat (*Triticum dicoccoides*) at Evolution Slope (Tabigha)

Allozymic variation in proteins encoded by 47 loci was analyzed electrophoretically in 1983–84 and 1984–85 in 356 individual plants of WEW,

Triticum dicoccoides, from a microsite at Tabigha, north of the Sea of Galilee, Israel, now dubbed Evolution Slope [21]. Each year the test involved two 100-m transects, each equally subdivided into basalt and abutting terra rossa soil types, and comparisons were based on 16 common allozyme polymorphic loci. Significant genetic differentiation, genetic phase disequilibria, and genome organization *according to soil type* were found over very short distances. Our results suggested that allozyme polymorphisms in WEW are partly adaptive, and that they differentiate at both single and multilocus structures primarily due to environmental stress of such ecological factors as soil type, topography, and temporal changes, most probably through aridity stress.

This study, together with the wild barley in the early 1980s [23] opened our current extensive studies in both wild cereals across "omics" and incipient SS at ES (Tabigha).

Evolution of wild emmer wheat avenin-like proteins at Evolution Slope (Tabigha)

Recently [111], we identified divergent avenin-like proteins, *TdALP*, in WEW abutting soil populations at ES (Tabigha): in *dry* terra rossa *only* alleles *bx-7AS-g*, *ay-7AS-N*, *AY-7as-c*, and in *humid* basalt, *only* alleles bx-7AS-n, ay-7AS-c, and ax-7AS-b, occurred. The absence of a significant relationship between geographic separation and genetic distances attests to a *sharp local ecological divergence* rather than a gradual change in allele frequencies across the range of WEW in ES specifically, and in Israel generally. Similar ecological correlates were identified in 21 WEW populations [111].

Adaptive evolution and sympatric speciation of the crucifer *Ricotia lunaria* at EC I

The crucifer *Ricotia lunaria* [112] showed highly denser populations and phenotypically smaller plants on the AS than on the ES. Plants were highly polymorphic for AFLP displaying two divergent intersopes AFLP clusters. The unbiased AFLP estimate of Nei genetic distances (D) indicated significantly higher ***inter***slope ($D = 0.124 \pm 0.011$) than ***intra***slope ($D = 0.076 \pm 0.015$) differences ($p < 0.001$, t-test). Correspondingly, mean *intraslope* gene flow was significantly higher than the *interslope* gene flow (2.9 ± 0.6 vs 1.9 ± 0.2). Furthermore, the slope divergent ecotypes were also divergent by *Arabidopsis* whole-genome tiling array data. Natural selection appears to adaptively diverge the plant ecotypes on the opposite slope, both *phenotypically* and *genotypically*. This includes significant divergence in flowering time likely to initiate incipient SS [112].

Indeed, a follow-up study unfolded SS transcriptomically in *Ricotia lunaria* at EC I [113]. Most remarkably a common garden glasshouse experiment showed that flowering time was earlier almost three months in bottom African population 3 (tropical) on AS, than population 5 (temperate) on ES, separated by only ~100 m. Moreover, slope ecotypes were divergent by frequencies of ~1064 unigenes [113]. Transcriptome analysis showed that 1064 unigenes were differentially expressed between the opposite ecotypes, which enriched in the four main pathways involved in abiotic and/or biotic stress responses, including flavonoid biosynthesis, alpha-linolenic acid metabolism, plant–pathogen interaction and linoleic acid metabolism. Furthermore, based on Ka/Ks analysis, nine genes were involved in the ecological divergence between the two ecotypes, whose homologs functioned in RNA editing, ABA signaling, photoprotective response, chloroplasts protein-conducting channel, and carbohydrate metabolism in *Arabidopsis thaliana*. Among them, four genes, namely, *SPDS1, FCLY, Tic21* and *BGLU25*, also showed adaptive divergence between *R. lunaria* and *A. thaliana*, suggesting that these genes could play an important role in plant speciation, at least in Brassicaceae. The dramatic interslope divergent flowering time provides a uniquely important premating isolating mechanism between the opposite slope populations due to the sharp interslope microclimatic divergence. Sympatric ecological speciation is still controversial and EC provides a hot spot of incipient and full SS *across life* from viruses (unpublished) and bacteria to mammals [4] (Fig. 10.1).

Evolution of fruit flies (Drosophilidae) in fitness, and incipient sympatric speciation at Evolution Canyon I, Mount Carmel

Cosmopolitan fruit fly *Drosophila melanogaster*, and eight other drosophilids are major model organisms at EC I [74], highlighting both adaptive evolution [114,115], and incipient ecological SS [26,116], in the contrasting *microclimates* of EC I, *tropical* AS versus *temperate* ES. Adaptive *interslope* divergence has been found in multiple genetic components and complexes some of which are described below. These involve mutation rate [117], recombination [118] fluctuating symmetry [119] thermoregulation (consistent across several years, 1997, 1999, 2000, and more developed on hot, tropical AS than on cool, temperate ES) [120]. Many adaptive complexes were intimately associated with their African or

European biome for years in the laboratory, indicating their genetic basis. These include oviposition temperature preferences in the lab following the slope origin of females, tolerance to high temperature, drought stress and starvation, habitat selection, and different longevity patterns [115]. DNA repair efficiency proved higher on AS, and better expressed by thermotolerant flies [121]. Likewise, heat shock proteins [122], microsatellites, and divergence in the regulatory region of *hsp70Ba*, which encodes the major inducible heat shock protein of *Drosophila* [123], were also higher on AS.

Incipient SS was identified in interslope divergence by *nonrandom*, assortative mate choice, single and multiple mate choice tests with *D. melanogaster* from the opposite slopes displaying highly significant assortative mating, with preference for sexual partners from the same slope [124,125]; sexual and reproductive behavior were also slope specific [126,127]; *D. melanogaster*, shows a clinal pattern in sperm size [128] associated with drought, wing size and shape [129]. Courtship song [130] aggression and courtship behavior [131], all showed interslope divergence. Dramatically biased interslope migration, tenfold higher from AS to ES, possibly caused by global warming affecting more the AS, with ongoing gene flow [132,133]; candidate behavioral genes [134], suggested that repeat length/composition may influence the functional features of flies, that is, habitat choice, *nonrandom* mating, and temperature choice. Genome and repeatome analysis of flies from the opposite slopes, separated by only 250 m, was remarkable. *Coding* genome [135], unfolded a total of 572 genes that were significantly different interslopes in allele frequency, 106 out of which were associated with 74 significantly overrepresented gene ontology (GO) terms, particularly so with response to stimulus, developmental, and reproductive processes. Thus they corroborate previous observations of interslope divergence in stress response, life history, and mating functions. There were at least 37 chromosomal "islands" of interslope divergence and low sequence polymorphism, plausible signatures of selective sweeps, more abundant in flies derived from ES. Correlation between local recombination rate and the level of nucleotide polymorphism was also found [135]. The no*ncoding* genome, or regulatory repeatome, affected genes associated with cognition, olfaction, and thermoregulation [136], affecting both structure and function. It is outstanding, that *Drosophila melanogaster* that can fly daily up to 16 km, and species like, *D. simulans, D. buzzati*, and the recent Indian colonizer *Zaprionus tuberculata*, evolve *incipient interslope genetic divergence and even* *incipient* new species, at an average short distance of 250 m. This is particularly impressive for the colonizing *Zaprionus*, that during a short time of some 40 years (appearing in EC and Israel at large in the early 1980s) displaying interslope adaptive allozyme evolution [74], and early incipient speciation, currently under study, unfolding the early stages of incipient SS after 40 years from its invasion Israel and EC. The interslope remarkable divergence of *Drosophila melanogaster* and *D. simulans* [124] in morphological, physiological, behavioral and extensive genomic and repeatomic structure and function, is not only evolving interslope divergence, but also partly convergent adaptive evolution patterns. It also unfolds at a microscale across hundreds of meters, incipiently new sympatrically evolving species with gene flow, in several drosophilids (*D. melanogaster, D. simulans, D. buzatti*), and recent invader *Zaprionus tuberculata* due to the sharp microclimatic, ecological interslope divergence between the AS and ES. Populations are expected, and confirmed, to adapt to divergent local *tropical* versus *temperate microclimates* and to sympatrically incipiently speciate with gene flow [9]. Clearly, adaptation to the contrasting microclimates overrules gene flow and is responsible for the genetic and phenotypic interslope divergence. *D. melanogaster* population divergence was established at EC involving habitat choice, mate choice, thermal and drought tolerances, adaptive genes, and mobile elements. Parallel patterns of stress tolerance, habitat choice, and mate choice were demonstrated in *Drosophila simulans* at EC, although on a smaller scale divergence [26,116]. This evidence and that in other taxa described later, substantiates Darwin's vision that the SS model of species origin is *common*, due to numerous ECs across planet Earth. *These include geologic, edaphic, climatic, biotic and abiotic microsites whose divergent ecologies, with ongoing gene flow, are cradles of SS unfolding numerous "Galapagos islands," but without geographical marine barriers separating the Galapagos Islands.*

Rodent genotypic and phenotypic interslope divergence at EC I

Genotypic and phenotypic divergence in two rodents, the African spiny mouse *A. cahirinus* and European wood mouse *Apodemus mystacinus* from 6 stations (populations) (*tropical* 1–3, and *temperate* 5–7) were studied at EC I [137]. ***Inter***- and ***intraslope*** allozyme, RAPD diversities, and morphological ***inter***slope divergence were found in both rodents. Local variation derived by solar radiation, temperature and aridity stress on the AS = SFS caused ***interslope*** and ***intraslope*** adaptive genotypic variation (proteins and DNA), in multilocus allozyme and RAPD polymorphisms, heterozygosity, gene diversity, unique alleles, and gametic phase disequilibria, across 250 m, indicate microsite genetic divergent patterns caused by microclimatic natural selection. Phenotypic (morphological, physiological and behavioral) differences at local *EC I* parallel *regional* phenotypic patterns across Israel in *Acomys* and in northern and central Israel in *Apodemus*. Higher genotypic variation characterized the microclimatically tropical hot and dry harsher AS. Phenotypic variation followed the ecological rules. Morphologically, in general, both rodents (excepting *Apodemus* males) showed smaller size and longer extremities on the warmer SFS, displaying *locally* the Bergmann and Allen ecological rules, respectively, promoting better thermoregulation, as is true across the species ranges *regionally* in Israel. Physiological comparison indicated higher activity pattern and metabolism (20% higher Oxygen consumption), in NSF than in SFS *Acomys, adaptively saving activity energy on the harsher AS.* Sharp physical and biotic interslope biome divergence results in the high predominance of tropical *Acomys*, and temperate *Apodemus* on the SFS and NFS, respectively. The local divergence, both genotypically and phenotypically, reflects their original regional—global biogeographical origins, African for *Acomys*, and European for *Apodemus*. Higher genetic diversity was observed in local and regional xeric environments in *Acomys* [24], while lower genetic diversity was observed in *Apodemus* [138]. This parallelism between ***local*** and ***regional*** patterns suggest that the genetic, morphologic, and physiologic diversities, both *locally* and *regionally*, represent networked adaptive complex syndrome contributing to fitness across ***micro***- and ***macro-scales***, across transects of increasing aridity (for details see Ref. [137]). This suggests that, at both the *micro*- and *macroscales*, diversifying natural microclimatic selection appears to be the major evolutionary driving force causing *inter*- and primarily AS intra*slope* adaptive genotypic and phenotypic divergence. "EC" proved in small rodents, as in other organisms, an optimal model for unraveling *evolution in action across life* [137].

Evolution caused by environmental stress

The ECs unfold numerous cases of adaptive evolution due to their abutting but contrasting dry and hot tropical versus humid and cool ***temperate*** microclimates. Increased male recombination rate in *D. melanogaster* is correlated with population adaptation to the AS stressful conditions [118]. Likewise, mutation rate in *Drosophila melanogaster* at EC I was correlated with stress [117], as were differences in fluctuating environments [119]. All could be considered adaptive to the hot and drought stresses on the AS. A remarkable case is that of the soil fungus *Sordaria fimicola* described below.

Fungal soil mutation, crossing over, and gene conversion in soil fungus *Sordaria fimicola*

Inherently and environmentally induced differences in mutation frequencies in the soil fungus *Sordaria fimicola* from the harsher AS had higher frequencies of new spontaneous mutations and of accumulated mutations for ascospore pigmentation and ascospore germination-resistance to acriflavine between slopes [139]. The mutation frequency in the hot-dry tropical harsher slope, AS, was 1.9%, more than twice than in the milder ES, 0.8%. Such inherited variation provides a basis for natural selection for optimum mutation rates in each environment [139]. Similarly, inherited interslope differences in crossing over and gene conversion frequencies between wild strains of the soil fungus *Sordaria fimicola* from EC I often differed significantly interslopes [140]. First- and second-generation descendants from selfing of the original strains from the harsher, more variable, AS = SFS had higher frequencies of crossing over in locus-centromere intervals and of gene conversion levels than those from the lusher NSF. Differences were more divergent ***interslope*** than ***intraslope***. Such inherited variation could provide a basis for natural selection for optimum recombination frequencies in the opposite biomes. Clearly, recombination generates new combinations from existing genetic variation; hence it is important in adaptive evolution.

Adaptive mutations in RNA-based regulatory mechanisms: computational and experimental investigations in soil bacteria at Evolution Canyon III, Negev

Recent discoveries of RNA-based regulatory mechanisms raised interest in the involvement of environmental stress in their evolution. Riboswitch in bacteria are regulating the biosynthesis of certain vitamins by an RNA genetic control element that senses small molecules and responds with a structural change that affects transcription termination or translation initiation without participation of proteins. We took the thiamin pyrophosphate (TPP)-riboswitch in *Bacillus subtilis* in EC III, Saharuth, in the southern Negev Desert, 50 km north of Eilat (Fig. 10.2), as a model system, aiming to examine whether beneficial mutations may exist at the level of RNA that will cause an improvement in organism fitness [141]. By computationally analyzing the difference in primary and secondary structure of the *B. subtilis* TPP-riboswitch collected from the xeric AS of EC III, at Nahal Shaharut, Southern Negev Desert, increase covered with cyanobacteria, and the "European" "mesic" ES-NSF covered by lichens, we aimed to experimentally study the environmental effect on transcription termination in these RNA-based regulatory mechanisms that are believed to be of ancient origin in the evolutionary time scale. Computational results, so far, indicate that specific mutations affect the riboswitch conformation by causing a global rearrangement. Our aim was to check whether such mutations could be adaptive mutations that may have a beneficial fitness effect, taking the TPP-riboswitch as a model system at the ***microscale***. Empirical results so far indicated that in the promoter region of the TPP-riboswitch, all mutations increase nucleotide **GC** content in the xeric SFS, whereas in the mesic NFS they increase **AT** content. Preliminary empirical examination of termination efficiency of strains found exclusively on one slope or the other, reveal increased termination efficiency in the presence of TPP and at more moderate temperatures, but only a suggestion of greater termination efficiency from strains found on both slopes. We expect that further results will shed light on the mutational differences of TPP-riboswitch sequences found on the opposite slopes of "EC" III at Nahal Shaharut, potentially leading to interesting discoveries that relate to the topic of *adaptive, nonrandom, mutations and their role in current evolution.*

Retrotransposon BARE-1 evolution in wild barley, *Hordeum spontaneum, at EC I*

The replicative spread of retrotransposons in the genome creates new insertional polymorphisms, increasing retrotransposon numbers and GS [87]. The BARE-1 retrotransposon is a major, dispersed, active component of *Hordeum* genomes, positively correlated with GS [87]. We examined GS and BARE-1 insertion patterns and number in wild barley, *Hordeum spontaneum*, in EC I across divergent microclimates, AS and ES [87]. BARE-1 has been sufficiently active for its insertional pattern to resolve individuals and identifying their ecology in a way consonant with their microecological distribution in and outside EC I. On both slopes, but especially on the drier (AS = SFS), a simultaneous increase in the BARE-1 copy number and a decrease in the relative number lost through recombination, as measured by the abundance of solo long terminal repeats, increased the BARE-1 share of the genome upward with slope dryness. The lower recombinational loss would increase more full-length copies, increasing GS. *Local* data are consistent with *regional* trends for BARE-1 in *H. spontaneum* across Israel [87]. Hence, it may reflect adaptive selection for increasing GS with increasing drought through retrotransposon activity [87].

Genome size is higher on the hot and dry more stressful tropical AS-SFS at EC I

Higher GS, at EC I on the ***tropical*** hot and dry more stressful slope AS has also been recorded in the following organisms: evergreen carob **tree**, ***Ceratonia siliqua*** [142]; annual legume ***Lotus peregrinus*** [143]; saw-toothed grain beetle, ***Oryzaephilus surinamensis*** [144], in eight silo pest populations GS was significantly smaller than in eight wild populations at EC I. The ability of *O. surinamensis* to colonize widespread habitats globally could be connected with an unusually AT-rich (for an invertebrate) genome (AT-base content ranging from 68% to 76%). All the above species had a *significantly* higher GS on the hot and dry tropical Slope (AS), except the perennial bulbous ***Cyclamen persica***, suggesting that large GS evolution may be adaptive to environmental drought stress, as was also shown above in wild barley, ***Hordeum spontaneum***. The data on all the above species at "EC" I showed that local variability in the *C*-value exists in these species and that ecological stress might be a strong adaptive evolutionary driving force in shaping the amount of DNA, that is, GS.

Repeatome evolution in *Drosophila melanogaster*

Repeat sequences, especially mobile elements, make up large portions of most eukaryotic genomes and provide enormous, albeit commonly underappreciated, evolutionary potential. We analyzed repeatomes of *Drosophila melanogaster* that has been diverging in response to a microclimate contrast in EC I [144]. Flies inhabiting the colder and more humid ES carried about 6% more transposable elements (TEs) than those from the hot and dry AS, in parallel to a suite of other genetic and phenotypic differences between the two opposite slope populations. *Nearly 50% of all mobile element insertions were slope unique*, with many of them disrupting coding sequences of genes critical for cognition, olfaction, and thermotolerance, consistent with the observed patterns of thermotolerance differences and assortative mating [125,144]. In another study, putative adaptive interslope divergence of 28 TEs of *Drosophila melanogaster* at EC I was analyzed, and 11 TEs were found to be selected [145].

Developmental instability of vascular plants in contrasting microclimates at EC

Is fluctuating asymmetry a reliable indicator of a population's state of stress adaptation? To answer this question, we studied leaf asymmetry of 12 species of vascular plants growing under contrasting microclimates on the opposing slopes of EC I [146]. Leaves of the trees *Quercus calliprinos* and *Pistacia palaestina* were significantly more asymmetrical on the microclimatically more temperature fluctuating, tropical, drier, savannoid AS. Leaves of the shrub *Calicotome villosa* were significantly more asymmetrical on the microclimatically less variable, shadier, and maquis-like ES. Differences in fluctuating asymmetry were negatively correlated with differences in local abundance. Species displayed higher fluctuating asymmetry on the slope where they were less abundant hence, more stressed ecologically. The results suggest that the tropical hot-dry AS at EC I is severely stressful only for species on the margins of their adaptive zone. Similarly, the temperate cool and humid European slope (ES = NFS) is severely stressful only for species on the margins of their, different, adaptive zone. Interaction of slope and species for both growth rate and asymmetry of microfungi in a common environment is evidence of genetic differences between the AS and ES of EC [147].

Fluctuating helical asymmetry and morphology of snails (Gastropoda) in divergent microhabitats at Evolution Canyon I and II

Developmental instability of shelled land snails was measured as deviations from a perfect equiangular (logarithmic) spiral. We measured six species of land snails at EC I and EC II [147], for fluctuating asymmetry, rate of whorl expansion, shell height, and, number of rotations of the body suture. Only *Eopolita protensa jebusitica* showed marginally significant differences in fluctuating helical asymmetry between the two slopes. Contrary to expectations, asymmetry was marginally greater on ES. Shells of *Levantina spiriplana caesareana* at EC I were smaller and more asymmetric than those at EC II. Moreover, shell height and number of rotations of the suture were greater on the ES of both canyons. The data are consistent with a trade-off between drought resistance and thermoregulation in snails. *Levantina* was significantly smaller on the AS, increasing surface area for better thermoregulation, whereas *Eopolita* was larger on the AS adapted to better resist water evaporation. EC I was more stressful than EC II for *Levantina* [147], being 38 km south and warmer than EC II, as is also evident by its *Quercus calliprinos* forest at EC I, replaced by a Pontic, colder forest of *Acer syriacus* at EC II, on the ES (see later).

Parallel biodiversity evolution of plants and animals at EC I

Plants and animals pattern at ***local*** (EC I) and ***regional*** (Israel) scales display parallel evolution. Angiosperm plant diversity [59] and cyanophyta (=cyanobacteria) species diversity [148] at EC I are correlated with interslope microclimatic stresses. Clearly, selection impacts similarity at *local* and *regional* scales when similar ecological stresses are operating. Biodiversity evolution parallelism of vascular plants, and lichens, *Caloplaca aurantiaca*, caused by interslope microclimatic divergence at EC I highlight the remarkable linkage between divergent ecological stress and parallel evolution across plants [59]. EC I and EC II, at Nahal Keziv, western Upper Galilee, 38 km northeastern of EC I (Fig. 10.2) represent ecological replicate, at a microscale, with the tropical savanna on AS, and temperate forest on ES at EC I. However, the forest constituents at EC II involve primarily *Acer syriacus* and *Laurus nobilis* [60], rather than *Quercus calliprinos* and *Pistacia palestina*, predominant in the forest of ES at EC I [59]. In an area of 7000 m^2, we recorded in EC II 283 vascular plant species belonging to 200 genera and 56 families. Plant cover varied between 70-90% (AS) and 100% (ES). Annuals were the predominant life form (54.7%). AS and ES at EC II varied in species composition, sharing ***only 11.6%*** of species [60], versus **38.7%** at EC I [59]. Phytogeographical types at EC II vary both ***inter-*** and ***intraslopes*** in species composition due to differential northern microclimatic conditions, demonstrating the effect of cooler climate on species diversity at a microscale. However, the biodiversity biomes and interslope divergence of vascular plants caused by sharp microclimatic divergence at EC II, replicated the AS—ES *interslope* divergence of an African savanna at AS and south European forest at EC II, displaying astounding plant community similarities to those of EC I, despite the 38 km northward separating EC II from EC I (Fig. 10.2). Interslope tropical hot-dry versus temperate cool-humid microclimates prevail at EC II generating interslope intercontinental scale divergent biomes similar to EC I. The best comparison of adaptive evolution and ecological incipient SS between EC I and EC II is the soil bacterium *Bacillus simplex* [31] (see later).

Xeric versus mesic patterns in woody plants at EC I

Xeromorphic phenotypes of woody plants characterize AS, contrasting with mesomorphic phenotypes on ES at EC I [149]. Seven morphological leaves' traits of three woody species (*Olea europea*, *Ceratonia siliqua*, and *Pistacia lentiscus*) were quantified interslopes at EC I. Overall, parallel but species-specific leaf *xeromorphic* interslope microclimatic adaptations were identified. Leaves of the three species from xeric tropical AS were smaller than leaves from mesic temperate ES. *Leaves of O. europea*, and *C. siliqua* were significantly thicker on AS than on ES, due to increased thickness of the photosynthetic, palisade and spongy layers, but only in the spongy layer in *P. lentiscus*, and opposite trend in its palisade layer. Thicker epidermis appeared in AS than in ES in *O. europea* and *P. lentiscus*. The results suggest *locally* microscale adaptive xeromorphic morphological drought resistance adaptations, paralleling the *regional* divergence, caused at both scales, by natural microclimatic ecological selection [149]. The following case of *Bacillus simplex* unfolds adaptive evolution and incipient SS displaying similar evolutionary scenarios at EC I similarly repeated at the ecological replicate EC II, separated by 38 km, due to similar microclimatic contrasts.

Adaptation and incipient sympatric speciation of soil bacterium *Bacillus* simplex under microclimatic contrast at Evolution Canyons I and II, Israel

A remarkable comparison between biodiversity evolution in EC I and EC II has been conducted on soil bacterium, *Bacillus simplex*, soil fungi, beetles, and ants [31]. These results could be considered as similar ecological replicates, close repetitive experiments, since the basic contrast between the African **TROPICAL** hot and dry slope and the European **TEMPERATE** cool and humid slopes are similar (Figs. 10.6–10.8).

Among 131 strains of *B. simplex* studied, "African," AS strains grow better than "European" strains at a stressful high temperature (43.25°C). The results suggest that adaptation to the hotter and more stressful SFS is continuously ongoing. The patterns of heat adaptation override the phylogenetic history of individual lineages. A positive correlation of growth rates at 43.25°C and 20°C, more markedly among AS strains, reflects probably the broader temperature range on the SFS. Thus the hot

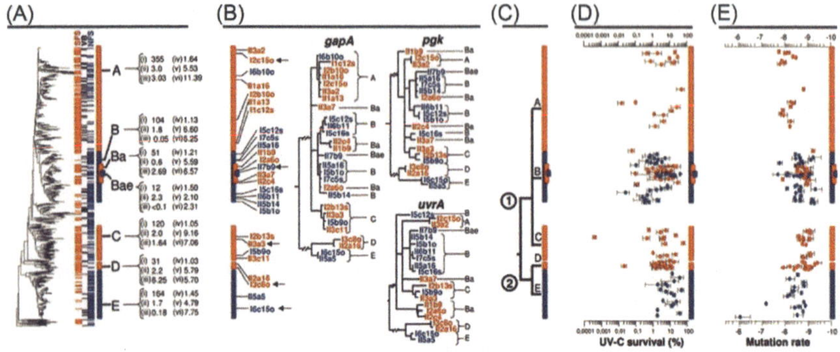

Figure 10.6 Evolutionary lineages in the metapopulation of *B. simplex* and their physiological properties of UV survival and mutation rate [31]. (A) Neighbor-joining tree (Jukes–Cantor distances) from RAPD data of all isolates. Thin horizontal lines to the right of the tree indicate origin from African-like slope (SFS = AS, red), valley bottom (VB, black), and European-like slope (NFS, blue). For each lineage, six characteristic values are given and denoted with the numbers of strains (i), the EC I/EC II ratio of strains (ii), the ratio of African-like to European-like strains (iii), the ratio of isolates to haplotypes (iv), the genetic diversity given as mean number of pairwise differences of the RAPD sequence (v), and the variance of RAPD pattern diversity (vi). The depicted RAPD clusters are supported by unweighted pair group method with arithmetic mean and maximum parsimony analyses (data not shown). (B) Representative RAPD cluster strains chosen from both canyons and their phylogenies of portions of the *gapA*, *pgk*, and *uvrA* genes. Names of the strains reflect their origin and are explained in *Supporting Materials and Methods.* The 16S sequence was determined from strains marked with an arrow. The phylogenetic trees were reconstructed from the results of neighbor-joining, minimum evolution, maximum parsimony, and maximum likelihood analyses of the gene sequences. Only nodes that were unambiguously resolved in the same way in all four reconstruction methods are shown. For divergence values see supporting information are given in Table 5 (from original paper) Outgroup sequences (which are not shown) were obtained from *Bacillus subtilis, Bacillus cereus, Bacillus halodurans, Bacillus licheniformis,* and *Oceanobacillus iheyiensis.* (C) Phylogenetic tree of ecological–genomic evolutionary lineages summarizing the results from the gene tree topologies and of the RAPD tree. (D and E) UVC survival (percentage) (D) and spontaneous mutation rates to rifampicin resistance of representative isolates from the evolutionary lineages A–E (E). Experiments were done twice, and results are given as mean and deviation from the mean. *From J. Sikorski, E. Nevo, Adaptation and incipient sympatric speciation of* Bacillus simplex *under microclimatic contrast at 'Evolution Canyons' I and II, Israel, Proc. Natl. Acad. Sci. U.S.A. 102 (44) (2005) 15924–15929, doi:10.1073/pnas.0507944102.*

temperature stress on the "AS" is a major environmental force driving the twin evolutionary processes of adaptation and speciation of *B. simplex* at "EC" [150].

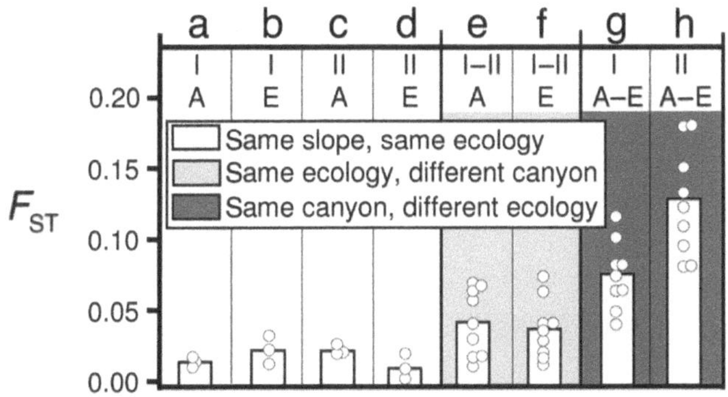

Figure 10.7 Genomic divergence of *B. simplex* across and within two ECs [31] as estimated from pairwise station comparisons of RAPD data within slopes ($n = 3$) (columns a–d), between slopes of same ecology but different canyon ($n = 9$) (columns e and f), and of interslope stations within canyons ($n = 9$) (columns g and h). Open circles indicate a pairwise station comparison; bars indicate the mean of all pairwise station comparisons. I and II denote EC I and EC II, respectively; and A and E denote African-like and European-like slopes, respectively. The p values (Mann–Whitney U test) between the indicated columns are as follows: a–e, $p = 0.100$; c–e, $p = 0.482$; b–f, $p = 0.372$; d–f, $p = 0.036$; a–g, b–g, c–g, d–g, a–h, b–h, c–h, and d–h, $p < 0.0001$; e–g, $p = 0.018$; f–g, $p = 0.003$; e–h and f–h, $p < 0.0001$ [31].

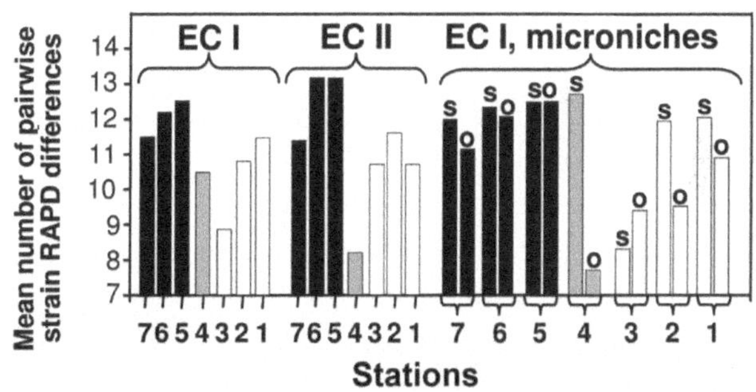

Figure 10.8 Genomic diversity of *B. simplex* in stations (populations) of EC I and EC II and of shady (s) and sunny (o) microniches on EC I. Black denotes stations 5–7 of European-like slopes; gray denotes valley bottom station 4; and white denotes stations 1–3 of the African-like slopes [31].

Microclimatic adaptive biodiversity interslope evolution of soil fungi across the four Evolution Canyons in Israel

Soil fungi have been major model organisms to assess adaptive evolutionary patterns at the four ECs (EC I–V) in Israel. They provide excellent organisms unfolding *local* and *regional* adaptive complexes selected by the environment, primarily, by edaphic and climatic stresses. Moreover, they also highlight the dynamics of protective melanin evolution due to differential solar radiation. The soil fungal studies in ECs in Israel have been led primarily by Dr. Bela Grishkan [53,55,151–153] revealing dramatic fungal interslope biodiversity divergence, as described below *locally* and *sequentially*.

Soil fungi in four Israeli Evolution Canyons

The environment selected dominant species and groups of soil microfungi in each canyon, determining spatial, interslope, seasonal variations, and diversity levels of the microfungal communities [53]. The hot-dry tropical, savannoid AS biomes of the northern ECs (EC I and EC II) with high interslope microclimatic stress supported abundant melanin-containing species with small, one-celled conidia, such as ***Aspergillus niger***, primarily on the AS. By contrast, located only 100 m apart, mild temperate, cool and humid, forested ES biome supported fast-reproducing ***Penicillium*** species as dominants [48,55,151]; In sharp contrast, ***desert*** environments, were overwhelmingly dominated by slow-reproducing, dark-colored species with *large,* ***multicelled conidia***. Both benign forest and strongly stressful desert territories were rather constant and stabilized in their spatial and seasonal structure of the microfungal communities. By contrast, fungal communities from the variable AS environments were subjected to remarkable spatiotemporal changes. Thus soil microfungal community structure can be a sensitive indicator of spatially contrasting and seasonally changing environmental stresses and demonstrate significant differences on a ***local*** scale and extreme differences on a ***regional*** scale. The microfungal communities in all localities and seasons at the extreme Negev desert EC III (Fig. 10.2) were characterized by a superdominance of *dark-colored melanin species with large multicelled conidia: Ulocladium atrum, U. botrytis, Alternaria alternata, and Al. chlamydospora* [151]. Following is the pattern and dynamics of the melanic soil fungus *Aspergillus niger*, as an experimental *melanic* model at EC I.

Solar radiation effects on adaptive melanin levels

Environmental stress, both *biotic* and *abiotic*, is a major driving force of evolution. I revisit our earlier studies [152,153]. The AS receives 200%–800% more solar radiation than the ES [9]. We measured conidial melanin concentration of 80 strains of *A. niger* from the opposite slopes and from sunny versus shady ***micro***niches in each slope. ***A. niger*** on savannoid AS had *threefold* more melanin than on the forested ES. As expected, AS strains of *A. niger* resisted UVA irradiation better than ES strains and the same was demonstrated by *A. niger* strains from *sunny* against abutting ***shady microniches*** on the ES. Remarkably, ***melanin*** effectively protects *A. niger*, and other melanic soil fungi, adaptively from solar radiation.

Subsequently, we reciprocally transplanted ***A. niger*** from AS to the ES and vice versa, in an attempt to highlight melanin evolutionary dynamics on the opposite slopes [154]. Noteworthy, the ES-to-AS soil transplantation ***increased tenfold*** the relative abundance of ***A. niger***, and significantly *increased* conidial melanin concentration in the transplanted population as compared to the original population ES (Fig. 10.9A). By contrast, the opposite AS-to-ES soil transplantation ***reduced tenfold*** the relative abundance of the fungus and highly significantly *decreased* the conidial melanin concentration in the transplanted population as in original ES. (Fig. 10.9B). Thus fungal populations ***adaptively responded to changing levels of solar radiation on the opposite slopes, high on AS and low on ES***. Genomic, transcriptomic and epigenomic studies will reveal in future studies the evolutionary mechanisms driving such fungal adaptations, whether Darwinian, Lamarckian, or both [154].

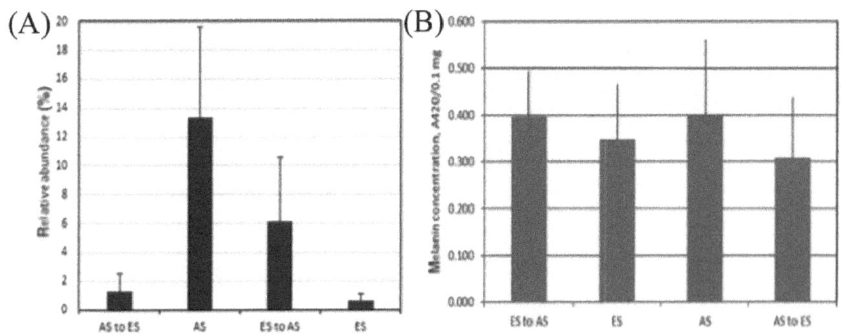

Figure 10.9 (A) Relative abundance of *Aspergillus niger* in original and transplanted soil at "Evolution Canyon" I, Mount Carmel, Vertical lines represent standard deviations (SD, $N = 6$). (B) Concentration of melanin in conidia of *Aspergillus niger* in original and transplanted soil at "Evolution canyon I, Israel." The vertical lines represent standard deviations (SD, $N = 10$).

Molecular-genetic biodiversity in a natural population of the yeast *Saccharomyces cerevisiae* from "Evolution Canyon": microsatellite polymorphism, ploidy, and controversial sexual status

Studies of natural populations of yeasts are very few. We isolated *S. cerevisiae* strains from EC I, and studied their genomic biodiversity. Analysis of 19 microsatellite loci revealed high allelic diversity and variation in ploidy level across the panel, from diploids to tetraploids, confirmed by flow cytometry. No significant differences were found in the level of microsatellite variation between strains derived from the major localities or microniches, whereas strains of different ploidy showed low similarity in allele content. Maximum genetic diversity was observed among diploids, and minimum among triploids. Phylogenetic analysis revealed clonal, rather than sexual, structure of the triploid and tetraploid strains populations. Viability tests in tetrad analysis also suggest that clonal reproduction may predominate in the polyploid strains [155].

Adaptive response of DNA-damaging agents in natural populations of yeast, *Saccharomyces cerevisiae* from "Evolution Canyon" I

UV radiation is one of the most important physical parameters which influences yeast growth in nature [156]. We examined 46 natural strains of *Saccharomyces cerevisiae* isolated from several natural populations at EC I. The interslope differences in solar radiation (200%–800% more on the "AS") caused the development of two distinct biomes: tropical hot-dry AS and temperate cool and humid ES. We studied the phenotypic response of the *S. cerevisiae* strains to UVA and UVC radiations and to methyl methanesulfonate (MMS) to evaluate the interslope effect on the strains' ability to withstand DNA-damaging agents, higher on AS. We exposed our strains to the different DNA-damaging agents and measured survival by counting colony forming units. The survival rates of the AS diploid and tetraploid strains were significantly higher than those from ES. On both slopes, the tetraploid strains were significantly more resilient to UVA than the neighboring diploid strains. Remarkably, the survival of the AS tetraploids from populations SFS1 and SFS 2 was significantly higher than the ES tetraploids, suggesting that they may have evolved specifically higher resistance to the higher solar radiation, and possibly even underwent SS on AS, which needs verification. The strains from the "AS" were more resilient to both UVA and MMS than the strains from the ES. We also found that the tetraploids strains from AS were more tolerant to all DNA-damaging agents than their neighboring diploid strains, and significantly better than ES tetraploid, suggesting that high ploidy level might be a mechanism of adaptation to high solar radiation, and AS tetraploids might have undergone an extra adaptation, or even SS, on AS, which awaits further verification. The above yeast results and those of parallel patterns in EC I in taxonomically distant organisms, for example, soil fungus *Aspergillus niger* [152] *Drosophila melanogaster* [121], and wild barley [157], highlight parallel adaptive responses against stressful UV radiation, suggesting that natural selection appears to select at a microscale, for different adaptive complexes that can tolerate the higher UV radiation on the AS [156].

Oxidative stress responses in yeast strains, *Saccharomyces cerevisiae*, from "Evolution Canyon"

We have also taken yeast strains from EC I, for an analysis by microarray hybridizations of the response of wild yeast accessions to environmental stress, in particular oxidative stress [158]. Strains were selected from the xeric, high irradiation AS, and compared with strains from the mesic, low irradiated ES, and from the valley bottom. H_2O_2-sensitive strains included a laboratory strain (S150-2B) and most strains from the NFS. We identified a statistical significant correlation between peroxide tolerance on the SFS, and microniches within a slope. Hierarchical clustering of regulated transcripts indicated maximum linkage of expression profiles between strains that showed the same phenotypic stress response. The analyses indicate strain-specific adaptive *interslope* and *micro-niche* adaptive evolution along the microclimatic gradient of "EC" that determine the yeast strain response to oxidative stress [158].

Continental biome *interslope* divergence across life at EC I

European and African biomes predominate on the ES and AS, respectively, due to the microclimatic interslope divergence, ***tropical*** hot-dry on AS versus ***temperate*** cool-humid on the ES [8] (Fig. 10.10).

We identified 2500 taxa on both slopes including 100, 500, 350, 1500, and 50 species of bacteria, soil fungi, plants, insects, and vertebrates, respectively. African species and European species characterize AS and ES, respectively, adapted to each continental origin, dry and hot Africa and humid and cold Europe (from Ref. [8]). The interslope opposite floras

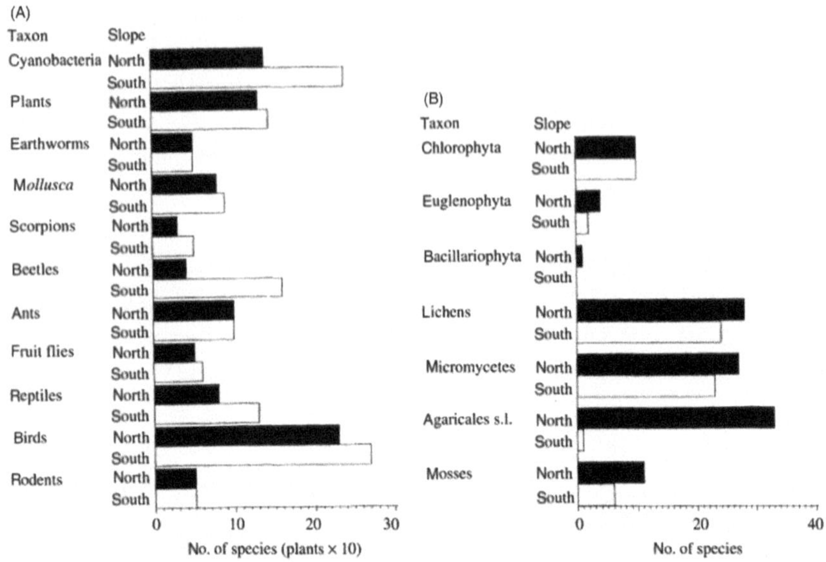

Figure 10.10 Biodiversity differential on the opposite S (=AS in *white bars*) and N (=ES in *black bars*) slopes at EC I of major plant and animal taxa (A) includes primarily animals and (B) includes cryptogamic, nonflowering plants [8].

appear in the caption of Fig. 10.1. African taxa are represented by the land snail *Sphincterochila carriossa*, and the lizards *Stelagama stellio* and *Ptyodactylus guttata*, by the birds *P. barbatus*, *N. ossea*, and by the spiny mouse *A. cahirinus*. European taxa are represented by the land snail *P. olivieri*, lizard *L. laevis*, birds *C. carduelis*, *T. merula*, and *T. troglodytes*, and the wood mice *Apodemus mystacinus* and *A. flavicollis* [8]. In European snail *P. olivieri*, whose southern border is Mount Carmel, snails of AS and ES differ in their susceptibility to hyperthermic desiccation and, genetic diversity increases with environmental stress on AS [159], as is paralleled by wild barley, *Hordeum spontaneum* [87]. Remarkably, earthworm species are differentially distributed on AS and ES [56].

Adaptive evolution and incipient sympatric speciation of spiny mouse, *Acomys cahirinus*, at Evolution Canyon I

Mitochondrial DNA

African-originated spiny mice, of the *A. cahirinus* complex, colonized Israel 30,000 years ago based on fossils. We showed by mtDNA [11], whole-genome analysis [12] and transcriptome, genetic editing, and microRNA [13] interslope evidence that *A. cahirinus*, incipiently speciates sympatrically, owing to sharp *microclimatic interslope divergence* at EC I, along with other organisms across life from bacteria to mammals [4]. Genotypically, we showed significantly higher genetic diversity of mtDNA and amplified fragment length polymorphism, DNA markers (AFLP) of *Acomys* on the stressful AS compared with the mild ES [11]. This is also true *regionally* across Israel. In complete mtDNA, 25% of the haplotypes at EC I were slope biased. Phenotypically, the opposite slopes' populations also showed adaptive morphology, physiology, and behavior divergence paralleling ***regional populations' pattern across Israel*** [24]. Preliminary tests indicate slope-specific mate choice. Colonization of *Acomys* at the EC first occurred on the AS and then moved to the ES (Fig. 10.11). Strong slope-specific natural selection (both positive and negative) overrules low interslope gene flow [3]. Both habitat slope selection and mate choice suggest ongoing incipient SS. We concluded, based on the above evidence that *Acomys* at EC I is ecologically and genetically adaptively, incipiently sympatrically speciating on the ES owing to divergent microclimatic natural selection. Similar results were found earlier and reported above by allozyme and RAPD variation [137].

Transcriptome analysis

Spiny mice, *A. cahirinus* (Fig. 10.11), colonized Israel 30,000 years ago from dry tropical Africa and inhabited rocky habitats across Israel. Earlier, we had demonstrated by mtDNA, that *A. cahirinus* possibly incipiently

Figure 10.11 Spiny mouse *Acomys cahirinus*: transcriptome analysis substantiating sympatric speciation.

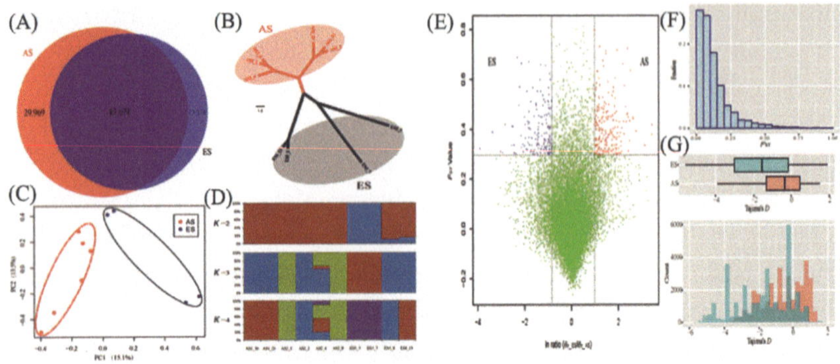

Figure 10.12 Transcriptome Analysis of *Acomys cahirinus* at Evolution Canyon I: Population divergence of spiny mouse, *A. cahirinus*, at EC I. (A) Venn diagram of SNPs unique to the AS and ES populations. (B) NJ tree of the AS (red) and ES (gray) populations. (C) PCA of the AS and ES populations. (D) Population genetic structure of the AS and ES populations when *K* was set to two, three, or four. Note that AS is the south-facing slope (SFS), whereas ES is the north-facing slope (NFS). The ecological divergence between the slopes is like between two continents, despite the very short distance separating them (on average 250 m). Natural selection on the two *A. cahirinus* opposite slope populations: (E) Distribution of ln ratio (π_ES/π_AS) and FST of each transcript. Red and blue dots (AS and ES, respectively) represent transcript under putative selection (corresponding to $p < 0.05$, where $F_{ST} > 0.295$, and ln ratio > 1). (F) F_{ST} distribution of the AS and ES populations. (G) Tajima's *D* distribution of the AS and ES populations. Tajima's *D* for the ES population is smaller than that for the AS population. The AS and opposite ES populations were marked in red and blue, respectively.

speciates sympatrically (SS) at EC I, in Mount Carmel, Israel, due to microclimatic interslope divergence. The EC I microsite consists of a dry and hot savannoid "AS" and an abutting humid and cool-forested "ES." We substantiated incipient SS in *A. cahirinus* at EC I based on the entire transcriptome [13], demonstrating that multiple slope-specific adaptive complexes across the transcriptome result in two divergent clusters (Fig. 10.12A–G). Tajima's *D* distribution of the abutting *Acomys* interslope populations demonstrates that the ES population is under strong positive selection (Fig. 10.12G), whereas the AS population is under balancing selection, harboring higher genetic polymorphism [13]. Considerable sites of the two populations differentiated with a coefficient of $F_{\mathrm{ST}} = 0.25–0.75$. Remarkably, 24 and 37 putatively adaptively selected genes were detected in the AS and ES populations, respectively (Fig. 10.12E). The AS selected genes involved DNA repair, growth arrest, neural cell differentiation, and heat shock proteins adapting to the local AS stresses of high solar radiation, drought, and high temperature. By contrast, the ES genes involved high ATP associated with energetics stress [13] (Fig. 10.12E). The sharp ecological interslope divergence led to strong slope-specific selection overruling the interslope gene flow [3]. Earlier tests suggested slope-specific mate choice. *Habitat interslope adaptive selection across the transcriptome and mate choice substantiates SS, suggesting its prevalence at EC and commonality in nature* (Fig. 10.11).

The figure shows EC I model in Mount Carmel, Israel. (A) The cross section of EC I. The sharp divergence of African savanna slope (AS) and forested, ES biomes are seen in the cross section of EC I. Samples were collected at station 2 of the tropical, hot, dry, savannoid, south-facing AS and station 6 of the abutting temperate, cool, humid, forested, north-facing ES. (B) The spiny mouse, *A. cahirinus* (Fig. 10.11B).

Evolution Canyon: a potential microscale monitor of global warming across life

Climatic change and stress is a major driving force of evolution. The effects of climate change on living organisms have been shown primarily on regional and global scales. I proposed the EC *microscale model* as a potential life monitor of global warming in Israel and the rest of the world [160]. The EC model reveals interslope and intraslope *evolution in action* at a microscale involving biodiversity divergence, adaptation, and incipient SS across life from viruses and bacteria through fungi, plants, and animals. The AS and ES exhibit drought and shade stress, respectively. Major adaptations on the AS are because of solar radiation, heat, and drought on AS, whereas those on the ES relate to shade stress and little light for photosynthesis. Preliminary evidence suggests the extinction of some European species on the ES, such as European lizard *Lacerta viridis*, on ES and the European wood mouse, *Apodemus flavicolis* on AS. In *Drosophila*, a remarkable tenfold higher migration was recorded in 2003 from the AS to ES, [133]. Some predictions were advanced that could be followed in diverse species in EC. The EC microclimatic model is optimal to track global warming at a microscale across life from viruses and bacteria to mammals in Israel, and in additional ECs across the planet [160].

Host–parasite interaction: Natural selection causes adaptive genetic disease resistance in wild emmer wheat against powdery mildew at "Evolution Canyon" I, Carmel

EC is also a model for an active cradle of **host–pathogen** interaction, driving disease resistance in both WEW and wild barley [161]. In WEW, we examined 278 single sequence repeats (SSRs) and the resistance phenotype diversity of WEW, *Triticum dicoccoides*, to powdery mildew between the opposite slopes of EC I. Furthermore, 18 phenotypes on the AS and 20 phenotypes on the ES, were inoculated by both *Bgt* E09 and a mixture of powdery mildew races. Very little polymorphism was identified *intra*slope in WEW accessions from both the AS or ES. By contrast, 148 pairs of SSR primers (53.23%) amplified polymorphic products between the phenotypes of AS and ES. There are some differences between the two WEW genomes and the *inter*slope SSR polymorphic products between genome A and B. Interestingly, all WEW growing on the AS were susceptible to a composite of the powdery mildew *Blumeria graminis*, while, remarkably, those growing on the ES were highly resistant to *Blumeria graminis* at both seedling and adult stages.

Remarkable *inter*slope evolutionary divergent processes occur in WEW at EC I, despite the short interslope average distance of 250 m. The AS, a dry and hot slope, did not develop resistance to powdery mildew, whereas the ES, a cool and humid slope, did develop resistance since the *disease stress* was strong there due to humidity. This is a remarkable demonstration of host–***pathogen interaction*** on how resistance develops when the pathogen stress causes an adaptive result at a *micro*scale distance. Noteworthy, similar results occur in wild barley, both in powdery mildew and rust (Unpublished), and ***regionally*** across Israel: high resistance in humid Galilee and susceptibility in the dry desert [162]. *Local* and *regional* humid ecologies highlight the evolutionary dynamics of host–parasite interaction and the origin of disease resistance.

Evolution in action: adaptation and incipient sympatric speciation with gene flow across life at "Evolution Canyon," Israel

The EC model is a microscale natural laboratory that can highlight some of the pending basic problems and mysteries of evolutionary biology requiring clarification (Nevo list of "EC" publications at http://evolution.haifa.ac.il). This is especially true if an interdisciplinary approach is practiced including ecological functional genomics, transcriptomics, proteomics, metabolomics, methylomic, and phenomics. Ref. [4] overviewed and reanalyzed the *incipient sympatric adaptive ecological speciation* of seven model organisms at EC, across life: the soil bacterium, *Bacillus simplex*; wild barley, the progenitor of cultivated barley, *Hordeum spontaneum*; the tiny beetle *Oryzaephilus surinamensis*; the cosmopolitan fruit fly, *Drosophila melanogaster*, and the African-originated spiny mouse, *A. cahirinus*. Additional taxa have been added recently, the crucifer *Ricotia lunaria* [112,113], and WEW, *Triticum dicoccoides* [108] (Fig. 10.1). All these seven model organisms display *evolution in action of microclimatic adaptation and incipient or full sympatric adaptive ecological speciation* on the ***tropical*** and ***temperate*** abutting slopes, separated from a few meters to hundreds of meters, and at EC I, in the study area, on average, by only ~250 m. Some distant species converge in their *micro*climatic adaptations to the hot and dry "African", SFS (or AS) and to the cool and humid "European", NFS (or ES), although in some taxa convergence is *partial* and other adaptations to the ecological stress are species specific as between *D. melanogaster*, and *D. simulans* [163], highlighting the adaptive richness and variation in nature. Natural selection overrules ongoing interslope gene flow (between the free interbreeding populations within and between slopes) [3] and leads to *adaptive incipient sympatric ecological speciation* on the dramatically opposite abutting xeric savannoid and mesic forested slopes.

Evolution Plateau: edaphic divergent microsite of incipient sympatric speciation in blind mole rat, and wild barley

Blind mole rats, S. galili: possible incipient SS unfolded by mitochondrial DNA

We reported on a possible incipient sympatric adaptive ecological speciation in *S. galili* ($2n = 52$) [11]. The study microsite (0.04 km^2) is sharply subdivided geologically, edaphically, and ecologically into abutting barrier-free ecologies divergent in rock, soil, and vegetation types. The Pleistocene Alma basalt abuts the Cretaceous Senonian Kerem Ben Zimra chalk. Only **28%** of 112 plant species on chalk and basalt were shared between the soils [11]. We examined mitochondrial DNA in the *control region* and *ATP6* in 28 mole rats from basalt and in 14 from chalk habitats. We also sequenced the complete mtDNA (16,423 bp) of four animals, two from each soil type. Remarkably, the frequency of all major haplotype clusters (HC) was highly soil biased. HC-I and HC-II are chalk biased. HC-III was abundant in basalt (36%) but absent in chalk; HC-IV was prevalent in basalt (46.5%) but was low (20%) in chalk. Up to 40% of the mtDNA diversity was edaphically dependent, suggesting constrained gene flow. We identified a homologous recombinant mtDNA in the basalt/chalk studied area. Phenotypically significant divergences differentiate the two populations, inhabiting different soils, in adaptive oxygen consumption [11] and in the amount of outside-nest activity. This identification of a possible incipient sympatric adaptive ecological speciation caused by natural selection indirectly refutes the allopatric alternative. Sympatric ecological speciation may be more prevalent in nature because of abundant and sharply abutting divergent ecologies. We next studied the genome and transcriptome of the chalk–basalt population substantiating the possible incipient SS from *possible* to *certain*. See below.

Blind mole rats, S. galili: incipient sympatric speciation unfolded by genomic analysis

Sympatric speciation (SS), that is, speciation within a freely breeding population or in contiguous populations, was first proposed by Darwin and is still controversial despite theoretical support and mounting empirical evidence. We showed in *S. galili* (Fig. 10.13A) [12] a case of genome-wide divergence analysis, demonstrating that SS in continuous populations, with gene flow, encompasses multiple widespread genomic adaptive complexes, associated with the sharply divergent ecologies, including soil plus vegetation (Fig. 10.13B) [12], based on geology of abutting chalk and basalt rocks and soils (Fig. 10.13C) [12]. The two abutting soil populations of *S. galili* in northern Israel involve the ancestral Senonian chalk and abutting derivative Pleistocene volcanic basalt populations. Population divergence originated approximately 0.2 million years ago based on nuclear genome analysis. Population structure analysis displayed two distinctly divergent clusters of chalk and basalt populations (Fig. 10.13D–F), with divergent LD (Fig. 10.14A) and population demography (Fig. 10.14B). GO enrichment analysis highlights strongly differential soil population adaptive complexes: in basalt, *sensory perception, musculature, metabolism and energetics*, and in chalk, *nutrition* and *neurogenetics* are outstanding (Fig. 10.13G). Population differentiation of chemoreceptor genes suggests inter–soil population's mate and habitat choice substantiating SS. Importantly, distinctions in protein degradation (Fig. 10.14C–E) [12] may also contribute to SS. Natural selection and natural genetic engineering overrule gene flow [3], evolving sympatric ecological adaptive speciation. Sharp ecological divergences abound in nature; therefore SS appears to be an important mode of speciation as first hypothesized by Darwin, and proved in the EC I *microclimatic model* and here at EP *edaphic model*.

Blind mole rats S. galili, incipient SS by transcriptome analysis

Incipient SS in blind mole rat, *S. galili*, in eastern upper Galilee, Israel, caused by sharp ecological divergence of abutting chalk–basalt ecologies,

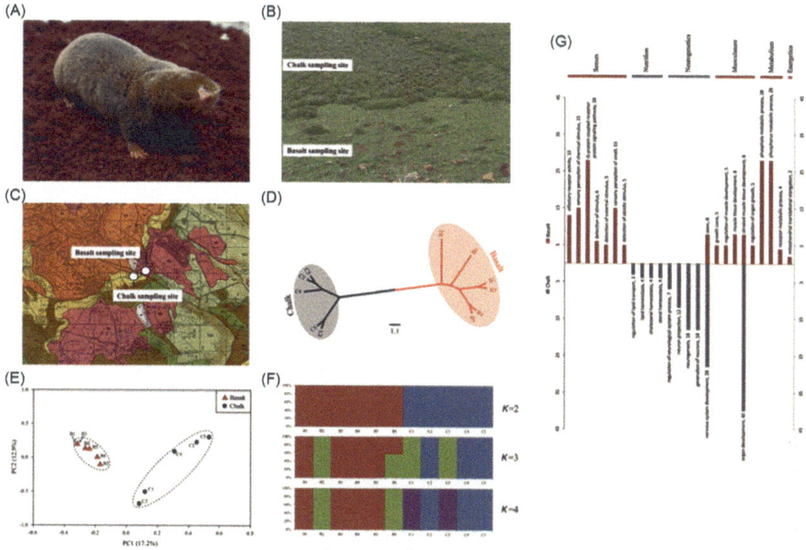

Figure 10.13 Sympatric speciation of *Spalax galili* ($2n = 52$) by ecological and genomic divergence, at "Evolution Plateau," eastern Upper Galilee. (A) The blind mole rat, *S. galili*. (B) The two divergent habitats: Senonian chalk formation, covered by dense bushlets of *Sarcopterium spinosum abuts with the abutting basalt. The two white circles represent the two mole rat soil populations studied on chalk (the progenitor), and on basalt the derivative new species called temporarily S. galili* basalt. (C) The geological map [164] displays chalk in yellow and basalt is in pink, separated by a geological fault. (D) Neighbor-joining tree based on all the SNPs. Red are basalt individuals, and black are chalk individuals. (E) Principle components analysis of two *S. galili* soil populations. Cluster with red triangles represents basalt individuals; black circles represent chalk individuals. (F) Population genetic structure of two *Spalax* soil populations when $K = 2$, 3, and 4. (G) Gene ontology (GO) of positively selected genes in chalk and basalt populations.

has been proposed previously based on mitochondrial DNA [11], and was verified by whole-genome nuclear DNA [12], and transcriptome [13] analyses. Here we present new evidence including transcriptome (Fig. 10.15A–H), microRNA (Fig. 10.16A and B), and DNA editing (Fig. 10.16C), which substantiated earlier evidence reinforcing adaptive divergence in the abutting chalk and basalt populations [13]. Genetic divergence, based on the previous and new evidence, is ongoing despite restricted gene flow between the two populations [3], *advancing SS* on the Pleistocene volcanic basalt. The principal component analysis, neighbor-joining tree, and genetic structure analysis of the transcriptome clearly show the *clustered divergent two mole rat populations* (Fig. 10.13). Gene-expression

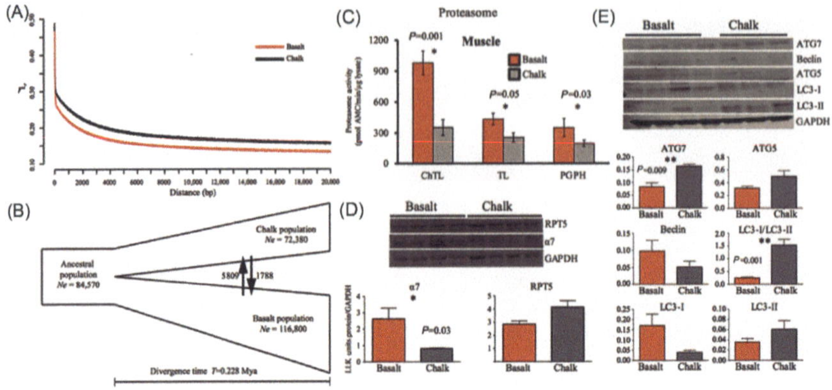

Figure 10.14 Genomic analysis of the two sympatric populations (species): Linkage disequilibrium and demographic structure of *Spalax galili* [12]. (A) Linkage disequilibrium decay of two continuous *S. galili* populations (species). *X* axis stands for physical distances (bp), while *Y* axis stands for r^2. (B) Inferred demographic history for the two abutting soil populations (chalk vs basalt) of *S. galili*. The extant and ancestral population sizes (*Ne*) of the chalk and basalt populations are indicated, and the migration rates between the two populations are provided. The divergence time (*T*) between two populations was inferred: $T = 0.228$ million years ago. (C) 20S proteasome activity measured by degradation of optimized peptides cleaved by the three active sites, chymotrypsin-like (ChTL), trypsin-like (TL), and postglutamyl, peptide hydrolyzing (PGPH) or caspase-like show higher levels of activity in the basalt population. (D) Higher levels of the constitutive 20S proteasome subunit $\alpha 7$ support the observation of the increase in activity of mole rats on the basalt [17]. (E) The chalk population protein degradation profile suggests more of a reliance on autophagy, with significantly higher levels of ATG7 and autophagic flux (LC3II/LC3I ratio) [12].

level analysis indicates that the population transcriptome divergence is also displayed by sex [13]. GO enrichment of the differentially expressed genes from the two abutting soil populations highlights reproductive isolation [13]. Alternative splicing variation of the two abutting soil populations displays two distinct splicing patterns. L-shaped F_{ST} distribution indicates that the two populations have undergone divergence with gene flow (Fig. 10.15H). Transcriptome divergent genes highlight neurogenetics and nutrition characterizing the poor food resource in the chalk population, and energetics, metabolism, musculature, and sensory perception characterizing the abutting basalt population [14]. The divergent soil ecologies evolved divergent behavioral phenotypes [17]. Remarkably, microRNAs (Fig. 10.16A and B) also diverge between the two populations. The GC content is significantly higher in chalk than in basalt, and stress—response genes mostly prefer *nonoptimal* codons. The multiple lines

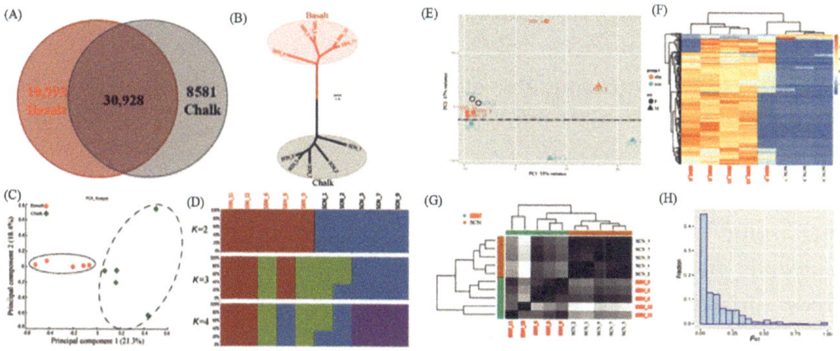

Figure 10.15 Transcriptome, genetic editing, and microRNA substantiate incipient sympatric speciation of *Spalax galili* and Genetic and expression divergence in sympatric speciation of *S. galili* [13]. (A) Venn diagram, note the larger number of SNPs in basalt than in chalk; (B) joint tree analysis (red denotes basalt animals; black denotes chalk animals); (C) PCA, red = basalt, blue = chalk; (D) Structure analysis ($K = 2, 3, 4$) red = basalt animals, blue = chalk animals. (E) PCA of gene-expression level of the 10 animals. Triangles denote males, and circles denote females. Animals from basalt were marked in red, and animals from chalk in blue. Note the separation of males and females; and the separation of chalk from basalt by a dashed line. Animals marked by a white circle in the basalt, are the two *recombinants* [13]. (F) Hierarchical clustering analysis of differential expressed transcript (DET) between animals of the chalk and abutting basalt populations. (G) Hierarchical cluster analysis of alternative splicing variation of the 10 animals. All of the individuals from chalk were clustered into one group, and the animals from basalt were subdivided into two subclusters. Red bar denotes animals from basalt, and blue bar represent animals from chalk. (H) The L-shaped F_{ST} distribution of the chalk and abutting basalt populations across the whole transcriptome. Note that in both F and G and, animals from basalt are denoted SBN in red, and animals from chalk are denoted SCN, in gray.

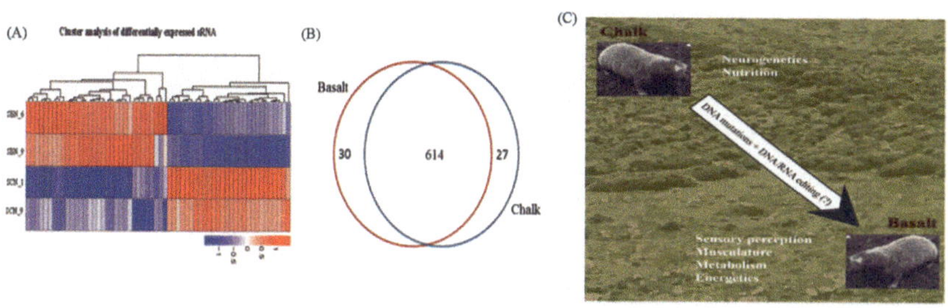

Figure 10.16 ***Transcriptome analysis of microRNA expression divergence of the chalk and abutting basalt populations*** [13]. (A) Cluster analysis of differentially expressed microRNA in the chalk and basalt mole rat populations. (B) Venn diagram displaying the number of microRNA shared by the two mole rat populations, and unique to basalt and chalk populations, respectively. (C) DNA-editing divergence of the chalk and basalt populations in sympatric speciation of *Spalax galili* [13].

of evidence of ecological–genomic and genetic divergence highlight that natural selection overrules the gene flow between the two abutting populations, substantiating the sharp ecological chalk–basalt divergence driving *primary* SS.

Adaptive methylation regulation of p53 pathway in SS of blind mole rats, S. galili

Epigenetic modifications play significant roles in adaptive evolution. The tumor suppressor *p53*, well known for cancer resistance, controlling cell fate and maintaining genomic stability, is much less known as a master gene in ecological adaptation involving methylation modifications. The blind subterranean mole rat *Spalax eherenbergi* superspecies in Israel consists of four chromosomal species that speciated peripatrically and regionally across Israel from north to south forming a speciation trend correlated with increasing aridity [165]. Remarkably, the northern Galilee species *S. galili* ($2n = 52$) underwent ***locally*** adaptive ecological SS, caused by the sharply divergent, but abutting, chalk and basalt ecologies. This was demonstrated by mitochondrial [11], nuclear genomic [12] and transcriptomic [13] evidence. We showed that the expression patterns of the *p53* regulatory pathway diversified between the abutting chalk and basalt ecologies, where the soil divergent populations of S. *galili* incipiently sympatrically speciate [16]. We identified higher methylation on several sites of the *p53* promoter in the chalk population. Site mutagenesis showed that methylation on these sites linked to the transcriptional repression of *p53* involving Cut-Like Homeobox 1 (Cux1), paired box 4 (Pax 4), Pax 6, and activator protein 1. Diverse expression levels of *p53* between the incipiently sympatrically speciating chalk–basalt abutting populations of *S. galili* selectively affected cell-cycle arrest but not apoptosis. We hypothesized that methylation modification of *p53* has adaptively shifted evolutionarily in supervising its target genes during SS of *S. galili* to cope with the contrasting environmental stresses of the abutting divergent chalk–basalt ecologies [16].

Conclusions and prospects

The genomic revolution, ecological stress, and the origin of species

The genomic era revolutionized evolutionary biology [6,7]. The enigma of genotypic–phenotypic diversity and biodiversity evolution of genes, genomes, phenomes, species, and biomes were central in the scientific programs of the Institute of Evolution, University of Haifa [166]. We explored the following questions. (1) How much of the genomic and phenome diversity in nature is adaptive and processed by natural selection? (2) What is the origin and evolution of adaptation and speciation processes under spatiotemporal variables and stressful ***macroscale*** and ***microscale*** ecologies? We advanced ecological genetics into *ecological functional genomics*, such as in WEW [106] and analyzed *globally* ecological, demographic, and life history variables in 1200 diverse species across life, and their effect on genome structure and organization in thousands of populations, and tens of thousands of individuals tested mostly for allozyme and partly for DNA diversity [167]. Likewise, we tested ecological stresses *thermal, chemical, climatic, biotic and atomic* stresses in several model organisms. Subterranean mole rats, *Spalax*, subjected to relatively underground constant microclimates coupled with multiple ecological stresses provided a major evolutionary model of adaptation and speciation [3,84,85,166,168], highlighting global regressive, progressive, and convergent evolution and the genome–phenome holistic model [165]. From 1975 to the present we started to explore wild cereal populations in edaphically divergent microsite ecologies [19,21]. These were followed by the long-term project of EC model, divergent *microclimatically*, and continued by *edaphically divergent* ecologies (1990 and continued) (see Nevo list of ECs in http://evolution@haifa.ac.il). The ECs diverge microclimatically into ***tropical*** (AS) slope abutting with ***temperate*** close abutting slopes (ES).

Our earlier results (1970–90) indicated abundant genotypic and phenotypic diversity in nature. The organization and evolution of molecular and organismal diversity in nature *at global, regional,* and local scales are *nonrandom* and *structured*; display *regularities across life*; and are *positively correlated* with, and *partly predictable* by, *abiotic and biotic environmental heterogeneity and stress* [166]. Biodiversity evolution, even in small isolated populations, is primarily driven by natural selection, including diversifying, balancing, cyclical, and purifying selective regimes, interacting with, but ultimately overriding, the effects of mutation, gene flow, and stochasticity. Plant genetic resources could be predicted by isozyme markers and ecology [166]. The EC model revealed that in fact it comprises several models of *evolution in action*: biodiversity evolution, adaptive evolution, SS with gene flow across life, from viruses and bacteria to mammals [4], monitoring global warming at a microscale [160], and host–parasite evolution of disease resistance [161]. The EC *microclimatic* model involves due to its *microsite*, a free interbreeding metapopulation, with differential degrees of gene flow between the opposite slopes, across life. However, selection overrules gene flow at ECs [3], generating new sympatric species on the opposite slopes [4]. Moreover, we expanded the model from a ***microclimatic*** *canyon microsite* to ***edaphically divergent microsites***, with chalk versus basalt at EP, and terra rossa versus basalt at ES, both in eastern Upper Galillee. These two ***edaphic evolution microsites*** strongly suggest that contrasting ecologies, abiotic like geological, and edaphic, or biotic, like divergent plant formations, could be fitting microsites harboring potential cradles of SS. The origin of new species may be associated with sharply divergent ecologies, *close* where *SS* prevails, and *distant*, where *allopatric* speciation exists. The EC model highlighted the ecological stress as a central driver of adaptation and speciation evolutionary processes. Moreover, it reinforced the case that adaptive ecological SS in a free interbreeding population, with gene flow, is a common world model of the origin of species as envisaged by Darwin, and highlighted by the EC model across life, and its extensions, geologic, edaphic, climatic, abiotic and biotic divergences in microsites which abound globally.

What next?

It will be exciting to expand the exploration of the EC model across the globe under diverse ecologies, in the tropics, subtropics, temperate, Arctic, and Antarctic climates. One could look into divergence, convergence and parallelism in phylogenetically distant organisms from microorganisms to mammals and explore how close and distant organisms may adapt not only to similar but also to divergent solutions to resist ecological stresses as was highlighted in the sibling species *Drosophila melanogaster* and *D. simulans* at EC [163]. A major mystery that could be unraveled at ecologically divergent microsites is regulation of genes by the *noncoding genome*. The repeatome, which was regarded in the past as "junk," is still the most mysterious element in regulating the genes [169]. We have now evidence that the *noncoding genome* undergoes parallel selective changes to those of the *coding genome* [170]. Another still mysterious issue that could be deciphered in ecologically divergent microsites is the level of *adaptive*, or *directed*, *nonrandom* mutations [141]. The genomic era opened up vast horizons to explore the genome as a dynamic read–write memory system undergoing natural selection and natural genetic engineering [6,7], ever changing under ecological stress as revealed by the adaptive epigenetic methylation regulation of p53 pathway in the SS of blind subterranean rodents in the *S. galili* complex [16] and in the whole genome in *Spalax* [170]. The great genomic revolution establishing the universal language of life enables now to revisit adaptation, speciation, convergence, parallelism, recombination, networking [171], symbiogenesis, interspecific hybridization, functional interaction of genomes in polyploidy, and between nuclear, mitochondrial, and chloroplast DNAs. The genome dynamics coupled with divergent stressful ecologies is a major key to decipher the enormous untapped mysteries of life's evolution. Ecological stress drives life's molecular-genomic diversity, mechanisms, and constant evolutionary change, genotypically, phenotypically, and phylogenetically in nature, accentuated **at diverse microsites evolution models**.

Acknowledgments

I'd like to thank Solomon Wasser, Ruth Ben David, Kexin Li, Xioaying Song, Esther Kabulanky, Olga Forman, Anat Marcovich, the late Avigdor Beiles, Andrei Galkin, and all my students, colleagues, and collaborators for their contributions to this overview of the Evolution Canyon model. I especially appreciate the dedicated collaboration of Dr. T. Pavlicek in this project. Special thanks are extended to Ancell-Teicher Research Foundation of Genetics and Molecular Evolution for supporting financially the Evolution Canyon Projects.

References

[1] A.A. Hoffmann, P.A. Parsons, Evolutionary Genetics and Environmental Stress, Oxford University Press, 1991.

[2] P.A. Parsons, Environments and evolution: interactions between stress, resource inadequacy and energetic efficiency (in English)Biol. Rev. Camb. Philos. Soc. 80 (4) (2005) 589–610. Available from: https://doi.org/10.1017/S1464793105006822.

[3] E. Nevo, Evolution under environmental stress at macro- and microscales (in English) Genome Biol. Evol. 3 (2011) 1039–1052. Available from: https://doi.org/10.1093/gbe/evr052.

[4] E. Nevo, Evolution in action: adaptation and incipient sympatric speciation with gene flow across life at "Evolution Canyon", Israel, Isr. J. Ecol. Evol. 60 (2–4) (2014) 85–98. Available from: https://doi.org/10.1080/15659801.2014.986879.

[5] E. Nevo, Evolution in action across life at "Evolution Canyons", Israel, Trends Evolut. Biol. 1 (1) (2009) e3. Available from: https://doi.org/10.4081/eb.2009.e3.

[6] J.A. Shapiro, Evolution: A View From the 21st Century, FT Press Science, 2011.

[7] J.A. Shapiro, No genome is an island: toward a 21st century agenda for evolution, Ann. N.Y. Acad. Sci. 1447 (1) (2019) 21–52. Available from: https://doi.org/10.1111/nyas.14044.

[8] E. Nevo, Asian, African and European biota meet at "Evolution Canyon" Israel: local tests of global biodiversity and genetic diversity patterns, Proc. R. Soc. London, Ser. B: Biol. Sci. 262 (1364) (1995) 149–155. Available from: https://doi.org/10.1098/rspb.1995.0189.

[9] T. Pavlíček, D. Sharon, V. Kravchenko, H. Saaroni, E. Nevo, Microclimatic interslope differences underlying biodiversity contrasts in "Evolution Canyon", Mt. Carmel, Israel, Isr. J. Earth Sci. 52 (2003) 1–9.
[10] A. Polyakov, A. Beharav, A. Avivi, E. Nevo, Mammalian microevolution in action: adaptive edaphic genomic divergence in blind subterranean mole-rats, Proc. R. Soc. London, Ser. B: Biol. Sci. 271 (Suppl. 4) (2004) S156–S159. Available from: https://doi.org/10.1098/rsbl.2003.0112.
[11] Y. Hadid, et al., Possible incipient sympatric ecological speciation in blind mole rats (*Spalax*), Proc. Natl. Acad. Sci. USA 110 (7) (2013) 2587–2592. Available from: https://doi.org/10.1073/pnas.1222588110.
[12] K. Li, et al., Sympatric speciation revealed by genome-wide divergence in the blind mole rat *Spalax*, Proc. Natl. Acad. Sci. USA 112 (38) (2015) 11905–11910. Available from: https://doi.org/10.1073/pnas.1514896112.
[13] K. Li, et al., Transcriptome, genetic editing, and microRNA divergence substantiate sympatric speciation of blind mole rat, *Spalax*, Proc. Natl. Acad. Sci. USA 113 (27) (2016) 7584–7589.
[14] M. Lövy, J. Šklíba, E. Hrouzková, V. Dvořáková, E. Nevo, R. Šumbera, Habitat and burrow system characteristics of the blind mole rat *Spalax galili* in an area of supposed sympatric speciation, PLoS One 10 (7) (2015) e0133157. Available from: https://doi.org/10.1371/journal.pone.0133157.
[15] M. Lövy, J. Šklíba, R. Šumbera, E. Nevo, Soil preference in blind mole rats in an area of supposed sympatric speciation: do they choose the fertile or the familiar? J. Zool. 303 (4) (2017) 291–300. Available from: https://doi.org/10.1111/jzo.12489.
[16] Y. Zhao, et al., Adaptive methylation regulation of *P53* pathway in sympatric speciation of blind mole rats, *Spalax*, Proc. Natl. Acad. Sci. USA 113 (8) (2016) 2146–2151. Available from: https://doi.org/10.1073/pnas.1522658112.
[17] J. Šklíba, et al., Activity of free-living subterranean blind mole rats *Spalax galili* (Rodentia: Spalacidae) in an area of supposed sympatric speciation, Biol. J. Linn. Soc. 118 (2) (2016) 280–291. Available from: https://doi.org/10.1111/bij.12741.
[18] K. Li, et al., Incipient sympatric speciation in wild barley caused by geological-edaphic divergence, Life Sci. Alliance 3 (12) (2020) e202000827. Available from: https://doi.org/10.26508/lsa.202000827.
[19] E. Nevo, A.H.D. Brown, D. Zohary, N. Storch, A. Beiles, Microgeographic edaphic differentiation in allozyme polymorphisms of wild barley (*Hordeum spontaneum*, Poaceae), Plant Syst. Evol. 138 (3) (1981) 287–292. Available from: https://doi.org/10.1007/BF00985193.
[20] V. Ivandic, et al., Phenotypic responses of wild barley to experimentally imposed water stress, J. Exp. Botany 51 (353) (2000) 2021–2029. Available from: https://doi.org/10.1093/jexbot/51.353.2021.
[21] E. Nevo, A. Beiles, T. Krugman, Natural selection of allozyme polymorphisms: a microgeographic climatic differentiation in wild emmer wheat (*Triticum dicoccoides*), Theor. Appl. Genet. 75 (3) (1988) 529–538. Available from: https://doi.org/10.1007/BF00276761.
[22] E. Nevo, T. Krugman, A. Beiles, Edaphic natural selection of allozyme polymorphisms in *Aegilops peregrina* at a Galilee microsite in Israel, Heredity 72 (2) (1994) 109–112. Available from: https://doi.org/10.1038/hdy.1994.16.
[23] X. Wang, et al., Genomic adaptation to drought in wild barley is driven by edaphic natural selection at the Tabigha Evolution Slope, Proc. Natl. Acad. Sci. USA 115 (20) (2018) 5223–5228. Available from: https://doi.org/10.1073/pnas.1721749115.
[24] Y. Hadid, T. Pavlíček, A. Beiles, R. Ianovici, S. Raz, E. Nevo, Sympatric incipient speciation of spiny mice *Acomys* at "Evolution Canyon," Israel, Proc. Natl. Acad.

Sci. USA 111 (3) (2014) 1043–1048. Available from: https://doi.org/10.1073/pnas.1322301111.

[25] K. Li, et al., Sympatric speciation of spiny mice, *Acomys,* unfolded transcriptomically at Evolution Canyon, Israel, Proc. Natl. Acad. Sci. USA 113 (29) (2016) 8254–8259. Available from: https://doi.org/10.1073/pnas.1608743113.

[26] A. Korol, E. Rashkovetsky, K. Iliadi, E. Nevo, *Drosophila flies* in "Evolution Canyon" as a model for incipient sympatric speciation, Proc. Natl. Acad. Sci. USA 103 (48) (2006) 18184–18189. Available from: https://doi.org/10.1073/pnas.0608777103.

[27] O.N. Vinogradova, O.V. Kovalenko, S.P. Wasser, E. Nevo, O. Kislova, O. Belikova, Biodiversity of cyanophyta in Israel. Preliminary studies at "Evolution Canyon", lower Nahal Oren, Mt. Carmel natural preserve, Hydrobiol. J. 33 (6–7) (1997) 109–123.

[28] O.N. Vinogradova, O.V. Kovalenko, New representatives of Chroococcophyceae (Cyanophyta) from the Mount Carmel National Park and Dead Sea area, Israel, Int. J. Algae 3 (3) (2001) 95–99. Available from: https://doi.org/10.1615/InterJAlgae.v3.i3.100.

[29] O.N. Vinogradova, O.V. Kovalenko, S.P. Wasser, E. Nevo, M. Weinstein-Evron, Species diversity gradient to darkness stress in blue-green algae/cyanobacteria: a microscale test in a prehistoric cave, Mount Carmel, Israel, Isr. J. Plant Sci. 46 (3) (1998) 229–238. Available from: https://doi.org/10.1080/07929978.1998.10676732.

[30] O.N. Vinogradova, M. Poem-Finkel, E. Nevo, S.P. Wasser, Diversity of *Cyanoprocaryota* in Israel. First data about blue-green algae of dry limestones of Upper Galilee, Int. J. Algae 2 (3) (2000) 27–45. Available from: https://doi.org/10.1615/InterJAlgae.v2.i3.40.

[31] J. Sikorski, E. Nevo, Adaptation and incipient sympatric speciation of *Bacillus simplex* under microclimatic contrast at "Evolution Canyons" I and II, Israel, Proc. Natl. Acad. Sci. U.S.A. 102 (44) (2005) 15924–15929. Available from: https://doi.org/10.1073/pnas.0507944102.

[32] J. Sikorski, E. Nevo, On the necessity to study natural bacterial populations-the model of *bacillus simplex* from "Evolution Canyons" I and II, Israel, Isr. J. Ecol. Evol. 52 (3–4) (2006) 527–542. Available from: https://doi.org/10.1560/IJEE_52_3-4_527.

[33] S. Timmusk, et al., Bacterial distribution in the rhizosphere of wild barley under contrasting microclimates, PLoS One 6 (3) (2011) e17968. Available from: https://doi.org/10.1371/journal.pone.0017968.

[34] L. Couch, L. Blaustein, D.W. Duszynski, G. Shenbrot, E. Nevo, A new coccidian from *Acomys cahirinus* Desmarest, 1819, from Evolution Canyon, Lower Nahal Oren, Mount Carmel, Israel, J. Parasitol. 83 (2) (1997) 276–279. Available from: https://doi.org/10.2307/3284455.

[35] M. Temina, M.P. Andreev, S. Barinova, E. Nevo, The diversity and ecology of epiphytic lichens in "Evolution Canyon" II, Lover Nahal Keziv, Upper Western Galilee, Israel, Turkish J. Bot. 33 (4) (2009) 263–275.

[36] O. Vinogradova, et al., Algae of the Mount Carmel National Park (Israel), Algologia 5 (2) (1995) 178–192.

[37] S.P. Wasser, et al., Contribution to study of algae, fungi and mosses in "Evolution Canyon" at Nahal Oren, Mount Carmel Natural Preserve, Israel, Ukr. Bot. J. 52 (1995) 354–371.

[38] S.P. Wasser, et al., Diversity of cryptogamic plants and fungi in "Evolution Canyon", Nahal Oren, Mount Carmel, natural preserve, Israel, Isr. J. Plant Sci. 43 (4) (1995) 367–383. Available from: https://doi.org/10.1080/07929978.1995.10676623.

[39] S.Y. Kondratyuk, I.L. Navrotskaya, S.D. Zelenko, S.P. Wasser, E. Nevo, (in English) The First Checklist of Lichen-Forming and Lichenicolous Fungi of Israel, Peledfus Publishing House, Haifa, 1996p. 136.

[40] I.L. Navrotskaya, S.Y. Kondratyuk, S.P. Wasser, E. Nevo, S.D. Zelenko, Lichens and Lichenicolous fungi new for Israel and other countries, Isr. J. Plant Sci. 44 (2–3) (1996) 181–193. Available from: https://doi.org/10.1080/07929978.1996.10676646.

[41] P.A. Volz, N. Rosenzweig, R.B. Blackburn, S.P. Wasser, E. Nevo, Cobalt 60 radiation and growth of eleven species of micro-fungi from Evolution Canyon, Lower Nahal Oren, Israel, Microbios 91 (368–369) (1997) 191–201. 1/1/pubmed 4/2/ medline 1/1/entrez.

[42] S. Kondratyuk, E. Nevo, S. Wasser, New and rare lichen-forming and lichenicolous fungi from the Carmel Mountains, Israel, Ukr. Bot. J. 61 (1) (2005) 100–110.

[43] S.P. Wasser, E. Nevo, M. Temina, S.Y. Kondratyuk, S.D. Zelenko, (in English) Lichen-Forming, Lichenicolous, and Allied Fungi of Israel (Biodiversity of Cyanoprokaryotes, Algae and Fungi of Israel), A.R.G. Gantner Verlag K.G., Ruggell, Liechtenstein, 2005.

[44] S.S. Barinova, O.V. Anissimova, E. Nevo, S.P. Wasser, Diversity and ecology of phytoplankton and periphyton of the Nahal Oren, Mt Carmel Natural Park, Northern Israel, Algol. Stud. 116 (2005) 171–199. Available from: https://doi.org/ 10.1127/1864-1318/2005/0116-0171.

[45] Y. Ofer, Y. Harris, A. Beiles, E. Nevo, Biodiversity of ants at "Evolution Canyon", Nahal Oren, Mt. Carmel, Israel, Isr. J. Entomol. 30 (1996) 115–119.

[46] M. Finkel, J. Ofer, A. Beharav, E. Nevo, Species interslope divergence of ants caused by sharp microclimatic stresses at 'Evolution Canyon' II, Lower Nahal Keziv, western Upper Galilee, Israel, Isr. J. Entomol. 44-45 (2015) 63–73.

[47] I.A. Ellanskaya, P.A. Volz, E. Nevo, S. Wasser, E. Sokolova, Soil micromycetes of Evolution Canyon, Lower Nahal Oren, Mt Carmel National Park, Israel, Microbios 92 (1997) 19–33.

[48] I. Grishkan, E. Nevo, S.P. Wasser, T. Pavlíček, Spatiotemporal distribution of soil microfungi in "Evolution Canyon", lower Nahal Oren, Mount Carmel, Israel, Isr. J. Plant Sci. 48 (4) (2000) 297–308. Available from: https://doi.org/10.1560/LAJV-KKNY-F0HD-59KG.

[49] K.G. Savchenko, L.M. Carris, L.A. Castlebury, V.P. Heluta, S.P. Wasser, E. Nevo, Revision of *Entyloma* (Entylomatales, Exobasidiomycetes) on *Eryngium*, Mycologia 106 (4) (2014) 797–810. Available from: https://doi.org/10.3852/13-317.

[50] K. Savchenko, V. Heluta, S. Wasser, E. Nevo, Rust fungi (Pucciniales) of Israel. II. The genus *Uromyces*, Nova Hedwig. 98 (2014) 393–407.

[51] K.G. Savchenko, V.P. Heluta, S.P. Wasser, E. Nevo, Rust fungi (Pucciniales) of Israel. I. All genera except *Puccinia* and *Uromyces* with *Caeoma origani* sp. nov, Nova Hedwig. 98 (1–2) (2014) 163–178. Available from: https://doi.org/10.1127/0029-5035/2013/0122.

[52] I. Grishkan, E. Nevo, Vertical divergence of microfungal communities in soil profiles of "Evolution Canyon" I, Lower Nahal Oren, Mount Carmel, Israel, Plant Biosyst. 142 (1) (2008) 51–59. Available from: https://doi.org/10.1080/11263500701872325.

[53] I. Grishkan, E. Nevo, Soil microfungal communities of 'Evolution Canyons' in Israel – extreme differences on a regional scale, Biol. J. Linn. Soc. 93 (1) (2007) 157–163. Available from: https://doi.org/10.1111/j.1095-8312.2007.00918.x.

[54] I. Grishkan, E. Nevo, Soil microfungi of Nahal Meitzar, "Evolution Canyon" IV, Golan Heights, Plant Biosyst. 138 (1) (2004) 21–26. Available from: https://doi.org/10.1080/11263500410001684080.

[55] I. Grishkan, E. Nevo, S.P. Wasser, A. Beharav, Adaptive spatiotemporal distribution of soil microfungi in 'Evolution Canyon' II, Lower Nahal Keziv, western Upper Galilee, Israel, Biol. J. Linn. Soc. 78 (4) (2003) 527–539. Available from: https:// doi.org/10.1046/j.0024-4066.2002.00164.x.

[56] T. Pavlíček, C. Csuzdi, G. Smooha, A. Beiles, E. Nevo, Biodiversity and microhabitat distribution of earthworms at "Evolution Canyon", a Mediterranean microsite, Mount Carmel, Israel, Isr. J. Zool. 42 (4) (1996) 449–454. Available from: https://doi.org/10.1080/00212210.1996.10688868.
[57] S.S. Nagornaya, T.V. Babich, V.S. Podgorsky, A. Beharav, E. Nevo, S.P. Wasser, Yeast interslope divergence in soils and plants of "Evolution Canyon", Lower Nahal Oren, Mount Carmel, Israel, Isr. J. Plant Sci. 51 (1) (2003) 55–57. Available from: https://doi.org/10.1560/RMNC-XLYU-456X-V6E5.
[58] D. Gotman, S.P. Wasser, E. Nevo, Biodiversity of yeast species isolated from soil and fallen leaves in Israel (biodiversity of cyanoprokaryotes, algae and fungi of Israel), Koeltz Sci. Books (2013) 162.
[59] E. Nevo, O. Fragman, A. Dafni, A. Beiles, Biodiversity and interslope divergence of vascular plants caused by microclimatic differences at "Evolution Canyon", Lower Nahal Oren, Mount Carmel, Israel, Isr. J. Plant Sci. 47 (1) (1999) 49–59. Available from: https://doi.org/10.1080/07929978.1999.10676751.
[60] M. Finkel, O. Fragman, E. Nevo, Biodiversity and interslope divergence of vascular plants caused by sharp microclimatic differences at "Evolution Canyon II", Lower Nahal Keziv, Upper Galilee, Israel, Isr. J. Plant Sci. 49 (4) (2001) 285–296. Available from: https://doi.org/10.1560/H8TG-HL9L-2D2P-TAGJ.
[61] G. Bretfeld, D. Poliakov, M. Broza, *Collembola symphypleona* (Insecta, Entognatha) from Israel, Isr. J. Zool. 46 (4) (2000) 313–341. Available from: https://doi.org/10.1560/0UQA-DF2T-M3U8-V80Y.
[62] T. Pavlíček, V. Chikatunov, I. Lopatin, E. Nevo, New records of leaf beetles from Israel, Phytoparasitica 25 (4) (1997) 337–338. Available from: https://doi.org/10.1007/BF02981098.
[63] V. Chikatunov, M. Lillig, T. Pavlíček, L. Blaustein, E. Nevo, Biodiversity of insects at a microsite, 'Evolution Canyon', Nahal Oren, Mt. Carmel, Israel. Coleoptera: Tenebrionidae (in English)J. Arid Environ. 37 (2) (1997) 367–377. Available from: https://doi.org/10.1006/jare.1997.0278.
[64] T. Pavlíček, V. Chikatunov, V. Kravchenko, J. Dorchin, E. Nevo, *Xylotrechus stebbingi* Gahan a new species for Israeli beetle fauna, Mitt. Internationalen Entomologischen Ver. 23 (1–2) (1998) 73–74.
[65] V.I. Chikatunov, T. Pavlíček, E. Nevo, Coleoptera of "Evolution Canyon," Lower Nahal Oren, Mount Carmel, Israel I, in: Pensoft series faunistica, 14th ed., Pensoft, Sofia, Bulgaria, 1999.
[66] V.I. Chikatunov, T. Pavlíček, I. Lopatin, E. Nevo, Biodiversity and microclimatic divergence of chrysomelid beetles at 'Evolution Canyon', Lower Nahal Oren, Mt Carmel, Israel, Biol. J. Linn. Soc. London 69 (2) (2000) 139–152. Available from: https://doi.org/10.1006/bijl.1999.0322.
[67] M. Volkovitsh, T. Pavlíček, V. Chikatunov, E. Nevo, Species diversity and microsite divergence of insects at "Evolution Canyon", Lower Nahal Oren, Mt. Carmel, Israel (Coleoptera: Buprestidae), Zool. Middle East 20 (1) (2000) 121–124. Available from: https://doi.org/10.1080/09397140.2000.10637822.
[68] M. Finkel, V.I. Chikatunov, E. Nevo, Coleoptera of "Evolution Canyon" II: Lower Nahal Keziv, Western Upper Galilee, Israel, Pensoft, Sofia/Moscow, 2002.
[69] V.I. Chikatunov, T. Pavlíček, E. Nevo, Coleoptera of "Evolution Canyon" Lower Nahal Oren, Mount Carmel, Israel Part II (Series Faunistica), Pentsoft, Sofia/Moscow, 2004, p. 192.
[70] V. Melamud, A. Beharav, T. Pavlíček, E. Nevo, Biodiversity interslope divergence of oribatid mites at "Evolution Canyon", Mount Carmel, Israel, Acta Zool. Acad. Sci. Hung. 53 (4) (2007) 381–396.

[71] T. Pavlíček, S. Vivanti, L. Fishelson, E. Nevo, Biodiversity and microsite divergence of insects at "Evolution Canyon", Nahal Oren, Mt. Carmel, Israel. II. Orthoptera: Acrididae, J. Entomol. Res. Soc. 4 (1) (2004) 1–39.
[72] A. Eitam, T. Pavlíček, L. Blaustein, E. Nevo, Effects of inter- and intra-slope contrasts at 'Evolution Canyon', Israel, on a gall aphid and its natural enemies, J. Mediterr. Ecol. 3 (2002) 47–52.
[73] T. Pavlíček, H.K. Mienis, S. Raz, V. Hassid, A. Rubenyan, E. Nevo, Gastropod biodiversity at the 'Evolution Canyon' microsite, lower Nahal Oren, Mount Carmel, Israel, Biol. J. Linn. Soc. 1 (2008) 147–155. Available from: https://doi.org/10.1111/j.1095-8312.2007.00920.x. https://doi.org/10.1111/j.1095-8312.2007.00920.x93.
[74] M. Harry, et al., Fine-scale biodiversity of Drosophilidae in "Evolution Canyon" at the Lower Nahal Oren microsite, Israel, Biologia 54 (6) (1999) 685–706.
[75] T. Pavlíček, V. Kravchenko, E. Nevo, V. Chikatunov, Seasonal and spatial distribution of butterflies (Lepidoptera Rhopalocera) in "Evolution Canyon", Lower Nahal Oren, Mt. Carmel, Israel, Ecol. Mediterr. (2002) 99–112.
[76] M.D. Zerova, L.Y. Seregina, New species of chalcidoid wasps of the family Eurytomidae (Hymenoptera, Chalcidoidea) from arid Palaearctic regions, Entomol. Rev. 88 (8) (2008) 960–966. Available from: https://doi.org/10.1134/S0013873808080101.
[77] M.D. Zerova, L.Y. Seryogina, G. Melika, T. Pavlíček, E. Nevo, New genus and new species of cynipid gall inducing wasp (Hymenoptera: Cynipidae) and new species of chalcid wasps (Hymenoptera: Chalcidoidea) from Israel (in English) J. Entomol. Res. Soc. 5 (1) (2003) 35–49.
[78] E. Nevo, S. Raz, A. Beiles, Biodiversity of reptiles at "Evolution Canyon", lower Nahal Oren, Mount Carmel, Israel (in English)Isr. J. Ecol. Evol. 42 (4) (1996) 395–402. Available from: https://doi.org/10.1080/00212210.1996.10688863.
[79] L. Blaustein, B.P. Kotler, E. Nevo, Rodent species diversity and microhabitat use along opposing slopes of Lower Nahal Oren, Mount Carmel, Israel, Isr. J. Zool. 42 (4) (1996) 327–333. Available from: https://doi.org/10.1080/00212210.1996.10688856.
[80] E. Nevo, A. Beiles, E. Ivanitskaia, Adaptive Radiation of Blind Subterranean Mole Rats: Naming and Revisiting the Four Sibling Species of the *Spalax ehrenbergi* superspecies in Israel: *Spalax galili* (2n = 52), *S. golani* (2n = 54), *S. carmeli* (2n = 58) and *S. judaei* (2n = 60), Backhuys Publishers (2001).
[81] Y. Gutterman, E. Nevo, Temperatures and ecological—genetic differentiation affecting the germination of *Hordeum spontaneum* caryopses harvested from three populations: the negev desert and opposing slopes on Mediterranean Mount Carmel (in English) Isr. J. Plant Sci. 42 (3) (1994) 183–195. Available from: https://doi.org/10.1080/07929978.1994.10676571.
[82] E. Nevo, Molecular evolution and ecological stress at global, regional and local scales: the Israeli perspective, J. Exp. Zool. 282 (1–2) (1998) 95–119. doi:10.1002/(SICI)1097-010X(199809/10)282:1/2 < 95::AID-JEZ12 > 3.0.CO;2-F.
[83] E. Nevo, M.G. Filippucci, A. Beiles, Genetic polymorphisms in subterranean mammals (*Spalax ehrenbergi* superspecies) in the Near East revisited: patterns and theory, Heredity 72 (5) (1994) 465–487. Available from: https://doi.org/10.1038/hdy.1994.65.
[84] E. Nevo, M.G. Filippucci, C. Redi, A. Korol, A. Beiles, Chromosomal speciation and adaptive radiation of mole rats in Asia Minor correlated with increased ecological stress, Proc. Natl. Acad. Sci. USA 91 (17) (1994) 8160–8164. Available from: https://doi.org/10.1073/pnas.91.17.8160.
[85] E. Nevo, M.G. Filippucci, C. Redi, S. Simson, G. Heth, A. Beiles, Karyotype and genetic evolution in speciation of subterranean mole rats of the genus *Spalax* in Turkey, Biol. J. Linn. Soc. 54 (3) (1995) 203–229. Available from: https://doi.org/10.1111/j.1095-8312.1995.tb01034.x.

[86] T. Krugman, N. Satish, O. Vinogradova, A. Beharav, Y. Kashi, E. Nevo, Genome diversity in the cyanobacterium *Nostoc linckia* at 'Evolution Canyon', Israel, revealed by inter-HIP1 size polymorphisms, Evolut. Ecol. Res. 3 (8) (2001) 899–915.
[87] R. Kalendar, J. Tanskanen, S. Immonen, E. Nevo, A.H. Schulman, Genome evolution of wild barley (*Hordeum spontaneum*) by BARE-1 retrotransposon dynamics in response to sharp microclimatic divergence, Proc. Natl. Acad. Sci. USA 97 (12) (2000) 6603–6607. Available from: https://doi.org/10.1073/pnas.110587497.
[88] N. Satish, T. Krugman, O.N. Vinogradova, E. Nevo, Y. Kashi, Genome evolution of the cyanobacterium *Nostoc linckia* under sharp microclimatic divergence at "Evolution Canyon," Israel, Microb. Ecol. 42 (3) (2001) 306–316. Available from: https://doi.org/10.1007/s00248-001-0013-0.
[89] V. Dvornyk, O. Vinogradova, E. Nevo, Long-term microclimatic stress causes rapid adaptive radiation of kaiABC clock gene family in a cyanobacterium, *Nostoc linckia*, from "Evolution Canyons" I and II, Israel, Proc. Natl. Acad. Sci. USA 99 (4) (2002) 2082–2087. Available from: https://doi.org/10.1073/pnas.261699498.
[90] V. Dvornyk, O. Vinogradova, E. Nevo, Origin and evolution of circadian clock genes in prokaryotes, Proc. Natl. Acad. Sci. USA 100 (5) (2003) 2495–2500. Available from: https://doi.org/10.1073/pnas.0130099100.
[91] V. Dvornyk, E. Nevo, Genetic polymorphism of cyanobacteria under permanent natural stress: a lesson from the "Evolution Canyons", Res. Microbiol. 154 (2) (2003) 79–84. Available from: https://doi.org/10.1016/S0923-2508(02)00015-3.
[92] E. Nevo, Evolution in action across phylogeny caused by microclimatic stresses at "Evolution Canyon" (in English)Theor. Popul. Biol. 52 (3) (1997) 231–243. Available from: https://doi.org/10.1006/tpbi.1997.1330.
[93] G. Bedada, A. Westerbergh, E. Nevo, A. Korol, K.J. Schmid, DNA sequence variation of wild barley *Hordeum spontaneum* (L.) across environmental gradients in Israel, Heredity 112 (6) (2014) 646–655. Available from: https://doi.org/10.1038/hdy.2014.2.
[94] Z. Yang, T. Zhang, A. Bolshoy, A. Beharav, E. Nevo, Adaptive microclimatic structural and expressional dehydrin 1 evolution in wild barley, *Hordeum spontaneum*, at 'Evolution Canyon', Mount Carmel, Israel, Mol. Ecol. 18 (9) (2009) 2063–2075. Available from: https://doi.org/10.1111/j.1365-294X.2009.04140.x.
[95] Z. Yang, T. Zhang, G. Li, E. Nevo, Adaptive microclimatic evolution of the dehydrin 6 gene in wild barley at "Evolution Canyon", Israel, Genetica 139 (11) (2011) 1429–1438. Available from: https://doi.org/10.1007/s10709-012-9641-1.
[96] T. Zhang, G. Li, Z. Yang, E. Nevo, Adaptive evolution of duplicated hsp17 genes in wild barley from microclimatically divergent sites of Israel, Genet. Mol. Res. 13 (1) (2014) 1220–1232. Available from: http://www.funpecrp.com.br/gmr/year2014/vol13-1/pdf/gmr2826.pdf.
[97] Z. Yang, E. Nevo, Heat shock proteins in wild barley at "Evolution Canyon", Mount Carmel, Israel, in: A. Asea, P. Kaur, S.K. Calderwood (Eds.), Heat Shock Proteins and Plants, 10, Springer International Publishing, Cham, 2016, pp. 79–102.
[98] E. Nevo, et al., Genomic microsatellite adaptive divergence of wild barley by microclimatic stress in 'Evolution Canyon', Israel, Biol. J. Linn. Soc. 84 (2) (2005) 205–224. Available from: https://doi.org/10.1111/j.1095-8312.2005.00425.x.
[99] S. Timmusk, et al., Sfp-type PPTase inactivation promotes bacterial biofilm formation and ability to enhance wheat drought tolerance, Front. Microbiol. 6 (2015) 387. Available from: https://doi.org/10.3389/fmicb.2015.00387.
[100] Z. Yang, T. Zhang, T. Lang, G. Li, G. Chen, E. Nevo, Transcriptome comparative profiling of barley eibi1 mutant reveals pleiotropic effects of HvABCG31 gene on cuticle biogenesis and stress responsive pathways, Int. J. Mol. Sci. 14 (10) (2013) 20478–20491.

[101] J.K. Cronin, P.C. Bundock, R.J. Henry, E. Nevo, Adaptive climatic molecular evolution in wild barley at the *Isa* defense locus, Proc. Natl. Acad. Sci. USA 104 (8) (2007) 2773–2778. Available from: https://doi.org/10.1073/pnas.0611226104.
[102] Y. Shen, K. Lebold, E.P. Lansky, M.G. Traber, E. Nevo, 'Tocol-omic' diversity in wild barley, short communication, Chem. Biodivers. 8 (12) (2011) 2322–2330. Available from: https://doi.org/10.1002/cbdv.201000363.
[103] P.K. Gupta, et al., Polymorphism at rDNA loci in barley and its relation with climatic variables, Theor. Appl. Genet. 104 (2) (2002) 473–481. Available from: https://doi.org/10.1007/s001220100690.
[104] E. Nevo, Evolution of wild barley at "Evolution Canyon": adaptation, speciation, pre-agricultural collection, and barley improvement, Isr. J. Plant Sci. 62 (1-2) (2015) 22–32. Available from: https://doi.org/10.1080/07929978.2014.940783.
[105] E. Nevo, "Evolution Canyon": a microcosm of life's evolution focusing on adaptation and speciation, Isr. J. Ecol. Evol. 52 (3–4) (2006) 501–506. Available from: https://doi.org/10.1560/IJEE_52_3-4_485.
[106] E. Nevo, Evolution of wild emmer wheat and crop improvement, J. Syst. Evol. 52 (6) (2014) 673–696. Available from: https://doi.org/10.1111/jse.12124.
[107] T. Krugman, E. Nevo, A. Beharav, H. Sela, T. Fahima, The Institute of Evolution Wild Cereal Gene Bank at the University of Haifa (in English)Isr. J. Plant Sci. 65 (3-4) (2018) 129–146. Available from: https://doi.org/10.1163/22238980-00001065.
[108] H. Wang, et al., Sympatric speciation of wild emmer wheat driven by ecology and chromosomal rearrangements, Proc. Natl. Acad. Sci. USA 117 (11) (2020) 5955–5963. Available from: https://doi.org/10.1073/pnas.1920415117.
[109] R. Avni, et al., Wild emmer genome architecture and diversity elucidate wheat evolution and domestication, Science 357 (6346) (2017) 93–97. Available from: https://doi.org/10.1126/science.aan0032.
[110] J. Dvorak, et al., Structural variation and rates of genome evolution in the grass family seen through comparison of sequences of genomes greatly differing in size, Plant J. 95 (3) (2018) 487–503. Available from: https://doi.org/10.1111/tpj.13964.
[111] Y. Zhang, et al., New insights into the evolution of wheat avenin-like proteins in wild emmer wheat (*Triticum dicoccoides*), Proc. Natl. Acad. Sci. USA 115 (52) (2018) 13312–13317. Available from: https://doi.org/10.1073/pnas.1812855115.
[112] O. Kossover, Z. Frenkel, A. Korol, E. Nevo, Genetic diversity and stress of *Ricotia lunaria* in "Evolution Canyon," Israel, J. Hered. 100 (4) (2009) 432–440. Available from: https://doi.org/10.1093/jhered/esp014.
[113] C. Qian, et al., Transcriptomes divergence of *Ricotia lunaria* between the two micro-climatic divergent slopes at "Evolution Canyon" I, Israel (in English)Front. Genet. 9 (506) (2018). Available from: https://doi.org/10.3389/fgene.2018.00506.
[114] E. Nevo, E. Rashkovetsky, T. Pavlíček, A. Korol, A complex adaptive syndrome in *Drosophila* caused by microclimatic contrasts, Heredity 80 (1) (1998) 9–16. Available from: https://doi.org/10.1046/j.1365-2540.1998.00274.x.
[115] E. Rashkovetsky, K. Iliadi, E. Nevo, A. Korol, Fitness related traits in *Drosophila melanogaster* subpopulations from the opposite slopes of 'Evolution Canyon', lower Nahal Oren, Mount Carmel, Israel, Drosoph. Inf. Serv. 83 (2000) 138–140.
[116] A. Korol, E. Rashkovetsky, E. Nevo, Enigmatic flies: is *Drosophila* in the "Evolution Canyon" a model for incipient sympatric speciation? Isr. J. Ecol. Evol. 52 (3–4) (2006) 507–525. Available from: https://doi.org/10.1560/IJEE_52_3-4_507.
[117] E. Derzhavets, A. Korol, T. Pavlíček, E. Nevo, Adaptation to stressful environment and mutation rate: a case study in *Drosophila melanogaster*, Drosoph. Inf. Serv. 80 (1997) 53.

[118] E. Derzhavets, A. Korol, E. Nevo, Increased male recombination rate in *D. melanogaster* correlated with population adaptation to stressful conditions, Drosoph. Inf. Serv. 77 (1996) 92–94.
[119] E. Derzhavets, A. Korol, E. Nevo, Differences in fluctuating asymmetry in *Drosophila melanogaster* caused by microclimatic contrasts, Drosoph. Inf. Serv. 80 (1997) 73.
[120] E. Rashkovetsky, et al., Adaptive differentiation of thermotolerance in *Drosophila* along a microclimatic gradient, Heredity 96 (5) (2006) 353–359. Available from: https://doi.org/10.1038/sj.hdy.6800784.
[121] A. Lupu, A. Pechkovskaya, E. Rashkovetsky, E. Nevo, A. Korol, DNA repair efficiency and thermotolerance in *Drosophila melanogaster* from 'Evolution Canyon', Mutagenesis 19 (5) (2004) 383–390. Available from: https://doi.org/10.1093/mutage/geh045.
[122] J. Carmel, E. Rashkovetsky, E. Nevo, A. Korol, Differential expression of small heat shock protein genes Hsp23 and Hsp40, and heat shock gene Hsr-omega in fruit flies (*Drosophila melanogaster*) along a Microclimatic Gradient, J. Hered. 102 (5) (2011) 593–603. Available from: https://doi.org/10.1093/jhered/esr027.
[123] P. Michalak, et al., Genetic evidence for adaptation-driven incipient speciation of *Drosophila melanogaster* along a microclimatic contrast in "Evolution Canyon," Israel, Proc. Natl. Acad. Sci. 98 (23) (2001) 13195–13200. Available from: https://doi.org/10.1073/pnas.231478298.
[124] A. Korol, E. Rashkovetsky, K. Iliadi, P. Michalak, Y. Ronin, E. Nevo, Nonrandom mating in *Drosophila melanogaster* laboratory populations derived from closely adjacent ecologically contrasting slopes at "Evolution Canyon", Proc. Natl. Acad. Sci. 97 (23) (2000) 12637–12642. Available from: https://doi.org/10.1073/pnas.220041397.
[125] S.R. Singh, E. Rashkovetsky, K. Iliadi, E. Nevo, A. Korol, Assortative mating in *Drosophila* adapted to a microsite ecological gradient, Behav. Genet. 35 (6) (2005) 753. Available from: https://doi.org/10.1007/s10519-005-6119-2.
[126] K. Iliadi, N. Iliadi, E. Rashkovetsky, I. Minkov, E. Nevo, A. Korol, Sexual and reproductive behaviour of *Drosophila melanogaster* from a microclimatically interslope differentiated population of "Evolution Canyon" (Mount Carmel, Israel), Proc. R. Soc. London, Ser. B: Biol. Sci. 268 (1483) (2001) 2365–2374. Available from: https://doi.org/10.1098/rspb.2001.1822.
[127] K.G. Iliadi, N.N. Iliadi, E.L. Rashkovetsky, S.V. Girin, E. Nevo, A.B. Korol, Sexual differences for emigration behavior in natural populations of *Drosophila melanogaster*, Behav. Genet. 32 (3) (2002) 173–180. Available from: https://doi.org/10.1023/A:1016017028041.
[128] D. Joly, A. Korol, E. Nevo, Sperm size evolution in *Drosophila*: inter- and intraspecific analysis, Genetica 120 (1) (2004) 233–244. Available from: https://doi.org/10.1023/B:GENE.0000017644.63389.57.
[129] V. Debat, R. Cornette, A.B. Korol, E. Nevo, D. Soulet, J.R. David, Multidimensional analysis of *Drosophila* wing variation in Evolution Canyon, J. Genet. 87 (4) (2008) 407–419. Available from: https://doi.org/10.1007/s12041-008-0063-x.
[130] K.G. Iliadi, et al., Peculiarities of the courtship song in the *Drosophila melanogaster* populations adapted to gradient of microecological conditions, J. Evolut. Biochem. Physiol. 45 (5) (2009) 579. Available from: https://doi.org/10.1134/S0022093009050041.
[131] C.B. Palavicino-Maggio, S. Trannoy, K.M. Holton, X. Song, K. Li, E. Nevo, Aggression and courtship differences found in *Drosophila melanogaster* from two different microclimates at Evolution Canyon, Israel, Sci. Rep. 9 (1) (2019) 4084. Available from: https://doi.org/10.1038/s41598-019-40701-8.

[132] T. Pavlíček, Z. Frenkel, A.B. Korol, A. Beiles, E. Nevo, *Drosophila* at the "Evolution Canyon" Microsite, MT. Carmel, Israel: selection overrules migration (in English) Isr. J. Ecol. Evol. 54 (2) (2008) 165–180. Available from: https://doi.org/10.1080/15659801.2008.10639620.
[133] E. Nevo, Selection overrules gene flow at "Evolution Canyons," Israel, Adv. Genet. Res. 5 (3) (2011) 67–89.
[134] I. Zamorzaeva, E. Rashkovetsky, E. Nevo, A. Korol, Sequence polymorphism of candidate behavioural genes in *Drosophila melanogaster* flies from 'Evolution canyon', Mol. Ecol. 14 (10) (2005) 3235–3245. Available from: https://doi.org/10.1111/j.1365-294X.2005.02616.x.
[135] S. Hübner, et al., Genome differentiation of *Drosophila melanogaster* from a microclimate contrast in Evolution Canyon, Israel, Proc. Natl. Acad. Sci. USA 110 (52) (2013) 21059–21064. Available from: https://doi.org/10.1073/pnas.1321533111.
[136] Y.B. Kim, et al., Divergence of *Drosophila melanogaster* repeatomes in response to a sharp microclimate contrast in Evolution Canyon, Israel, Proc. Natl. Acad. Sci. USA 111 (29) (2014) 10630–10635. Available from: https://doi.org/10.1073/pnas.1410372111.
[137] E. Nevo, et al., Genotypic and phenotypic divergence of rodents (*Acomys cahirinus* and *Apodemus mystacinus*) at "Evolution Canyon": micro-and macroscale parallelism, Acta Theriol. 43 (1998) 9–34.
[138] M.G. Filippucci, S. Simson, E. Nevo, Evolutionary biology of the genus *Apodemus* Kaup, 1829 in Israel. Allozymic and biometric analyses with description of a new species: *Apodemus hermonensis* (Rodentia, Muridae), Boll. Zool. 56 (4) (1989) 361–376. Available from: https://doi.org/10.1080/11250008909355663.
[139] B.C. Lamb, M. Saleem, W. Scott, N. Thapa, E. Nevo, Inherited and environmentally induced differences in mutation frequencies between wild strains of *Sordaria fimicola* from "Evolution Canyon", Genetics 149 (1) (1998) 87–99.
[140] M. Saleem, B.C. Lamb, E. Nevo, Inherited differences in crossing over and gene conversion frequencies between wild strains of *Sordaria fimicola* from "Evolution Canyon", Genetics 159 (4) (2001) 1573–1593.
[141] D. Barash, J. Sikorski, E.B. Perry, E. Nevo, E. Nudler, Adaptive mutations in RNA-based regulatory mechanisms: computational and experimental investigations (in English)Isr. J. Ecol. Evol. 52 (3-4) (2006) 263–279. Available from: https://doi.org/10.1560/IJEE_52_3-4_263.
[142] P. Bures, T. Pavlíček, L. Horova, E. Nevo, Microgeographic genome size differentiation of the carob tree, *Ceratonia siliqua*, at 'Evolution Canyon', Israel, Ann. Bot. 93 (5) (2004) 529–535. Available from: https://doi.org/10.1093/aob/mch074.
[143] N. Gasmanova, et al., Genome size variation of *Lotus peregrinus* at evolution canyon I microsite, Lower Nahal Oren, Mt. Carmel, Israel, Acta Biol. Crac. Ser. Bot. 49 (1) (2007) 39–46.
[144] K. Sharaf, Y. Hadid, T. Pavlíček, E. Nevo, Local genetic population divergence in a saw-toothed grain beetle, *Oryzaephilus surinamensis* (L.) (Coleoptera, Cucujidae), J. Stored Prod. Res. 53 (2013) 72–76. Available from: https://doi.org/10.1016/j.jspr.2013.03.002.
[145] A. Beiles, S. Raz, Y. Ben-Abu, E. Nevo, Putative adaptive inter-slope divergence of transposon frequency in fruit flies (*Drosophila melanogaster*) at "Evolution Canyon", Mount Carmel, Israel, Biol. Direct 10 (1) (2015) 58. Available from: https://doi.org/10.1186/s13062-015-0074-5.
[146] S. Raz, J.H. Graham, H. Hel-or, T. Pavlíček, E. Nevo, Developmental instability of vascular plants in contrasting microclimates at 'Evolution Canyon', Biol. J. Linn. Soc. 102 (4) (2011) 786–797. Available from: https://doi.org/10.1111/j.1095-8312.2011.01615.x.

[147] S. Raz, J.H. Graham, A. Cohen, B.L. de Bivort, I. Grishkan, E. Nevo, Growth and asymmetry of soil microfungal colonies from "Evolution Canyon," Lower Nahal Oren, Mount Carmel, Israel, PLoS One 7 (4) (2012) e34689. Available from: https://doi.org/10.1371/journal.pone.0034689.

[148] O. Vinogradova, O.V. Kovalenko, S.P. Wasser, E. Nevo, Cyanophyta: Checklist of Continental Species From Israel. International Centre for Cryptogamic Plants and Fungi, Institute of Evolution, University of Haifa, Haifa, Israel, 1996.

[149] E. Nevo, et al., Drought and light anatomical adaptive leaf strategies in three woody species caused by microclimatic selection at "Evolution Canyon", Israel, Isr. J. Plant Sci. 48 (1) (2000) 33–46. Available from: https://doi.org/10.1560/RNPF-9HJE-8J3L-B5F1.

[150] J. Sikorski, E. Nevo, Patterns of thermal adaptation of *Bacillus simplex* to the microclimatically contrasting slopes of 'Evolution Canyons' I and II, Israel, Environ. Microbiol. 9 (3) (2007) 716–726. Available from: https://doi.org/10.1111/j.1462-2920.2006.01193.x.

[151] I. Grishkan, A. Beharav, V. Kirzhner, E. Nevo, Adaptive spatiotemporal distribution of soil microfungi in 'Evolution Canyon' III, Nahal Shaharut, extreme southern Negev Desert, Israel, Biol. J. Linn. Soc. 90 (2) (2007) 263–277. Available from: https://doi.org/10.1111/j.1095-8312.2007.00722.x.

[152] N. Singaravelan, I. Grishkan, A. Beharav, K. Wakamatsu, S. Ito, E. Nevo, Adaptive melanin response of the soil fungus *Aspergillus niger* to UV radiation stress at "Evolution Canyon", Mount Carmel, Israel, PLoS One 3 (8) (2008) e2993. Available from: https://doi.org/10.1371/journal.pone.0002993.

[153] E. Nevo, Evolution under ecological stress: fungal divergent adaptive melanization at Evolution Canyons in Israel, Fungal Genom. Biol. 7 (2017) 149–154.

[154] I. Grishkan, K. Wakamatsu, T. Perl, K. Li, E. Nevo, Adaptive response of a soil fungus, *Aspergillus niger*, to changed environmental conditions in a soil transplant experiment at 'Evolution Canyon' I, Mount Carmel, Israel, Biol. J. Linn. Soc. 125 (4) (2018) 821–826. Available from: https://doi.org/10.1093/biolinnean/bly138.

[155] T.K. Ezov, et al., Molecular-genetic biodiversity in a natural population of the yeast *Saccharomyces cerevisiae* from "Evolution Canyon": microsatellite polymorphism, ploidy and controversial sexual status, Genetics 174 (3) (2006) 1455–1468. Available from: https://doi.org/10.1534/genetics.106.062745.

[156] G.A. Lidzbarsky, T. Shkolnik, E. Nevo, Adaptive response to DNA-damaging agents in natural *Saccharomyces cerevisiae* populations from "Evolution Canyon", Mt. Carmel, Israel, PLoS One 4 (6) (2009) e5914. Available from: https://doi.org/10.1371/journal.pone.0005914.

[157] A. Lupu, E. Nevo, I. Zamorzaeva, A. Korol, Ecological–genetic feedback in DNA repair in wild barley, *Hordeum spontaneum*, Genetica 127 (1) (2006) 121–132. Available from: https://doi.org/10.1007/s10709-005-2611-0.

[158] S. Miyazaki, E. Nevo, I. Grishkan, U. Idleman, D. Weinberg, H.J. Bohnert, Oxidative stress responses in yeast strains, *Saccharomyces cerevisiae*, from "Evolution Canyon", Israel, Monatsh. Chem. Chem. Mon. 134 (11) (2003) 1465–1480. Available from: https://doi.org/10.1007/s00706-003-0072-7.

[159] D. Rankevich, B. Lavie, E. Nevo, A. Beiles, Z. Arad, Genetic and physiological adaptations of the prosobranch landsnail *Pomatias olivieri* to microclimatic stresses on Mount Carmel, Israel (in English)Isr. J. Ecol. Evol. 42 (4) (1996) 425–441. Available from: https://doi.org/10.1080/00212210.1996.10688866.

[160] E. Nevo, "Evolution Canyon," a potential microscale monitor of global warming across life, Proc. Natl. Acad. Sci. USA 109 (8) (2012) 2960–2965. Available from: https://doi.org/10.1073/pnas.1120633109.

[161] H. Yin, Y. Ben-Abu, H. Wang, A. Li, E. Nevo, L. Kong, Natural selection causes adaptive genetic resistance in wild emmer wheat against powdery mildew at "Evolution Canyon" microsite, Mt. Carmel, Israel, PLoS One 10 (4) (2015) e0122344. Available from: https://doi.org/10.1371/journal.pone.0122344.

[162] J. Yan, Y. Wang, Y. Gutterman, E. Nevo, C. Jian-ping, Caryopsis dormancy and seedling drought tolerance of wild barley, *Hordeum spontaneumat* at "Evolution Canyon" Microsite, Israel, J. Sichuan Agric. Univ. (2011) 01.

[163] L. Kang, et al., Genomic divergence and adaptive convergence in *Drosophila simulans* from Evolution Canyon, Israel, Proc. Natl. Acad. Sci. USA 116 (24) (2019) 11839–11844. Available from: https://doi.org/10.1073/pnas.1720938116.

[164] D. Levitte, Geological Map of Zefat 1:50,000, Geological Survey of Israel, Jerusalem, 2001.

[165] E. Nevo, Mosaic Evolution of Subterranean Mammals: Regression, Progression, and Global Convergence, Oxford University Press, Oxford, 1999.

[166] E. Nevo, Evolution of genome–phenome diversity under environmental stress, Proc. Natl. Acad. Sci. 98 (11) (2001) 6233–6240. Available from: https://doi.org/10.1073/pnas.101109298.

[167] E. Nevo, A. Beiles, R. Ben-Shlomo, The evolutionary significance of genetic diversity: ecological, demographic and life history correlates, In: G.S. Mani (Ed.) Evolutionary Dynamics of Genetic Diversity, Springer, Berlin, Heidelberg, 1984, pp. 13–213.

[168] E. Nevo, Evolutionary theory and processes of active speciation and adaptive radiation in subterranean mole rats, *Spalax ehrenbergi* superspecies, in Israel, Evolut. Biol. 25 (1991) 1–125.

[169] E. Nevo, Plant genetic resources: prediction by isozyme markers and ecology (in English) Isozymes 16 (1987) 247–267.

[170] K. Li, et al., Genome evolution of blind subterranean mole rats: adaptive peripatric versus sympatric speciation, Proc. Natl. Acad. Sci USA 117 (51) (2020) 32499–32508. Available from: https://doi.org/10.1073/pnas.2018123117.

[171] A.K. Conley, J.L. Neuwald, A.R. Templeton, (New Horizons in Evolution) Network Analyses of the Impact of Visual Habitat Structure on Behavior, Demography, Genetic Diversity, and Gene Flow in a Metapopulation of Collared Lizards *(Crotaphytus Collaris Collaris)*, Elsevier, 2021.

Index

Note: Page numbers followed by "*f*" and "*t*" refer to figures and tables, respectively.

M

N

O

P

T

U

V

W

X

Y

Z

Kolibri 360 GmbH
Ferdinand-Jühlke-Straße 7
99095 Erfurt, Deutschland
produktsicherheit@kolibri360.de